Challenges and Risks Involved in Deploying 6G and NextGen Networks

A.M. Viswa Bharathy
GITAM University, Bengaluru, India

Basim Alhadidi
Al-Balqa Applied University, Jordan

A volume in the Advances in Wireless Technologies and Telecommunication (AWTT) Book Series

IGI Global
PUBLISHER of TIMELY KNOWLEDGE

Published in the United States of America by
IGI Global
Information Science Reference (an imprint of IGI Global)
701 E. Chocolate Avenue
Hershey PA, USA 17033
Tel: 717-533-8845
Fax: 717-533-8661
E-mail: cust@igi-global.com
Web site: http://www.igi-global.com

Library of Congress Cataloging-in-Publication Data

Names: Bharathy, A. M. Viswa, 1986- editor. | Alhadidi, Basim, 1972- editor.
Title: Challenges and risks involved in deploying 6G and NextGen networks / A M Viswa Bharathy and Basim Alhadidi, editors.
Description: Hershey, PA : Information Science Reference, an imprint of IGI Global, 2022. | Includes bibliographical references and index. | Summary: "This book focuses on AI-enabled NextGen (6G and beyond) networks making readers aware of the challenges waiting in NextGen networks, so that proper steps could be taken to either minimize or eliminate issues on 6G and NextGen networks and incorporating AI in NextGen networks for privacy and security policies"-- Provided by publisher.
Identifiers: LCCN 2021061499 (print) | LCCN 2021061500 (ebook) | ISBN 9781668438046 (h/c) | ISBN 9781668438053 (s/c) | ISBN 9781668438060 (ebook)
Subjects: LCSH: Wireless communication systems--Technological innovations.
Classification: LCC TK5103.2 .C44 2022 (print) | LCC TK5103.2 (ebook) | DDC 621.384--dc23/eng/20220307
LC record available at https://lccn.loc.gov/2021061499
LC ebook record available at https://lccn.loc.gov/2021061500

This book is published in the IGI Global book series Advances in Wireless Technologies and Telecommunication (AWTT) (ISSN: 2327-3305; eISSN: 2327-3313)

British Cataloguing in Publication Data
A Cataloguing in Publication record for this book is available from the British Library.

Advances in Wireless Technologies and Telecommunication (AWTT) Book Series

Xiaoge Xu
University of Nottingham Ningbo China, China

ISSN:2327-3305
EISSN:2327-3313

Mission

The wireless computing industry is constantly evolving, redesigning the ways in which individuals share information. Wireless technology and telecommunication remain one of the most important technologies in business organizations. The utilization of these technologies has enhanced business efficiency by enabling dynamic resources in all aspects of society.

The **Advances in Wireless Technologies and Telecommunication Book Series** aims to provide researchers and academic communities with quality research on the concepts and developments in the wireless technology fields. Developers, engineers, students, research strategists, and IT managers will find this series useful to gain insight into next generation wireless technologies and telecommunication.

Coverage

- Telecommunications
- Broadcasting
- Mobile Communications
- Grid Communications
- Wireless Broadband
- Digital Communication
- Mobile Web Services
- Virtual Network Operations
- Wireless Technologies
- Network Management

IGI Global is currently accepting manuscripts for publication within this series. To submit a proposal for a volume in this series, please contact our Acquisition Editors at Acquisitions@igi-global.com or visit: http://www.igi-global.com/publish/.

The Advances in Wireless Technologies and Telecommunication (AWTT) Book Series (ISSN 2327-3305) is published by IGI Global, 701 E. Chocolate Avenue, Hershey, PA 17033-1240, USA, www.igi-global.com. This series is composed of titles available for purchase individually; each title is edited to be contextually exclusive from any other title within the series. For pricing and ordering information please visit http://www.igi-global.com/book-series/advances-wireless-technologies-telecommunication/73684. Postmaster: Send all address changes to above address.

Titles in this Series

For a list of additional titles in this series, please visit: www.igi-global.com/book-series

Modelling and Simulation of Fast-Moving Ad-Hoc Networks (FANETs and VANETs)
T.S. Pradeep Kumar (Vellore Institute of Technology, India) and M. Alamelu (B.S. Abdur Rahman University, India)
Information Science Reference • © 2022 • 300pp • H/C (ISBN: 9781668436103) • US $240.00

Industrial Applications of the Internet of Things and 5G and 6G Networks
Hadi Alasti (Purdue University Fort Wayne, USA)
Information Science Reference • © 2022 • 300pp • H/C (ISBN: 9781799892663) • US $250.00

Achieving Full Realization and Mitigating the Challenges of the Internet of Things
Marcel Ohanga Odhiambo (Mangosuthu University of Technology, South Africa) and Weston Mwashita (Vaal University of Technology, South Africa)
Engineering Science Reference • © 2022 • 263pp • H/C (ISBN: 9781799893127) • US $240.00

Handbook of Research on Design, Deployment, Automation, and Testing Strategies for 6G Mobile Core Network
D. Satish Kumar (Nehru Institute of Engineering and Technology , India) G. Prabhakar (Thiagarajar College of Engineering, India) and R. Anand (Nehru Institute of Engineering and Technology, India)
Engineering Science Reference • © 2022 • 490pp • H/C (ISBN: 9781799896364) • US $360.00

Implementing Data Analytics and Architectures for Next Generation Wireless Communications
Chintan Bhatt (Charotar University of Science and Technology, India) Neeraj Kumar (Thapar University, India) Ali Kashif Bashir (Manchester Metropolitan University, UK) and Mamoun Alazab (Charles Darwin University, Australia)
Information Science Reference • © 2022 • 227pp • H/C (ISBN: 9781799869887) • US $240.00

Handbook of Research on Advances in Data Analytics and Complex Communication Networks
P. Venkata Krishna (Sri Padmavati Mahila University, India)
Engineering Science Reference • © 2022 • 275pp • H/C (ISBN: 9781799876854) • US $380.00

Emerging Materials and Advanced Designs for Wearable Antennas
Vinod Kumar Singh (SR Group of Institutions, India) Vikas Dubey (Bhilai Institute of Technology, India) Anurag Saxena (SR Group of Institutions, India) Ratnesh Tiwari (Bhilai Institute of Technology, India) and Himani Goyal Sharma (Department of Electrical Engineering, Poornima College of Engineering, Jaipur, India)
Engineering Science Reference • © 2021 • 210pp • H/C (ISBN: 9781799876113) • US $250.00

IGI Global
PUBLISHER of TIMELY KNOWLEDGE

701 East Chocolate Avenue, Hershey, PA 17033, USA
Tel: 717-533-8845 x100 • Fax: 717-533-8661
E-Mail: cust@igi-global.com • www.igi-global.com

Table of Contents

Detailed Table of Contents

As most of the countries in the world have moved on to 5G networks and the remaining are already in the process, everyone's attention is on the NextGen networks (6G and further). Security and privacy issues have always been the concern in the communication networks, where more technology and measures have been deployed to enhance user experience and protect critical information of potential applications. As the number of users is growing rapidly, the number of applications deployed has also grown. In this chapter, the authors discuss the upcoming risks and challenges associated with the NextGen networks with respect to security and privacy policies and the supporting steps to be taken to eliminate threats and attacks. The chapter also focuses on an artificial intelligence-enabled communication network and a feasibility study on the same.

Fifth-generation (5G) communication will be formally released very soon, with several more characteristics than fourth-generation communication. Between 2027 and 2030, a revolutionary wireless transmission paradigm, the 6G network, with full artificial intelligence integration, is projected to be implemented. Faster ultimate bearing ability, higher bandwidth speed, reduced dormancy, and enhanced quality of service (QoS) compared to 5G systems are among some of the key challenges that need to be resolved throughout 5G. This chapter discusses the variations of 6G data transmission and its system architectures. It gives a thorough look at the 6G network's specifications, including the wireless medium and various technologies. More specifically, this chapter emphasizes the different features of the planned 6G networks using ultra-dense heterogeneous network (UD HetNets) scenarios and implementations. This chapter explores the 5G/6G network's key performance indicators (KPI), difficulties, and research directions.

access for optimizing parameters of network and automatic network adjustment. In this chapter, the authors explore the various applications of AI/ML in sixth-generation and next generation networks and provide detailed explanation on how AI/ML may be implemented in 6G network effectively. Moreover, they provide possible issues while implementing AI-empowered 6G networks.

Ramya R., SRM Institute of Science and Technology, India
Ramamoorthy S., SRM Institute of Science and Technology, India

The internet of things (IoT) technology, which connects internet-connected devices, continues to extend the current internet by allowing communication and interactions across the physical and cyber worlds. The IoT generates big data that is characterized by its velocity in terms of time and place dependency, with a diversity of diverse modalities and fluctuating data quality, in addition to increased volume. The key to designing smart IoT applications is intelligent data processing and analysis. This chapter first describes the details of how cloud services and edge technology work and support the internet of things with many challenges and limitations in the overall internet services. Second, it describes the support of different machine learning algorithms (MLA) in the different fields of internet of things applications. Finally, there is a description of the future research scopes and open issues in the field of the internet of things with machine learning algorithms for further research work initiation.

Sivaramakrishnan S., Dayananda Sagar University, India
Rathish C. R., New Horizon College of Engineering, India
Lingasamy V., Sterlite Technologies Limited, India
Premalatha S., KSR Institute for Engineering and Technology, India

The communication technology has taken a long journey starting from telegraphy to high-definition video transmission and moving toward the high-density hologram. It is noticed that the data rate has raised linearly with reduced latency as the technology has progressed from first generation to the sixth generation. 5G technology has its limitation in meeting the challenge in creating the high-density hologram and future artificial intelligence application and that's one of the reasons for the focus in the 6G technology. The terahertz frequency that will be used in 6G communication will offer a wideband channel which can support multiple gigahertz channel. Terahertz communication in 6G can support data rate of terabits per second and the latency in 5G can be reduced to one tenth compared to the 6G. This chapter focuses on architecture, necessary hardware requirement, use cases, and deployment of machine learning/artificial intelligence algorithms for certain applications of interest.

Shikha Garg, Punjabi University, Patiala, India
Sonia Goyal, Punjabi University, Patiala, India
Abhinav Bhandari, Punjabi University, Patiala, India

Over the last decade, the demand for faster data rates has fueled the rapid growth of wireless communication systems that introduce the concept of new wireless technologies such as 6G or beyond 6G. The world is going through an ultimate transformation with the arrival of the intelligent information era. It integrates

Wireless cellular communication systems are witnessing tremendous improvement in QoS, new features, and new technologies from the past four decades. This leads to realization of advanced mobile phone systems (AMPS/1G), 2G, 3G, 4G, and 5G networks. Each generation, networks have faced potential challenges to cater large coverage, huge connectivity, high data rates, etc. Even though 5G has provided high standard infrastructures, the demand for a huge volume of data is exploding due to billions of IoT devices and multimedia services over mobile networks. Because of remarkable expansion in data traffic and improved digital makeover of industry and society, 6G is focused to provide ultra-broadband and ultra-low latency connectivity for wide range of digital devices and to address several 5G constraints. The implementation of 6G networks faces many technological and non-technical issues and challenges. Some anticipated challenges and potential solutions are discussed in this chapter.

Chapter 12
Defending IoT Security Infrastructure with the 6G Network, and Blockchain and Intelligent Learning Models for the Future Research Roadmap 177
Janani K., SRM Institute of Science and Technology, India
Ramamoorthy S., SRM Institute of Science and Technology, India

The battle for wireless networks' 6G has started. Many review articles expressed their views and critically evaluated the overall state of the art in order to give readers a clear picture of current advances. Most of the studies, though, are focused on broad views and a big-picture perspective, and do not address the real issues arising from internet of things (IoT) use cases. The authors give a sample of IoT use cases that are indicative of a wide range of IoT implementations, which is a novel approach in the review. The use cases chosen represent the most research-intensive industries that potentially profit from 6G and its enabling technologies. Healthcare, smart grid, transportation, and blockchain learning techniques are among these sectors. They also identified some of the actual hurdles and lessons encountered in putting these use cases into action. The evaluation focuses on the major requirements of the scenarios and how they intersect with the primary drivers for the next generation of wireless networks.

Chapter 13
Wireless Brain-Computer Interface (WBCI) and 6G Technology Security Issues, Safety Mechanisms 204
Saravana Kumar Ganesan, Karpagam College of Engineering, India
Parthasarathy V., Karpagam Academy of Higher Education, India
Arunachalam M., Sri Krishna College of Engineering and Technology, India
Viswa Bharathy A. M., GITAM School of Technology, GITAM University, India

A brain-computer interface (BCI) is a communication system that does not depend upon the brain's normal output pathways of peripheral nerves and muscles. Wireless brain-computer interface (WBCI) systems are a branch of BCI systems with an exclusive method to acquire the electrical activities of the brain, that is, electroencephalogram (EEG) using an effective non-invasive, implantable electrode scheme and employment of wireless communication schemes to transfer the acquired EEG for further processing. The five paramount security and privacy issues are authentication, access control, malicious behavior, encryption, and communication. With the appropriate implementation of the wireless BCI in the context of 6G technology, this chapter presents a comprehensive overview of WBCI and 6G technology and outlines artificial intelligence-based schemes' utility to address security and privacy issues arising out of 6G network deployment to the contexts revolving around WBCIs.

It is really a very challenging task for the private sector, public sector, and the government to make use of this blockchain technology for various applications in smart cities. The task is not going to be easy for the system for providing better and the best services for the people in the country. The solution for all the policies to be implemented is to be by making use of blockchain technology. This technology will create trust on the policymakers and will also help in reducing the cost of data storing along with security for the data stored. The technology provides security by decentralization process and will allow anyone to verify the records. As we all know, trust is going to be a big question when it comes to finance and any other kind of dealing, but blockchain will help to regain and maintain the factor of trust as it is a transparent system allowing the people to verify and check what is happening behind the policy. This banking concept is considered as it plays a vital role in every person's life. 6G will change remote correspondence by utilizing AI.

Foreword

Dear Readers,

Greetings in the name of Almighty. I would like to thank the editors of the book Dr A M Viswa Bharathy and Dr Basim Alhadidi for giving me this opportunity to write the foreword for this book. Its an immense pleasure for me to be a part of this book in one or the other way. I am Dr Hong Jer Lang, working at Taylors University, Malaysia, having a vast experience in research and teaching. I have published more than 100 papers in international journals and conferences, besides being reviewer and chair for more than 150 conferences. When the editors first approached me for the foreword, I was little reluctant as I had a couple of projects in my folio and was completely busy in that. But the editors sent me the copies of extracts of the book chapters, then I decided I have to write the foreword for this book.

Prof Viswa is known to me from 2017, where he approached me to evaluate his PhD thesis and that's how the journey started. The way his research findings were documented impressed me and of course his work. From then on, he would approach me for any kind of research and book proposals. An ardent researcher and reader of upcoming technologies, ha slot more in his kitty to offer.

I would like to thank and congratulate the editors for coming up with such a topic for a book, which would open gates to the research and development of future communication. This one-of-a-kind early research is a necessity of the today's dynamic world. As the world is slowly moving towards 5G communication, and we can see majority of the countries are struggling towards this transition, this book on 6G will help the governments, researchers and others in knowing the challenges and risks in moving towards the 6G and further. The title of the book is interesting enough the capture the attention of the technologists, researchers and communication experts.

The way the editors, called for the chapter proposals of the book was definitive and constructive. The editors have attempted to classify the issues in different areas and the challenges need to be addressed before proceeding towards establishing NextGen communication systems.

As Artificial Intelligence is taking up the world, editors have made the best effort to explore how Artificial Intelligence can be coupled with NextGen communication systems to provide service to the stake holders. This can be a small step in achieving the Sustainable Development Goals of the United Nations.

On the whole, the book is a go to material for every technology experts, scientists and service providers to get hold of the future and glimpse of the challenges it has for them to be tackled. My best wishes for the editors to come up with more books of this type and help technology enthusiasts like me to learn and explore more.

Hong Jer Lang
Taylor's University, Malaysia

Preface

As most of the countries in the world has moved onto 5g networks and remaining are already on the process, now everyone's attention is on the NextGen networks (6G and further). Security and privacy issues have always been the concern in the communication networks, where in more technology and measures have been deployed to enhance user experience and protect critical information of potential applications. As the users are growing rapidly, number of applications deployed has also become manifold.

The Ultra-dense Heterogeneous Networks (UD HetNets) are preparing to hit the future Mobile Communication Generation (6G) Smartphone market with extensive development of considering the increase in wireless devices, service complexity and requirements needed to access mobile services. The networking trademarks move towards the 6G communication by considering the main features like Smart usage, faster bit rate and less power consumption than 4G/5G communication. We are in need to develop the 6G UD HetNets with essential network elements, allocation of framework, performing parallel operation at Heterogeneous frequency bands and formative spectrum allocation. The main intention of framing 6G with UD HetNets small cells deployment scenario is to concern the network optimization with connecting zillions of wireless devices through Internet. The ultra-dense small cell deployment scenario benefits both the indoor and outdoor region mobile users.

AI has been playing major role in wireless communication. AI enabled wireless communication system has enhanced performance and diverse applications. The advent of computationally efficient hardware which supports the demands of AI has accelerated the use of AI in variety of applications. In the 6G networks, AI is used to provide intelligent solutions for better management and optimization of network resources. The major concerns of 6G network which includes resource management, interference management, spectrum sensing, channel estimation, signal detection, decision making, mobility management, privacy, and security and main more are better addressed using AI techniques.

In modern times, smart applications such as Augmented Reality (AR), Virtual Reality (VR), Mixed Reality (MR), Extended Reality (XR) and holography, massive robotics, Internet of Everything, autonomous driving can be accelerated by integration of AI in 6G. All these present and future applications have diverse requirements like handling large volumes of data, intelligence, low latency, massive and reliable connectivity. These growing demands can be satisfied with the aid of AI and 6G. Such being the importance of AI in 6G, it becomes indispensable to study some of the recently developed AI techniques and their possible use in 6G. Topics such as Deep unfolding, AI enabled Mobile Edge computing, Explainable AI and Federated learning are given special attention in this book

Machine learning (ML) allows users to create models that analyze data fast and offer results based on historical and real-time data. Healthcare providers can use machine learning to make better decisions about patient diagnoses and treatment alternatives, resulting in an overall improvement in healthcare services. ML is revolutionizing education and reshaping teaching, learning, and research. ML is being used by educators to identify difficult kids earlier and take steps to increase their achievement and retention. ML assists teachers in raising their game by providing them with all of the information they require. It also enables teachers to create content that is tailored to the needs of their students, ensuring personalized learning. Machine learning, a form of AI that employs algorithms and data to provide automated insights to healthcare practitioners, is a subset of AI that identifies patterns. ML is significant because it provides businesses with insight into customer behavior trends in the healthcare and education sectors, as well as assisting in the development of new goods. The application of machine learning in medicine will open up a slew of new options for improving patient care, providing real-time data analytics, and enabling continuous patient monitoring. Machine learning should be learned by clinicians and health informatics.

A Brain-Computer Interface (BCI) is a communication system that does not depend upon the brain's normal output pathways of peripheral nerves and muscles. With the appropriate implementation of the wireless BCI in the context of 6G technology, this chapter presents a comprehensive overview of WBCI and 6G technology and outlines Artificial Intelligence-based schemes' utility to address security and privacy issues arising out of 6G network deployment. The chapter focus is on Utilizing AI based 6G technology for EEG transmission, Optimization of EEG transmission through an appropriate antenna design scheme and CNN, Highlighting security threats in interlinking WBCI and 6G, Scheme for building a trustworthy 6G network with WBCI. The Researchers, Industries associated with Assistive Systems Development, Collaborative research organizations stemming out of deployment of 6G technology for BMI networks shall be benefitted from this book.

Many practical answers to real-world technical, scientific, and social challenges can be driven by blockchain and its implementations in 6G networks. Many security solutions are being proposed for the security and privacy of 6G networks using state-of-the-art techniques that include statistical methods, machine learning, deep learning techniques, theory-based information approaches, etc. Among all technologies, blockchain technology will be proven to be the most effective.

Major part of earth's surface is covered with water still a large portion of the underwater areas are unexplored despite the abundant advantages of the ocean. Marine operations such as environmental monitoring, underwater exploration, and scientific data collection depend primarily on underwater wireless communications. Due to the unique and harsh conditions that constitute underwater channels, underwater wireless communications remains relatively tough. Advancement in technology and invention of smart terminals has resulted in tremendous increase in Wireless data traffic. Conventional cellular technology cannot meet the rapidly increasing requirements. 6th generation (6G) mobile network will be able to meet the requirements and attain a high degree of efficiency.

The book focuses on AI enabled NextGen (6G and beyond) networks on the whole. We need to be aware of the challenges waiting in NextGen networks, so that proper steps could be taken to either minimize or eliminate issues on 6G and NextGen networks. Incorporating AI in NextGen networks for Privacy and Security policies will serve the purpose. The four key components of NextGen networks will be, Real-Time Intelligent Edge, Distributed AI, Intelligent Radio, and 3D Intercoms. The book will focus on these key areas identifying the challenges and risks in each, therefore increasing the scope and viability of future potential applications such as Multi-sensory XR, Autonomous Wireless Robotics and Wireless Brain – Computer communication. As per current research the technologies involved are AI-based software, Molecular and Quantum Communications, Blockchain, TeraHertz (THz) technology, and the Visible Light Communication (VLC) technology. The feasibility of all these technologies in connection with NextGen networks, the strategies for design and deployment of 6G and beyond, identification of challenges and risks will also form the core theme of the book.

THE CHALLENGES

The challenges and issues identified in the book are

Technical challenges and risks in deploying the 6G and NextGen networks
The political, social, and geographical challenges involved in realizing the 6G networks
Strategies for design and deployment of a more secured and user centered NextGen (6G and further) networks through artificial intelligence to enrich user experience
Ways to change security and privacy policies, to protect user data from illegal access
Increase the security of future potential applications such as multi-sensory XR, autonomous wireless robotics, and wireless brain-computer communication through artificial intelligence.

ORGANIZATION OF THE BOOK

Chapter 1

In this chapter, we discuss the upcoming risks and challenges associated with the NextGen (6G and further) networks with respect to security and privacy policies and the supporting steps to be taken to eliminate threats and attacks. The chapter also focuses on an Artificial Intelligence enabled communication networks and the feasibility study on the same.

Chapter 2

This chapter aims on (UD HetNets) - Network Architecture scenarios and Performance metrics, by considering 6G networks under Diverse Tier (D- Tier) UD HetNets with macro-cells and few low power node cells. The deployments of these cells will increase the cell density with better network performance than the existing conventional UD HetNets. The Perceptive aspects on 6G UD HetNets performance are analyzed by estimating various key performance metrics like mobility, peak data rate, Ergodic rate, spectral efficiency, channel access probability and throughput. This chapter reading helps the Engineering Students, Research scholars, faculties and academicians for enhancing their research interest in the field of HetNets and Mobile computing. This chapter was appraised in covering few tools and applications that are employed in 6G UD HetNets communication and about the potential barriers, research directions in 6G and Next Gen Networks with UD HetNets.

Chapter 3

The Internet of Everything (IOE) is the goal of the 6th generation network, which unites processing, navigation, and sensing with communications. 6G networks aims at creating a global, pervasive, and integrated network to enable device-to-device communication. An intelligent network that combines artificial intelligence and machine learning technology. Network has built-in security as it uses real-time dynamic analysis. Recently, there has been a lot of innovation in the field of marine and undersea communication. Purpose of this chapter is to provide an overview of traditional acoustic and optical communication technologies as well as the obstacles they may confront. It continues with the progress in marine communication with the support of the 6G network.

The article emphasises the increased capabilities of 6G-based undersea communications. The paper focuses on all unique hybrid techniques and their advantages over traditional techniques. The summarised approaches described shed light on the smart world that can be attained through the unification of all modalities of communication via 6G. Marine and underwater communication is one of the untapped fields that could be used for a variety of applications. The following advancements in the field of networking and communication can assist both the military and civilians.

Chapter 4

This chapter is intended to cover role of AI in different layers of 6G network and some important research challenges. The target audience of this content would be those who are interested to research and work on integrating AI in 6G and future wireless communication system. This chapter motivates researches to do some break through developments in the field of technology.

Chapter 5

This chapter has been written with a view to provide necessary solutions to tackle the challenges posed while implementing Sixth-Generation and next generation networks with Artificial Intelligence /Machine Learning.

This chapter is primarily intended to present the challenges that can be arrived when implementing 6G technology with Machine Learning. With respect to each type of issue, solutions as well as recommendations are given which can be found useful in the near future when implementing the technology. Our primary aim is to convey various solutions to the problems while combining Machine Learning with the 6G technology.

This chapter contains introduction section to machine learning and its types as well as various algorithms, then explained how Ml can be implemented with the 6G technology and discussed its applications in various fields. This chapter also provides the issues that can be emerged while implementing 6G technology with AI/ML along with the solutions and recommendations to overcome the issues which can be found useful to various researches.

Chapter 6

Machine learning algorithms play an important role in many fields of technology. This chapter explains the importance of machine learning algorithms in edge cloud services in the field of the internet of things. The internet of things is defined as a thing made by connecting large numbers of heterogeneous devices for getting services like storing and computing with the use of cloud and edge computing technology through the internet. There are many difficulties in handling and providing services for large numbers of heterogeneous devices. Many machine learning algorithms have become solutions for different applications, which are described in this chapter. The machine learning-based optimisation algorithms for efficient edge optimization with the help of machine learning trained models are discussed. There are some benefits discussed from the invention of the architecture for the formation of intelligent edge management with AI/ML. The discussion about why and how to run machine learning algorithms on edge devices provides the details of all popular edge devices and their supporting software. Finally, the discussion is about the future implementation of 5G with an edge in intelligence. The findings in this chapter help to choose the right algorithm for any application in the IoT and will help to choose further research work in the required field of machine learning algorithms.

Chapter 7

In 5G, computing and intelligence functions are add-on functionalities that are implemented incrementally on top of the connectivity. However, in 6G, these functionalities are implemented as an integral part along with the very high speed (in Tbps) and very low latency. In this context, this chapter covers various unique features of 6G over 5G along with application of Machine Learning in 6G.

Unlike traditional wireless standards that have been catering only voice and data, the 5G and 6G have revolutionized the telecommunication by employing various technologies like AI/ML driven RAN, Mobile Edge Computing (MEC), Network slicing etc… to support various possible use cases in eMBB, MMTC, and URLLC.

This chapter covers uniqueness and need for 6G over 5G along with specialized benefits, portrayed 6G empowering the modern applications, presented the premise of AI key innovations and summed up important use of AI in various situations. It also provides an overview 5G and 6G along with the comparison in terms of various aspects/key features. Also, this covers the transition of 6G from 5G, functional block diagrams, deployment scenarios, few possible use cases, and finally role of ML in 6G.

Chapter 8

This chapter presents an overview of blockchain technology and how it can secure 6G networks. This chapter will help envision and comprehend the requirements, problems, and case studies of numerous 6G-based applications. The focus is on understanding the significance of blockchain technology to secure the 6G networks, to provide insights into the development, research directions, issues, challenges, and ways to enhance the security of 6G networks using blockchain. In the future, 6G wireless networks will be widely used, with the integration of Artificial Intelligence and blockchain technology. 6G Architecture, Intelligent Mobility Management, Dynamic Resource Management, Decentralized Network Management, and Smart Contracts in 6G are just a few of the many areas of study currently being pursued.

This chapter provides a more comprehensive understanding of the 6G networks based on AI and distributed ledgers like blockchain. AI training necessitates a large amount of computing power, making 6G more expensive. Each miner in a blockchain has many computational resources, which might use for AI training, and the blockchain's smart contract services would lower the overall cost of smart applications. In a 6G network, AI and blockchain can manage dynamic resource management and mobility management.

Chapter 9

The proposed chapter intends to bring together key scholars and practitioners working on machine learning and networking strategies in this sector. The suggested book will undoubtedly be a useful resource for students working in any of these fields, with the goal of promoting, presenting, and discussing current research in the field. The proposed book will provide an overview of machine learning's role in

6G technologies in the healthcare and education sectors, as well as case studies, so that readers may understand the methodologies and their relevance.

Chapter 10

Data security is a prominent need in wireless communication in the era of sixth generation 6G. Although 6G network has many features when compared to 4G or 5G, the Confidentiality, Integrity and Availability (CIA) of the network needs to be secured. And the role of Unmanned Aerial Vehicles (UAV) in high-risk military and surveillance operations are increasing due to its deployment and coverage of areas restricted to human intervention. So, the era of 6G in the deployment of UAV will enhance the navigational and network operations but still it will be more prone to security attacks and the data security is not secured. Since 6G network is still evolving, the security features need to be focused to incorporate the new technologies and it should overcome the drawbacks in the 4G and 5G. This chapter discusses on how the use of blockchain can ensure the data security and privacy and allows only the authorized UAV nodes to participate in the network. And the support of blockchain in UAV swarm in 6G can make the military and surveillance operations more secure and resilient. The blockchain can make the 6G aided UAV swarm network to prevent from security attacks such as denial-of-service attack and signal jamming attacks. This chapter mainly focuses on the support of 6G in UAV high profile military surveillance operations and how the data security and privacy can be provided by incorporating blockchain technology in 6G and how the blockchain supported 6G network can reduce the security attacks in UAV swarm.

Chapter 11

6G mobile communication network enables Internet of Everything by 2030. Even though 5G has provided high standard infrastructures, the demand for huge volume of data is exploding due to billions of IOT devices and multimedia services over mobile networks. Because of remarkable expansion in data traffic and improved digital makeover of industry and society, 6G is focused to provide ultra broadband and ultra low latency connectivity for wide range of digital devices and to address several 5G constraints. Hence researchers started to plan, vision and gather requirements of 6G. It promises connection of every smart device to internet, the device can rage from smart phones to intelligent vehicles. 6g aims to provide sophisticated high Quality of Service (QoS) such as holographic communication, augmented reality/ virtual reality and many more. The Quality of Experience (QoE) provides rich experience to users such as high speed and high quality. Research on 6G has already started, which requires understanding of issues and challenges associated with 6G. Various alternatives are explored by researchers to achieve desired quality parameters. Starting from devices to softwarization huge number of challenges are faced by researchers. Hence some of these technical and non technical challenges and issues are discussed in this chapter. Important technological and non technical issues and challenges and potential solutions are discussed in this chapter.

Chapter 12

Widely deployed 5G across several countries is now moving beyond 5G. This chapter discusses next generation 6G network with particular reference to AI and ML. The current position of 6G research is discussed. AI will play a critical role in optimizing future 6G networks by addressing the challenges that cannot be presented easily using closed-form models. This paper also presents a comprehensive study of 6G trends and technologies.

The rise of modern AI based on Machine Learning has significantly improved Predictive and Prescriptive Analytics. Deep learning need to be explored for the design and optimization of 6G architecture which can be deployed in smart cities, smart grid, autonomous vehicles, and industrial automation. We identify significant issues to towards an intelligent, efficient and secure 6G system. Networks needs to develop fully-user-centric architecture without much supervision from centralized controllers. 6G connects humans and machines in an unprecedented manner more than IoT started during 5G. Wireless applications in 6G needs higher speeds, greater security, new information technology laws to govern and flexibility to change. Also the challenges of 6G race needs to be properly used for benefit to the mankind. Thus the relevance of 6G is highly significant in country's development.

Chapter 13

This chapter delves into the concerns setting the groundwork for the 6G network's development. This chapter also looks at four major aspects of the 6G network in order to reveal security concerns about future technologies. The inquiry comes to a close with a consideration of the possible uses for WBCI that the 6G network will enable. This chapter provides a complete overview of WBCI and 6G technologies, as well as the applicability of AI-based strategies in addressing security and privacy challenges originating from the deployment of 6G networks in WBCI contexts

Chapter 14

6G-Transformation of Smart cities with Blockchain and AI technology will create trust on the policy makers and will also helps in reducing the cost of data storing along with security for the data stored. The technology provides security by decentralization process and will allow anyone to verify the records. The trust is going to be a big question when it comes to finance and any other kind of dealing, but blockchain will help to regain and maintain the factor of trust as it is transparent system allowing the people to verify and check what is happening behind the policy. The proposed method works in accordance to the procedure adapted right from the inception phase of capturing the data till the process of providing information as and when required by making using of the nitizen identification number which is generated using luhn algorithm followed by adding up a prefix in terms of name of the state. The process of maintaining the data for specified number of years is also maintained including the process of backup mechanism. AI will be crucial to effectively implementing the key technologies and intelligently optimize the 6G performance under vastly dynamic conditions and environments.

A. M. Viswa Bharathy
GITAM University, Bengaluru, India

Basim Alhadidi
Al-Balqa Applied University, Jordan

Acknowledgment

We would like to thank Dr P Karthikeyan, Post-Doctoral Researcher, National Chung Cheng University, Taiwan, for guiding us in the initial stages of the book with his expertise and willing to help us anytime.

Chapter 1
Security and Privacy Policies in Artificially Intelligent 6G Networks:
Risks and Challenges

A. M. Viswa Bharathy
GITAM University, Bengaluru, India

P. Karthikeyan
https://orcid.org/0000-0001-8977-5520
National Chung Cheng University, Taiwan

ABSTRACT

As most of the countries in the world have moved on to 5G networks and the remaining are already in the process, everyone's attention is on the NextGen networks (6G and further). Security and privacy issues have always been the concern in the communication networks, where more technology and measures have been deployed to enhance user experience and protect critical information of potential applications. As the number of users is growing rapidly, the number of applications deployed has also grown. In this chapter, the authors discuss the upcoming risks and challenges associated with the NextGen networks with respect to security and privacy policies and the supporting steps to be taken to eliminate threats and attacks. The chapter also focuses on an artificial intelligence-enabled communication network and a feasibility study on the same.

INTRODUCTION

5G technology is still not completely operational, but its shortcomings imply we must begin studying 6G networks immediately. The first ever 6G conference was held in Finland in March 2019, and Finland's leading communications specialists worked with the world's first 6G white paper. The foundation of study in the area of 6G networking was set with this one act. The proliferation of their activity has occurred

DOI: 10.4018/978-1-6684-3804-6.ch001

with the announcement of new initiatives in the 6G network from governments and organisations. For instance, the government of UK is announcing its desire to invest in technologies and technologies for the use of 6G technology and beyond, while the Finnish Academy announced the start of the "6 Genesis" research project, which focuses on basic research.

What's a 6G and What are the Differences Compared to 5G Networks?

Until yet, the 6G network lacks guidelines, common features, or standards, but there are plenty of opportunities. The enhancement of 5G technology rather than simply a more advanced 5G network, say others. This is an example of poor network coverage: The 5G network only covers the ground level. Instead, the system should provide comprehensive coverage of the ocean floor.

It is believed that the 6G network will be able to handle much more intelligent algorithms. Many academics see the 6G network as a "AI-empowered" infrastructure, implying AI would be both its primary engine and defining characteristic. The 5G network should not be the only one to employ AI in its infrastructure. A thorough integration of existing AI technologies and networking functionalities shall comprise the 6G network. As concerns about network security and privacy have grown in interest in recent times, contingency plan should be included in the design as an essential component. To this end, in the next article, we will address the many network components that may form the basis of a 6G network, together with the problems of privacy and security that each new technology raises.

The four critical features of a 6G network are real-time smart edges, distributed AI, smart radio, as well as 3D intercommunication systems. We selected these four points because they touch on the most important portion of the 6G research done so far. Additionally, they face the greatest security and privacy risks. The study involves a variety of current technologies, including AI-based software, molecular communication, quantum communication, blockchain, THz technology, and VLC.

The following technologies have tremendous potential for 6G network usage, including multi-sensory XR applications, linked robots, autonomous intelligent systems, mobile brain-computer interfaces, and blockchain and decentralized consensus technologies. The distributed AI, real-time intelligent edge, and intelligent radio domains are all made up of overlapping technologies.

AI, since it is a unique convergence of all three spaces, is the result of the assumption that the next generation of networks will be AI-powered. Security and privacy concerns are noted in the diagram, and the following list identifies the top five: Most of the elements in this illustration are at risk of unauthorised access, authentication problems, and malevolent conduct. Some technologies, however, are more vulnerable to specific concerns. One example of the device's shortcomings is its lack of ability to defend against malicious activity and data transfer. Intelligent radio benefits from the cell based communication and THz technology.

One big drawback of THz technology is authentication security, because it's known to be vulnerable to attack. Decentralized AI and IR (Intelligent Radio) have both intersected with the blockchain systems and quantum communication. Two of the major security and privacy issues in this area are the systems used for authentication, data access, and encryption. Applications built for the 6G standard also have certain risks. VLC and AI, which may enable both malevolent behavior and data encryption/transmission, are standard features of robots and automated systems.

Multi-sensory XR applications rely on molecular communication technology, THz technology, and quantum communication technology, thus they are vulnerable to assaults, data transfer exposure, and criminal behavior. The XR application's multi-sensory abilities can be adapted to use for wireless brain-computer connections, but that creates its own set of security and privacy concerns. Maliciousness and encryption are two of the biggest problems. Because 6G technology will soon be extinct, the block chain technology and decentralized consensus technologies (mostly based on blockchain) will become obsolete. Though they are somewhat secure, these tools may still be vulnerable to malevolent activity.

For these new places, five kinds of security and privacy vulnerabilities may be expected: authentication, access control, malevolent behavior, encryption, and data transfer, and we'll refer to them as S1, S2, S3, S4, and S5. Sections III, IV, and V are where the methods to minimize the risks are discussed, as well as how the risks impact the 6G network. This survey's main contributions are summarized as follows:

1. 6G network important issues have been identified and presented with their respective security and privacy concerns.
2. In addition, a thorough description of the user privacy and confidentiality problems that arise from these emerging essential technologies is provided.
3. The new privacy and security concerns that arise from 6G technology are outlined and addressed.

The chapter's concluding section is as follows: Section 2 offers a summary of the historical evolution of cellular phone networks, from 1G to 5G. Section 3 of our report goes on to discuss the 4 key aspects of the 6G technology. Section 4 is devoted to the core competencies of the 6G network, while Section 5 explores various 6G applications. The three sections after this one likewise provides an overview of the privacy and security vulnerability issues in the subjects they cover. Section 6 is where the paper ends.

THE EVOLUTION OF COMMUNICATION NETWORKS

This part gives a summary of the development of cellular network security and privacy from first-generation (1G) to fifth-generation (5G) as per technological standards.

1G

Voice service was the design intent of the 1G network, which was launched in the 1980s. The information is transferred by means of analogue signals, and it has no defined wireless standard. All of these issues make transfers between users very difficult. The mobile service really aren't protected, which means neither communications nor conversations can be private or safe. This means that the whole internet and its clients are facing many difficulties when it comes to security and privacy, such as duplication, spying, and clandestine access (Chandra et al., 2011; Union et al., 2022; Gupta et al., 2015)

2G

The 2nd Generation communication network is digital and relies on a pair of technologies: TDMA and CDMA, which allow both voice and brief messaging. GSM (Arunabha et al., 2010) is the most significant and extensively used 2G mobile communication standard. The GSM is intended to provide the system with the same level of security as a PSTN.

In addition to anonymity, authentication, signaling protection, and user data protection, the service provides privacy. The use of ephemeral identifiers makes it impossible to trace a user's actual identity, thereby achieving anonymity. After being powered on, a temporary ID is generated, however the device's true identification must be utilized at first. Network operators typically employ authentication to identify their users.

The authentication method is called "challenge and response" and is an encryption-based strategy. Furthermore, user data protection and encryption utilize the SIM, which also is essential to the keys used in the encryption. Two of the most popular ways to keep the personal information of users secure are radio path encryption and Temporary Mobile Subscriber Identity (TMSI) (Gindraux, 2002). Despite the considerable advances in privacy and security over 1G and 2G, both generations still face significant vulnerabilities.

The inability of users to verify the network is one of the key issues in network security, since the user can only be authenticated, not the network. Base stations may mask themselves as lawful network members and therefore dupe users, while pilfering information from them (Cattaneo et al., 2013). On top of that, the encryption isn't end-to-end. Adversaries have a target-rich environment since only a portion of the wireless medium is protected, and no encryption is provided for the fixed network (Toorani & Beheshti, 2008). The TMSI and radio path encryption have shortcomings when it comes to privacy and are vulnerable to many types of threats, such as eavesdropping.

3G

High-speed internet access and data transfer were developed by the 3G network, which appeared in 2000. Yet, increased capacity would allow for new services, like surfing the web, streaming television, and video on demand, which are not feasible with 1G and 2G networks (Sharma, 2013). The 3G system's security is based on the 2G network. While parts of GSM and other 2G components were vital, new security measures also had to be put in place. In addition to the 2G's security vulnerabilities, the 3G addresses 2-way authentication system, AKA, and other elements.

The 3rd Generation Partnership Project (3GPP) also offers a comprehensive security system that encompasses two aspects: the air interface security that helps protect customers and the signaling information sent via wireless links; and user network authentication to provide physical reliability for users and the network. To address privacy concerns, 3GPP included a number of measures for protecting the identities, locations, and footprints of 3G users. But, although 3G networks remain susceptible to attacks like IP traffic and cryptographic keys. In addition, the terminal equipment's radio interface with the service network offers the opportunity for a set of assaults. The types of threats linked to wireless interface assaults are as follows:

A lack of security may give rise to a number of things: 1) the illegal access to information; 2) a breach of data integrity; 3) a denial-of-service attack (DOS); and 4) illegal access to services.

Private or classified data is often compromised by to assaults, such as AKA error codes, which are intended to erase user identities.

4G

The downstream capacity was up to 1 Gbps and the upstream was up to 500 Mbps when LTE networks were launched in 2009. These networks can serve sophisticated applications like DVB, HDTV, and video chat, because to their higher spectrum efficiency and decreased latency. The LTE combines many current and new technologies, including CoMR, MIMO, or OFDM (Varshney, 2012).

Wireless core networks (WCN), wireless access networks (WAN), and smart mobile terminals (SMT) are all examples of 4G systems. In short, the major sources of security risks are as follows: network security flaws, hacking, eavesdropping, data manipulation, insertion, or deletion, as well as hardwired and wireless entity authentication problems (Cao et al., 2013). Users are more in contact with mobile devices than in the past, thus the 4G network is much more susceptible to privacy and security issues. 4G wireless interactions are more complicated, and dangers are more prevalent since 4G devices are used by all wireless networking standard protocols and support different wireless applications. An even greater threat is the rise in the amount of malware capable of infecting mobile devices. A typical example of cybercrime is viruses, hardware tampering, and OS vulnerabilities, etc. Also, because of issues with key management protocols, WiMAX has vulnerabilities at the MAC layer, such as denial-of-service (DoS), spying, and impersonation attack. According to (Seddigh et al., 2010), both LTE and 3G networks are susceptible to DoS and DDoS attacks, file integrity assaults, illicit usage of equipment and users, and MAC layer location monitoring.

5G

As the 5G network approaches, we may expect higher speeds, more comprehensive applications, and more secure designs (Gupta et al., 2015). One of the major benefits of 5G networks is that they will be able to connect more devices and offer high-quality services to all of them at the same time. Other devices like Internet of Things (IoT) technology may also access the network, so smartphones aren't the only ones that will function.

Five-generation (5G) wireless networks face a number of security and privacy problems, which may be split into three categories based on the network architectural style: the access networks, backhaul networks, and the network infrastructure. The variety of endpoints and access methods in access networks creates additional security problems since switching between various access technologies raises the threat of an attack. There is connection between the backhaul network's base station and its core network through backhaul communication, which may be carried out by wireless channels, microwaves, and even satellite links (Jaber et al., 2016). These networks have less security and privacy risks than access networks since there are no links between the devices. In addition, the GPRS Tunnel Protocol (GTP) may offer certain extra security assurances by adjusting access network components such as EUTRAN Node B (eNB) or the Mobility Management Entity (MME). NFV and Software-Defined Networking (SDN) methods are used to move the backhaul network into the data plane, transferring security risks to the core network as well (Prados-Garzon et al., 2016).

The core network in 5G networks is made up of several functionalities (Kim et al., 2017). Network functions virtualization (NFV), software-defined networking (SDN), and cloud computing all increase the dynamic nature of networks, opening them up to new vulnerabilities. Massive devices and services, for example, may exceed the signalling load, increasing the likelihood of DoS or resource attacks (Harel & Babbage, 2016). So far, two approaches for dealing with signal overloads have been established (Bisson & Waryet, 2017). The first is to enable communication for a large number of devices by using minimal key exchange protocol methods. Protocols that enable equipment to be clustered via different group-based AKA protocols are used in the second method of grouping. The new methods utilised to enhance the 5G network's performance also introduce security vulnerabilities (Ahmad et al., 2019) [16]. Massive MIMOs, for example, may be used to mask both passive and active listening. SDN deployment of OpenFlow raises the danger level from malicious apps or activities because of its open nature. NFV also presents security concerns when moving a system or functionality across resources (Pattaranantakul et al., 2018). Due to the wide range of business models and application possibilities in 5G networks, there are indeed new privacy concerns to be addressed. Because of the platform's openness, critical data about a user may go from being in a locked state to being in an open one quickly and often. When the contact status switches from offline to online, the danger of leakage increases dramatically.

As a result, the privacy concerns associated with 5G becomes a pressing issue that will need attention and resolution in the next years. Nevertheless, because to recent developments in machine learning and statistical, existing privacy protection measures are now more effective than ever.

6G

In the physical layer, channel coding and estimation have already adopted AI as its backbone, while in the MAC layer multiple access and different network applications have already used AI (Huang et al., 2019). The use of AI in 5G networks is restricted by the limitations of the conventional architecture that was accessible during the early phases of its development, and as a result, AI applications are rare. As a result, distributed AI and intelligent radio aren't supported since they're purely AI-based. Although 5G networks have already incorporated the real-time intelligent edge, such as car networks, emergency situations cannot be handled in "real-time" because of latency problems. With the advent of 6G, this may be possible, however.

A good example of this is 6G networks, which have radio latency of just 0.1 ms, a tenth the speed of 5G networks. Furthermore, 5G coverage is currently limited to the surface of the earth; connections in space or beneath the sea are not yet feasible at certain levels of 3D intercoms. We'll go through how these four categories may progress in 6G networks now. On top of everything else, we go through potential security and privacy concerns. Table 1 provides an overview of the part.

6G Prime

Real time Intelligent Edge networks can't be deployed in 5G since managing the entire network needs low transmission delay and real strategy, and our present technology can't do it. Especially for vehicle networks. While the 5G network currently supports autonomous driving and certain AI-based cloud services already utilise it, networking objects with self-awareness, self-adaptation, or forecasting could not be enabled (Kibria et al., 2018). So a new network with these capabilities is needed. Intuitive AI-powered services should be supported by a 6G network Real-time intelligent edge may enable autonomous driving in unfamiliar environments (Huang et al., 2019). Also, emerging technologies like VLC will be helpful in vehicular communications (Tariq, et al., 2019).

Unlike 5G networks, which comply with the Internet of Things (IoT), 6G networks must comply with the Internet of Everything (IoE) (IoE). 6G is designed to be a large, decentralised system of creating intelligent choices at many levels. Decentralized AI will be required to meet various criteria in IoE since everything is interconnected through the internet. Edge devices should have access to and control over a portion of a training data set that is spread unevenly among most of the devices.

The gadgets transmitter and receiver techniques used to be developed jointly in earlier network versions. Hardware capabilities of networks like decoder processing and the total no. of antennas have remained relatively unchanged as a result. Nevertheless, new circuit and antenna advancements have made it possible to decouple hardware from transceiver algorithms, meaning intelligent radio may function inside a single framework. Instead of configuring and updating themselves based on equipment information, the transceiver algorithms may use this paradigm.

3D Intercoms

The ability to analyse, design, and optimise networks in 3D will be required in 6G networks, up from 2D in earlier generations (Dang et al., 2020). hence it is necessary to have 6G network's ability to enable satellite communications in 3D space, unmanned aerial vehicles (UAVs) and communications beneath the sea This may be provided via a 3D intercom. service based on the customer's location and the time available for instance, in distant areas, deployment and reaction are required when an emergency arises than 4G or 5G networks, the use of UAV networks is more cost efficient. because of the generational infrastructure that is fixed It's immovable. Including altitude and its corresponding in order to obtain more degrees of freedom, new approaches to network optimization are required planning and control of transportation, route selection, and resource allocation Context of 3D-intercom.

SECURITY AND PRIVACY ISSUES OF 6G

AI

AI is generally regarded as one of the essential components of the future network architecture for 6G networks, compared to all other anticipated technologies. Putting it mildly would be an understatement of There has been a lot of discussion about artificial intelligence in the area of network. With that focus comes a growing lot of new security mechanisms. and there are new concerns about privacy (Zhu & Philip, 2019). Despite the fact that 5G will include AI, network is purportedly run at remote locations where large amounts of data are large quantities of training examples and powerful yet private computer centers are essential As AI becomes more widely accessible, it will play a larger role in the 6G network. For the second time, artificial intelligence (AI) technologies may be split into two categories: architectural they put up a show Devices such as these are found in the physical layers (Zhang et al., 2019) as well as network infrastructure, such as data connections, and the layers of computers, NFV (Network Function Virtualization), SDN (Software-Defined Networks), and other emerging technologies and computing in the cloud, on the edge, in the fog, etc. We have a conversation about each one in turn below.

DNN (Deep Neural Networks), K-Means, and supervised/unsupervised learning were proposed by (Zhang et al., 2019) as AI technologies that might be used at various physical levels. As well as the fact that they the physical layers are made to function better by optimizing connection, but they can also anticipate traffic patterns and improve. Nawaz et al. (2019) pointed out that quantum cryptography and machine learning are also important topics. may be used to safeguard 6G communication connections from being compromised networks.

Loven et al. (2019) suggested that AI may enhance edge security through security measures and quite well controls in terms of network design. This idea, on the other hand, needs further research. Additionally, AI is a component of the different layers and the network architecture. it's also helpful for things like large data processing and decentralized artificial intelligence. administration as well as the optimization of the network. Dang et al. (2020) claims that AI It would aid in the detection of network abnormalities and the provision of proactive warning methods to enhance the 6G security Further. Note that the edge of a 6G network may benefit from distributed and federated AI. Network performance is improved since devices do not have to share data. security. Certain ML algorithms that use correlation may see an improvement in accuracy (Zhang et al., 2020). Zhu et al. (2017) described various algorithms based on privacy breaches and the use of comprehensive privacy to address some of 6G's problems Concerns about confidentiality.

Molecular Communication

Living things with nanoscale structures communicate through MC (Molecular Communication) is a common occurrence (Akan et al., 2016). In the past, nanotechnology, bioengineering, and synthetic biology all advanced rapidly. By the end of the decade, micro- and nanoscale gadgets will be commonplace. In addition, the production and use of electricity consumes MC signal transmission is quite limited. Although despite the fact that this biological phenomenon has been researched for quite some time, until recently has, for the most part, been a study subject in the area of communication decade. There is a lot of promise in the field of MC technology 6G technology is the latest in wireless connectivity. It is, however, a multidisciplinary effort. it's a new method that's currently being developed. Molecular biology's fundamental concept Communication is the transmission of information through biological signals. MMC (Mobile Molecular Communication) was shown by Nakano et al. (2019), a procedure through which the sender, the recipient, and any other nodes connected to them may communicate move about and talk to each other while doing so.

Quantum Communication

6G networks also have tremendous practical applicability for QC (Quantum Communication), which uses photons instead of electrons. It has the potential to substantially improve data security and dependability, which is one of its major advantages. transmission. in the event that a foe listens in, measurements or copies The quantum status will be impacted by anything involving QC. As a result, the receiver cannot remain in the dark about the interfering with the work of others Assuming QC works as intended, it may provide perfect safety and, if certain advances are made, a great degree of suitability for transmission over a distant place It provides many brand-new options. It raises the bar for connections beyond what is possible with conventional methods (Gyongyosi et al., 2018) is out of reach for our systems.

Blockchain

Access control, authentication, and communication procedures are all intertwined in terms of blockchain security and privacy. Blockchain radio access network design is discussed by Ling et al. (2019) to secure such that trustless parties may efficiently control access to the network and authentication whereas Kiyomoto et al. (2017) offer a new conceptual framework for network entities design of a mobile service architecture based on the authorisation The blockchain is a new technology. In addition, Kotobi et al. (2018) presented a technique for enhancing media access methods' security using the block chain technology and cognitive radio to get access to the licenced spectrums that are currently not in use. Aside from that, despite 6G's decentralised design network implies that records may only be altered if there are Hacker had control over 51% of nodes, indicating that There isn't a dependable third party in charge of safeguarding the data that's secure enough. When a data breach occurs, storing and managing data is necessary and pointing out that validating a cryptographic key requires hashing capabilities.

TeraHertz Technology (THz) and Visible Light Communication (VLC)

Despite mm-wave bands were extensively employed in 5G networks, they are insufficient for 6G due to high data rate. Regardless, the RF band is almost filled and cannot be utilised for future technologies (Elmeadawy & Shubair, 2019). These considerations fueled terahertz technology development. The 0.1–10 Terahertz band has larger available spectrum in comparison to the mm-wave range. It also uses electromagnetic and optical waves (Zhang et al., 2019). THz has several advantages, according to Huang et al. First, THz communication technology may enable data rates up to 100 Gbps. Second, due to short length in pulse and slender beam utilised in THz transmission limits external hacking, increasing communications security. Third, THz waves have a low attenuation through specific materials, allowing for a wide range of applications. THz communication may also be extremely directed, reducing intercell influence (Zhang et al., 2019).

Unlike RF, which has severe delay and congestion, VLC provides greater bandwidths (Khan, 2017). The SSL (Solid-state lighting) advanced VLC technology. Chen et al. (2018) developed a VLC system called LiFi that enables multiple access and might possibly achieve superior services to many mobile users.

However, there are significant flaws in the VLC technology. For instance, intense natural light may impact broadcasts, therefore use VLC mostly inside. Some of VLC's security and privacy vulnerabilities include hostile conduct or communication. Pathak et al. (2017) emphasised that an attacker should be in alignment of the target to assault an ongoing VLC operation. Clearly, this makes the attacker simpler to identify. Ucar et al. (2016) proposed a SecVLC protocol for securing data transfer in a vehicle network.

CHALLENGES IN IMPLEMENTING 6G

Multi-Sensory XR Applications

5G's data transfer rate and reduced dealy have already enhanced VR/AR users' experiences. However, there are several issues with limiting VR on 5G networks that need to be addressed in 6G networks. For instance, cloud VR/AR services now provide certain interactive elements, but latency is an issue, and the ambiguity that results causes extra issues. Cloud-based VR/AR makes technology more transportable and accessible, however 5G bandwidths need pictures to be reduced, therefore delivering large amounts of lossless photos or films in realtime will have to wait until 6G. The realistic VR/AR performance will be enhanced with 6G networks. To gather data and give input to consumers, many sensors will be employed

Autonomous Robotics Systems

Although 5G network restrictions have prevented complete deployment of autonomous drone systems, 6G networks may unleash their full capabilities. But these systems are also attacked. The SDN controllers used to monitor and control UAV systems are obvious targets, say Li et al. A Tiro opponent might deploy a WiFi-based attack, according to Hooper et al. Fotouhi et al. (2019) emphasise that eavesdropping, hijacking, spoofing, and denial-of-service attacks may occur in autonomous drone systems. As a result, security protection techniques are required. For autonomous drone systems, Challita et al. (2019) offer an AI-based method.

Self-driving car security and privacy concerns include a wide range of challenges, including difficulties associated with system confidentiality, location privacy, and power utilisation systems that are particularly susceptible.

To safeguard the safety and privacy of its passengers, Xu et al. (2019) suggest an Efficient and Privacy-Preserving Truth Discovery (EPTD) technique for self-driving cars. Self-driving car protocol by Ni et al. (2019) mentions two-factor authentication with reciprocal tracking for lowering auto theft risk and also for mitigating disclosure of sensitive information. Using Ding et al. (2017) unique fuel-efficient route planning paradigm, autonomous driving systems may be able to reduce their energy usage.

Wireless Brain-Computer Interactions

In spite of its age, Brain-Computer Interaction (BCI) is a well-developed technology. With wireless BCI, the main goal is to establish a link between the brain and a machine or equipment. There are two possible locations for the device: within the body (like the visual cortex) or outside the body (such as an artificial limb). There are generally 4 processes involved in this process: acquiring the signal, extracting the features, translating the features, and providing feedback. Malicious conduct and encryption are the main security and privacy concerns highlighted by radio BCI systems. A major difficulty in wireless BCI is data security, according to Mccullagh et al. (2014), however certain hacking apps might be built to get access to super delicate neurologic data, according to Ramadan et al. (2017). One solution developed by Svogor et al. (2012) prevents reply attacks by making users reach a specific mental state before entering their secret key.

Blockchain and Distributed Ledger

When it comes to 6G network security, the introduction of blockchain technologies is expected to make a big difference since it makes data available to all parties (Ahmad et al., 2019). Some of these problems, on the other hand, are nonetheless very nefarious. Attacks on most vulnerable systems, leaking of transaction privacy, and double-spending are all mentioned by Nguyen et al. (2020). To address these challenges in 6G networks, they offer many alternatives related to blockchain, including incentive methods and cryptosystems, among others. Some forms of private block chain technology, have minimal security, according to Dai et al. (2019), while other types of blockchains, such as consortia blockchains, have high levels of security and may still be utilised for safe resource transactions. We suggest interested people to study (Li et al., 2020), which has a full analysis of blockchain security.

REFERENCES

Ahmad, I., Shahabuddin, S., Kumar, T., Okwuibe, J., Gurtov, A., & Ylianttila, M. (2019). Security for 5g and beyond. *IEEE Communications Surveys and Tutorials*, *21*(4), 3682–3722. doi:10.1109/COMST.2019.2916180

Akan, O. B., Ramezani, H., Khan, T., Abbasi, N. A., & Kuscu, M. (2016). Fundamentals of molecular information and communication science. *Proceedings of the IEEE, 105*(2), 306–318. doi:10.1109/JPROC.2016.2537306

Bisson & Waryet. (2017). *5GPPP phase 1 security landscape*. 5G PPPSecurity Group White Paper.

Cao, J., Ma, M., Li, H., Zhang, Y., & Luo, Z. (2013). A survey on security aspects for lte and lte-a networks. *IEEE Communications Surveys and Tutorials*, *16*(1), 283–302. doi:10.1109/SURV.2013.041513.00174

Cattaneo, G., De Maio, G., & Petrillo, U. F. (2013). Security issues and attacks on the gsm standard: A review. *J. UCS*, *19*(16), 2437–2452.

Challita, Ferdowsi, Chen, & Saad. (2019). Machine learning for wireless connectivity and security of cellular-connected UAVs. *IEEE Wireless Communications*, *26*(1), 28–35.

Chandra, P., Bensky, D., Bradley, T., Hurley, C., Rackley, S. A., Rittinghouse, J., Ransome, J. F., Cism, C., Stapko, T., & Stefanek, G. L. (2011). *Wireless Security: Know it All.* Newnes.

Chen, C., Bian, R., & Haas, H. (2018). Omnidirectional transmitter and receiver design for wireless infrared uplink transmission in lifi. *2018 IEEE International Conference on Communications Workshops (ICC Workshops),* 1–6. 10.1109/ICCW.2018.8403727

Dai, Xu, & Maharjan, Chen, He, & Zhang. (2019). Blockchain and deep reinforcement learning empowered intelligent 5g beyond. *IEEE Network*, *33*(3), 10–17.

Dang, S., Amin, O., Shihada, B., & Alouini, M.-S. (2020). What should 6g be? *Nature Electronics*, *3*(1), 20–29. doi:10.103841928-019-0355-6

Ding, C., & Zhang, G., Yu, & Wang. (2017). Greenplanner: planning personalized fuel-efficient driving routes using multi-sourced urban data. *2017 IEEE International Conference on Pervasive Computing and Communications (PerCom),* 207–216.

Elmeadawy, S., & Shubair, R. M. (2019). 6g wireless communications: future technologies and research challenges. *2019 International Conference on Electrical and Computing Technologies and Applications (ICECTA)*, 1–5. 10.1109/ICECTA48151.2019.8959607

Ferraro, P., King, C., & Shorten, R. (2018). Distributed ledger technology for smart cities, the sharing economy, and social compliance. *IEEE Access: Practical Innovations, Open Solutions*, *6*, 62728–62746. doi:10.1109/ACCESS.2018.2876766

Fotouhi, A., Qiang, H., Ding, M., Hassan, M., Giordano, L. G., Garcia-Rodriguez, A., & Yuan, J. (2019). Survey on uav cellular communications: Practical aspects, standardization advancements, regulation, and security challenges. *IEEE Communications Surveys and Tutorials*, *21*(4), 3417–3442. doi:10.1109/COMST.2019.2906228

Gindraux, S. (2002). From 2g to 3g: a guide to mobile security. *Third International Conference on 3G Mobile Communication Technologies*, 308–311.

Gupta, L., Jain, R., & Vaszkun, G. (2015). Survey of important issues in uav communication networks. *IEEE Communications Surveys and Tutorials*, *18*(2), 1123–1152. doi:10.1109/COMST.2015.2495297

Gyongyosi, L., Imre, S., & Nguyen, H. V. (2018). A survey on quantum channel capacities. *IEEE Communications Surveys and Tutorials*, *20*(2), 1149–1205. doi:10.1109/COMST.2017.2786748

Harel & Babbage. (2016). *5g security recommendations package #1*. NGMN Alliance.

Huang, T., Yang, W., Wu, J., Ma, J., Zhang, X., & Zhang, D. (2019). A survey on green 6g network: Architecture and technologies. *IEEE Access: Practical Innovations, Open Solutions*, *7*, 175758–175768. doi:10.1109/ACCESS.2019.2957648

Jaber, M., Imran, M. A., Tafazolli, R., & Tukmanov, A. (2016). 5g backhaul challenges and emerging research directions: A survey. *IEEE Access: Practical Innovations, Open Solutions*, *4*, 1743–1766. doi:10.1109/ACCESS.2016.2556011

Khan, L. U. (2017). Visible light communication: Applications, architecture, standardization and research challenges. *Digit. Commun. Network.*, *3*(2), 78–88. doi:10.1016/j.dcan.2016.07.004

Kibria, M. G., Nguyen, K., Villardi, G. P., Zhao, O., Ishizu, K., & Kojima, F. (2018). Big data analytics, machine learning, and artificial intelligence in next-generation wireless networks. *IEEE Access: Practical Innovations, Open Solutions*, *6*, 32328–32338. doi:10.1109/ACCESS.2018.2837692

Kim, J., Kim, D., & Choi, S. (2017). 3gpp sa2 architecture and functions for 5g mobile communication system. *ICT Express*, *3*(1), 1–8. doi:10.1016/j.icte.2017.03.007

Kiyomoto, S., Basu, A., Rahman, M. S., & Ruj, S. (2017). On blockchain-based authorization architecture for beyond-5g mobile services. *2017 12th International Conference for Internet Technology and Secured Transactions (ICITST),* 136–141. 10.23919/ICITST.2017.8356363

Kotobi, K., & Bilen, S. G. (2018). Secure blockchains for dynamic spectrum access: A decentralized database in moving cognitive radio networks enhances security and user access. *IEEE Vehicular Technology Magazine*, *13*(1), 32–39. doi:10.1109/MVT.2017.2740458

Li, Jiang, & Chen, Luo, & Wen. (2020). A Survey on the Security of Blockchain Systems. *Future Generation Computer Systems*, *107*, 841–853.

Ling, X., Wang, J., Bouchoucha, T., Levy, B. C., & Ding, Z. (2019). Blockchain radio access network (b-ran): Towards decentralized secure radio access paradigm. *IEEE Access: Practical Innovations, Open Solutions*, *7*, 9714–9723. doi:10.1109/ACCESS.2018.2890557

Loven, Leppanen, Peltonen, Partala, Harjula, Porambage, Ylianttila, & Riekki. (2019). Edge Ai: A vision for distributed, edge-native artificial intelligence in future 6g networks. *The 1st 6G Wireless Summit*, 1–2.

McCullagh, Lightbody, Zygierewicz, & Kernohan. (2014). Ethical challenges associated with the development and deployment of brain computer interface technology. *Neuroethics*, *7*(2), 109–122.

Nakano, T., Okaie, Y., Kobayashi, S., Hara, T., Hiraoka, Y., & Haraguchi, T. (2019). Methods and applications of mobile molecular communication. *Proceedings of the IEEE*, *107*(7), 1442–1456. doi:10.1109/JPROC.2019.2917625

Nawaz, S. J., Sharma, S. K., Wyne, S., Patwary, M. N., & Asaduzzaman, M. (2019). Quantum machine learning for 6g communication networks: State-of-the-art and vision for the fu-ture. *IEEE Access: Practical Innovations, Open Solutions*, *7*, 46317–46350. doi:10.1109/ACCESS.2019.2909490

Nguyen, Tran, Loven, Partala, Kechadi, & Pirttikangas. (2020). Privacy aware blockchain innovation for 6g: challenges and opportunities. *2020 2nd 6G Wireless Summit (6G SUMMIT)*, 1–5.

Ni, J., Lin, X., & Shen, X. (2019). Toward privacy-preserving valet parking in autonomous driving era. *IEEE Transactions on Vehicular Technology*, *68*(3), 2893–2905. doi:10.1109/TVT.2019.2894720

Pathak, P. H., Feng, X., Hu, P., & Mohapatra, P. (2015). Visible light communication, networking, and sensing: A survey, potential and challenges. *IEEE Communications Surveys and Tutorials*, *17*(4), 2047–2077. doi:10.1109/COMST.2015.2476474

Pattaranantakul, M., He, R., Song, Q., Zhang, Z., & Meddahi, A. (2018). Nfv security survey: From use case driven threat analysis to state-of-the-art countermeasures. *IEEE Communications Surveys and Tutorials*, *20*(4), 3330–3368. doi:10.1109/COMST.2018.2859449

Prados-Garzon, J., Adamuz-Hinojosa, O., Ameigeiras, P., Ramos-Munoz, J. J., Andres-Maldonado, P., & Lopez-Soler, J. M. (2016). Handover implementation in a 5g sdnbased mobile network architecture. *2016 IEEE 27th Annual International Symposium on Personal, Indoor, and Mobile Radio Communications (PIMRC)*, 1–6.

Puthal, D., Malik, N., Mohanty, S. P., Kougianos, E., & Yang, C. (2018). The blockchain as a decentralized security framework. *IEEE Consum. Electron. Mag.*, *7*(2), 18–21. doi:10.1109/MCE.2017.2776459

Ramadan & Vasilakos. (2017). Brain computer interface: Control signals review. *Neurocomputing*, *223*, 26–44.

Seddigh, N., Nandy, B., Makkar, R., & Beaumont, J.-F. (2010). Security advances and challenges in 4g wireless networks. *2010 Eighth International Conference on Privacy, Security and Trust*, 62–71. 10.1109/PST.2010.5593244

Sharma, P. (2013). Evolution of mobile wireless communication networks-1g to 5g as well as future prospective of next generation communication network. *Int. J. Comput. Sci. Mobile Comput.*, *2*(8), 47–53.

Svogor & Kisasondi. (2012). Two factor authentication using eeg_augmented passwords. *Proceedings of the ITI 2012 34th International Conference on Information Technology Interfaces*, 373–378.

Tariq, F., Khandaker, M., Wong, K.-K., Imran, M., Bennis, M., & Debbah, M. (2019). *A Speculative Study on 6g*. arXiv Preprint arXiv:1902.06700.

Toorani, M., & Beheshti, A. (2008). Solutions to the gsm security weaknesses. In *2008 The Second International Conference on Next Generation Mobile Applications, Services, and Technologies*. IEEE. 10.1109/NGMAST.2008.88

Ucar, S., Coleri Ergen, S., Ozkasap, O., Tsonev, D., & Burchardt, H. (2016). Secvlc: secure visible light communication for military vehicular networks. *Proceedings of the 14th ACM International Symposium on Mobility Management and Wireless Access*, 123–129. 10.1145/2989250.2989259

Varshney, U. (2012). 4g wireless networks. *IT Professional*, *14*(5), 34–39. doi:10.1109/MITP.2012.71

Xu, Li, & Liu, Wen, & Lu. (2019). Efficient and privacy-preserving truth discovery in mobile crowd sensing systems. *IEEE Transactions on Vehicular Technology*, *68*(4), 3854–3865.

Zhang, Zhu, Xiong, Huo, Tari, & Zhou. (n.d.). Correlated differential privacy: feature selection in machine learning. *IEEE Trans. Ind. Inf.*

Zhang, Z., Xiao, Y., Ma, Z., Xiao, M., Ding, Z., Lei, X., Karagiannidis, G. K., & Fan, P. (2019). 6g wireless networks: Vision, requirements, architecture, and key technologies. *IEEE Vehicular Technology Magazine*, *14*(3), 28–41. doi:10.1109/MVT.2019.2921208

Zhu, T., Li, G., Zhou, W., & Philip, S. Y. (2017). Differentially private data publishing and analysis: A survey. *IEEE Transactions on Knowledge and Data Engineering*, *29*(8), 1619–1638. doi:10.1109/TKDE.2017.2697856

Zhu, T., & Philip, S. Y. (2019). Applying differential privacy mechanism in artificial intelligence. *2019 IEEE 39th International Conference on Distributed Computing Systems (ICDCS)*, 1601–1609. 10.1109/ICDCS.2019.00159

Chapter 2
6G and Next Gen Networks With Ultra-Dense Heterogeneous Networks:
System Architecture, Performance Metrics

Rekha M.
Dhanalakshmi Srinivasan Engineering College (Autonomous), India

Bhuvaneswari Mariappan
University College of Engineering Bharathidasan Institute of Technology Campus, Trichy, India & Anna University, Tiruchirappalli, India

ABSTRACT

Fifth-generation (5G) communication will be formally released very soon, with several more characteristics than fourth-generation communication. Between 2027 and 2030, a revolutionary wireless transmission paradigm, the 6G network, with full artificial intelligence integration, is projected to be implemented. Faster ultimate bearing ability, higher bandwidth speed, reduced dormancy, and enhanced quality of service (QoS) compared to 5G systems are among some of the key challenges that need to be resolved throughout 5G. This chapter discusses the variations of 6G data transmission and its system architectures. It gives a thorough look at the 6G network's specifications, including the wireless medium and various technologies. More specifically, this chapter emphasizes the different features of the planned 6G networks using ultra-dense heterogeneous network (UD HetNets) scenarios and implementations. This chapter explores the 5G/6G network's key performance indicators (KPI), difficulties, and research directions.

INTRODUCTION

The Internet of Things (IoT), three-Dimensional (3D) media, Artificial Intelligence (AI) and Virtual Reality (VR) have all witnessed rapid rise in traffic (Alsharif & Rosdiadee Nordin, 2017). The traffic is8 EB/month in the earlier 2010, and it is anticipated to increase to 5000 EB/month by 2030 (Ser-

DOI: 10.4018/978-1-6684-3804-6.ch002

gey, Vitaly, Mischa, & Halim, 2019). The importance of improving communications infrastructure is mainly needed. The World is becoming a civilisation ruled by fully automated systems. Production, agriculture, commerce, the oceans, and aerospace are all areas where autonomous robots are gaining popularity. Thousands of scanners are installed to provide effective utilisation and autonomous systems. These programs will need high data throughput and consistent connectivity. Cellular systems of the 5th generation (5G) were in use in various parts of the world. By 2022, 5G is fully operational across the world.5G Wired connections will not be able to create a fully automated network that offers anything as a single, total immersion experience (Shanzhi, Ying Chang, Shaohui, Shaoli, Wenchi, & Mugen, 2020). Despite the fact that 5G fiberoptic cables will significantly outperform existing models, they will not meet the expectations of IoT and commercial processes until at least the next ten years (Chiaraviglio & Melazzi, 2018). The proposed system will offer advanced functionalities and a higher QoS than 4G communications (Yueyue, Du Xu & Zhuang, 2019; Ding & Poor, 2019; Giordani & Zorzi, 2019; Gu, 2018). 5G includes bandwidths (mm), better spectrum utilisation and supervision, and unlicensed band. Nonetheless, 5G wireless internet connectivity may be outstripped by the fast expansion of data-centric and automated technology. In 5G communication, the intersection of connection, intelligence, sensing, control, and processing functions was largely overlooked. Future Internet of Things (IoE) applications, on the other hand, will require this convergence. Virtual worlds, for example, need data rates of at least 10 Gbps (Alsharif & Rosdiadee Nordin, 2017), thus they must advance beyond 5G. (B5G). When a result, as 5G reaches its limits in 2030, academics are already looking into the design goals for the next phase.

I hulking beast interaction, (ii) a lot of advanced between many local source and the network, (iii) sensory help to analyze to generate multi-verse map of the area and type of foreign feelings, and (iv) correctness in crispness to manage the material world that require 6G (Han, 2018). To address this issue and overcome the 5G limitations, B5G transmitters need to be built. By launching innovative formation of future benefits such as acoustic detection intellectual capacity, a widespread AI initiation and innovative technology solutions like THz, 3D networking, subatomic connectivity, backscatter information exchange, smart total internal reflection surface (IRS), and proactive data storage, 6G infrastructure addresses the 5G system limitations (Huq & Busari, 2019). The confluence of all previous aspects, such as layer is important, maximum throughput, high dependability, reduced energy consumption, and enormous connection, will be the biggest factors of 6G. The 6G systems continue prior administrations' by introducing new goods and services and technologies. It includes AI, wearable technology, implants, and driverless automobiles, computing devices, and sensing (Jaber & Tukmano, 2019). The ability to manage massive amounts of data and a much higher connections per unit is the most important need for 6G signal repeaters (Alsharif & Rosdiadee Nordin, 2017).

The 5G framework is improved and expanded as 6G, in terms of strength and client QoS, the 6G system will exceed 5G, as well as offer some fascinating new features. It has the ability to protect user data, defend the system, and deliver useful services (Letaief & Zhang, 2019). 6G connection is expected to become a global communication network. In many scenarios, the per-user data throughput in 6G is projected (Alsharif & Rosdiadee Nordin, 2017; Li & Avestimehr, 2017). In comparison to 5G, 6G is expected to offer 1000 times more immediate wireless connectivity.

Furthermore, ultra-long-range transmission with latency of <1 ms is predicted. (Li & Wu,2014). The addition of completely implemented AI for driving automated systems is the most intriguing element of 6G. In 6G communications, video-type transmission is anticipated to be prominent across different data traffic network. The 3D networking, AI, UAV, THz band, IRS, OWC and wireless power transfer are the most significant innovations that will drive 6G. This study attempted to offer a comprehensive analysis on future 6G communication infrastructure with recent trends and research activities from around the world. Other articles touched on the subject as well. This article, on the other hand, shows an insight of technology, use cases, needs, and problematic challenges (Liu, 2017). Although it is difficult to discover every element of 6G in the time available, we think that our study will point research community in the correct path. This article's achievements may be summed up as follows:

- The growing trend of data roaming and WiFi connection is briefly explored.
- Approaches to the 6G connectivity system are discussed.
- Expected 6G connectivity needs are outlined.
- A quick comparison is made between the planned 6G communications network and the 4G and 5G networks.
- New 6^{th}Generation technologies are discussed.
- The functions of various expertises in the 5G and 6G networks are explored.
- A list of 6thGeneration application is provided, along with their specifications.
- Development on 6G is discussed with reference to other active projects.
- Potential roadblocks and research approaches for achieving the 6G target are highlighted.

SYSTEM ARCHITECTURE

Assume you have a thick distribution of homogeneous cells, with a macrocell layer which provide operations and penetration and a thick layer of small base stations (SBSs) with varying properties. Cross-tier interference is avoided since macrocells and small cells function in distinct frequency bands. SBSs, on the other hand, utilise the same spectrum resources. We take a look at two distinct instances:

- There are two types of users covered by the comparatively tiny and macrocell layers (i.e., users need secure communications by small cells and not by macrocell) (Liu & Vucetic, 2018). As a result, the attention is solely on the small-cell layer with micro, pico, and femto cells, all of which are referred to as small cells in the following;
- Consumers can be serviced by macrocells with no specialization.

In both cases, a Poisson point process (PPP) distribution is adopted to scatter infrastructure over the study area at arbitrary. The letter S represents the saturation of the PPP. Users are uniformly distributed in A, according to an isolated PPP probability with probability (Liu & Vucetic, 2018).The reference voltage emitted by the BSs all around mobile station is received by each user. Each sensor plus interfering ratio (SINR) received beyond the sensitivities barrier may subsequently be evaluated and ranked by the access point. The user u receives (SINR) from cell c is represented as

$$SINR_{u,c} = \frac{P_c^{TX} \left| h_{u,c} \right|^2 \rho_{u,c}}{\sigma_u^2 + \sum_{s=1}^{N_c} \left| h_{u,s} \right|^2 P_s^{TX}} \tag{1}$$

Figure 1. Ultra-dense network model

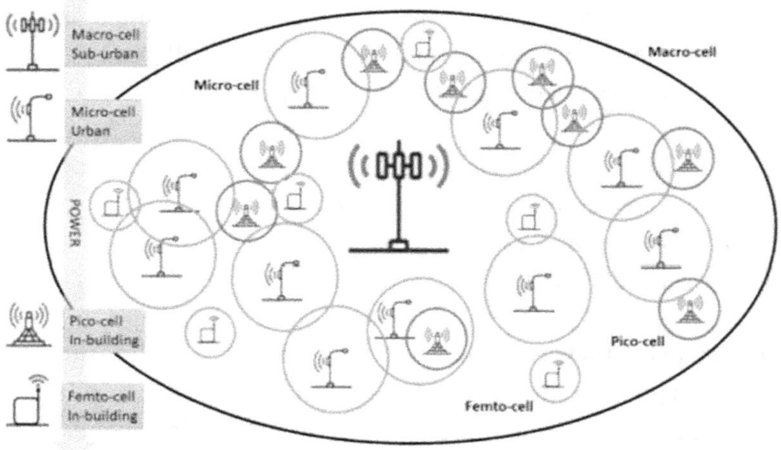

P_c^{TX} is the kind of designed transceiver transmitted on broadcast by s-th cell, h_c is the total number of SBSs, $p(u,s)$ is the pathway loss, and $h(u,s)$ is the route loss. The transmission noise intensity is u^2, and the streams efficiency among s and u is $|h_{u,c}|^2$, which follows an elliptic curve with unit mean. The channel loss models _(u,c) are thought it's double sloping, since they contain the LOS element BS and properly describe the rising BS density rising.

$$\rho_{u,c}\left(d_{u,c}\right)=\begin{cases}K_0 d_{u,c}^{-\alpha_1} & d_{u,c}\leq \bar{d}\\ K_1 d_{u,c}^{-\alpha_2} & d_{u,c}\geq \bar{d}\end{cases} \quad (2)$$

where $d*(u,c)$ is the distance among BS and user u's station, d specifies essential separation separating the close-in and lengthy path loss zones, K_0 is a catch-all best condition possible to the fading channel for unitarist distance, K_1 is a continuous to ensure the integrity among two path-loss regions, 1 specifies close-in propagation loss extension (equal to 2) (Mao, 2020).

BASIC REQUIREMENTS

Several concerns, including as performance, latency, energy efficiency, rollout costs, dependability, and processing capacity, are connected with 5G technology. After 2030, 5G is improbable to fulfil market needs. To handle the issues among 5G and market demand, 6G will be introduced. The major objectives for 6G systems are based on historical patterns and forecasts of future demands.

(i) Increased data rates per gadget, (ii) a huge proportion of mobile networks, (iii) global accessibility, (iv) extremely low variability, (v) lesser energy consumed with battery-less IoT devices, (vi) ultra-high credible gateways, and (vii) attached intellectual ability. A comparative of all the G' communications networks is shown in Table 1. Indeed of KPIs in 5G, it is expected that 6G need some added key KPI drivers. Many of the 5G program's KPIs will be applicable to 6G as well. The 5G KPIs, on the other hand, must be evaluated, and new KPIs for 6G must be explored. Numerous KPI groups are presently complex to describe for 6G, but further research are planned to complete them. When opposed to the earlier generation of communications networks, 5GB transceivers aim to enhance capabilities by a factor of 10–100 in most of the technological fields. Research and business research organisations have recommended the following KPIs for 6Process by providing: Peak data price of 1 Tbps, energy efficiency of ten times, optimal outage of one out of one millions, and high precision in positioning. The first list includes jitter, connectivity throughput; upgrade options, three dimensional, wireless modems, velocity profile, cost, and resource KPIs. Standardized, data protection, open-source technology and global application case are among KPIs in the second group. The essential criteria of all classic communication networks are KPIs relating to capacity, efficiency, fuel efficiency, data throughput, lag, and connection (Mao, 2020). The KPIs for intelligence gathering, on the other hand, are brand new for 6G. Along with the growth of 6G systems, all feasible KPIs will be realised.

SERVICE REQUIREMENTS

The following sorts of KPI-related services are projected to be available in 6G communication systems (Liu & Vucetic, 2018):

- Portable ultra-broadband is ubiquitous (uMUB)
- Networking at ultra-high speeds with little delay (uHSLLC)
- Large-scale machine-to-machine connectivity (mMTC)
- Extremely dense data (uHDD)

The 6G communication system will be defined by the following fundamental characteristics:

- Enhanced data protection
- AI-integrated interaction
- Physical network
- High energy good organization
- Low down backup and network admission backlog

The 6G is expected to provide 1k times the continuous connection of the 5G system. Unlike 5G's improved mobile broadband (eMBB), 6G is anticipated to provide partially or totally, such as uMUB. With E2E delay compared with fewer than 1 ms [14], upwards of 99.99999 percent durability (Mao, 2020), and a peak available bandwidth of 1 Tbps, ultra-reliable low-latency connections (uHSLLC), which was a major feature of 5G, will be a prominent driver again in 6Process by providing. A high number of connected objects (up to 10 million per km^2) will be facilitated by the 6G communications network (Mao, 2020).6G is anticipated to deliver Gbps connectivity throughout, with new settings like sky (10,000 km) and sea (20 nautical miles) covered (Mao, 2020). In 6G, volume spectral effectiveness will be significantly greater than the commonly utilised area transmission rate (Li & Wu, 2014). The

6G system will have an extremely extra battery lifespan and sophisticated resource collecting battery technologies. Smart phones will not need to be powered separate in 6G systems.

STANDARDIZATION

The 5G standards have been previously finalised, and while it has previously been realize in some areas of the planet, the entire chapter of 5G deployment will take place in 2020. The development on 6G is still in its early phases. A number of studies on the standardisation of 6G will be conducted internationally beginning in 2020; 6G transmission is still budding. Various scientists refer to 6G as 5G+. In the U.S.A, preliminary research efforts have indeed commenced. The implementation of 6G has been requested by the US president. In 2019, China has already begun an impression study for just beginning and standardising 6G infrastructure. In 2020, the Chinese want to conduct aggressive research on 6G. Several 6G projects are being planned in European nations. The 6G research programme is anticipated to begin in 2020 (Mozaffari & Debbah, 2018). We'll go through a few research projects and standardisation initiatives in this section.

Table 1. Comparison of various communication system

Issues	4G	5G	6G
Per device peak information charge	1 Gbps	10 Gbps	1 Tbps
E2E latency	100 ms	10 ms	1 ms
greatest spectral efficiency	15 bps/Hz	30 bps/Hz	100 bps/Hz
Mobility sustain	350 km/hr	500 km/hr	1000 km/hr
settlement mixing	No	No	Entirely
AI	No	Unbiased	Entirely
self-sufficient vehicle	No	Unbiased	Entirely
XR	No	Unbiased	Entirely
Haptic announcement	No	Unbiased	Entirely

The International Telecommunication Union's (ITU), ITU-R and 5G standardisation operations are based on IMT-2020. As a result, ITU-R is expected to produce IMT-2030, a document that summarises the potential needs for communications technology in 2030.

The first 6G wireless conference was conducted in March 2019, and was a huge success. At the conference, cutting-edge wireless communication experts were represented. The meeting also included representatives from the world's top telecom firms. This 6G summit will kick off talks on important topics including the reason for 6G, how to transition from 5G to 6G and innovative solutions (Nawaz, Sharma & Asaduzzaman, 2019).

The 6G research efforts at the University of Oulu began as part of Finland's flagship initiative., distributed compute, forces, Wireless connectivityand applications are the four unified intended research elements of the 6G Flagship. Significant technological components of 6G networks will be created using medical discoveries.

Samsung Electronics has traditional a research and expansion centre for the development of critical technologies for 6G wireless services. Samsung investigate is undertaking investigate work on cellular systems in order to expedite the establishment of products and the standardisation of 6G; they have elevated some next telecom study group to a centre.

PERFORMANCE METRICS

6G infrastructures are projected to accomplish a particular task and have more performance indicators than 5G (IMT-2020) (Ding & Poor, 2019). With the use of THz and laser frequencies bands, 5G networks may reach peak data rates of 20 Gbps, while 64 g can reach 1–10 Tbps. With these frequency spectrums, the user-experienced information rate can reach Gbps. The capacity of the area traffic might exceed 1 Gbps/m2. Better on assignments can climb by 3–5 times to provide for the 100-fold increasing data flow, but performance must gain compared to 5G. Other key performance indicators, including as reducing costs, encryption strength, penetration, and intelligence level (Nguyen & Pirttikangas, 2020), should also be added.

In March 2019, Bell Laboratories provided certain important performance indicators as a comparative. The maximum data rate is projected to exceed 100 Gbps. The data rate observed by the user is 1–10 Gbps. The connection density is 107 devices/km^2. The delay should be less than 0.3 milliseconds. In comparison to 5G, the fuel efficiency is predicted to be ten times higher (Pan, 2019). The capacity will be 10000 times that of 5G networks. Interior placement consistency of 10 cm and outdoor high precision of 1 metre, as well as six "9" constancy, are among the other measures. In comparison to 5G, the researchers proposed the following new key performance indicators: > 1 Tbps high bit rate, 1 Gbps user observed connection speed, latency, motion range, equipment connection liquid limit, traffic power, effectiveness, and times transmission power (Popovski, 2018).

While 5G networks primarily focus on 6G networks will greatly expand and expand potential uses. In February 2020, the ITU published the 6G first timetable. By 2023, it expects to have finished its 6G vision and technological trends research. During its meeting in Geneva, ITU-T market research formed a latest technology for network 2030. It intends to investigate broadband internet, whenever it is to support creative forward-thinking scenarios like visual transmission, rapid response, and the great response expectations of increasing industrial vertical.

The vision for 6G application areas contain 5G situations that have been improved, as well as additional situations. The effectiveness of the three enhanced 5G scenarios are much improved in 6G. 6G also enables new application areas, like healthcare, lengthy and elevated communication systems, extremely low-power communication systems, consolidation of correspondence, computing, control, classification, spatial interconnected connections, dispersed AI, and space-air-ground-sea. To mention a few, there are autonomous visual surveillance systems and the Internet of bio-nano-things (Letaief & Zhang. 2019).

KEY PERFORMANCE INDICATORS

This page discusses the main KPIs of 6G wireless communication technologies. Any of the important performance benchmarks are 100 mbps velocity, mobility demands, digital phones per km^2 , area congestion capability, latencies, dependability, network coverage, and power efficiency (Qi & Li, 2016).

Peak Data Rate

Enhanced mobile broadband,, which simply means fast data throughput, is one of the uses cases for this next wireless data transmission. As a result, we may download HD videos in a couple of moments. Since the inception of wireless communications, consumers' data rate needs have risen. 1G had download speeds (kbps) grew to gbps. For certain applications, these data rates are insufficient. As a result, we need to create certain standards and interoperability with data speeds of 10 to 100 gigabits per second.

Mobility

In next-generation communication networks, more movement robustness is also necessary. In highly mobile equipment, high bandwidth rates should be sustained (Rappaport, Xing, Kanhere, Ju, Madanayake, Mandal, & Trichopoulos, 2008). When travelling by airline or high-speed bullet train, for example, communications are not disrupted, and data speeds are maintained. ITU defines the mobility criteria for 6G as >1000 km/hr.

Massive Connectivity (devices/Km 2)

mMTC is another application for upcoming communication systems. This is where the IoT enters the picture, and it is machine-to-machine without the need for human intervention. The messages, calls,and orders are sent from one machine to another. A person does not carry out these acts (Saad, 2019). It is the algorithms, not the humans that communicate with one another. Next-generation wi - fi are projected to support 107 devices per square kilometre.

Area Traffic Capacity (Mbps/m 2)

Impact of increased bandwidth routes and backhauling rose as the number of internet - connected units per unit area increased. On a daily basis, a densely distributed sensor network creates more than a terabyte (TB) of data (Strinati, 2019). To handle the traffic, this data creation demands a significant back-hauling route.

Extremely-High Consistency and Low Latency With Security (eRLLCS)

Low down latency translates to lightning communication. In 6G, the maximum allowed delay is 10 seconds (Su & Chih-Lin, 2019). High dependability and ultra-low latency will be required in the network design of autonomous mobiles and robotics. Smart houses, smart automobiles, company's ability, smart school systems, and ability of firms will all be part of future cities. Airplanes, ships, bullet trains, and unmanned aerial vehicles (UAVs) will all need to be connected to smart cities. Health care, defence, tracking, and surveillance are just a few of the significant functions that will demand ultra-reliability and low latency. Table 2 depicts the critical analysis of diverse approaches proposed for 6G system.

Table 2. Critical analysis on various approaches for 6G systems

Technology	Pros	Cons	Use cases
Quantum Communication (QC)	Higher computational capabilities that is better	Costly multifaceted	Drug industry Radar industry Mathematics
Blockchain	Reliability in Multiple Locations Integrity Traceability Formal verification	Large memory inefficiency Concerns about privacy Decentralize	Procurement Healthcare Security Biometric - based Voting
Reconfgurable Intelligent Surfaces	Low level of complexity Technical feasibility in terms of power Cost-effective	Stage confguration is difficult.	Industry of communication and defence
Tactile Internet and Haptics	Gaming on the internet Make a difference in the lives of handicapped people.	Noise Extremely expensive because it is less energy efficient.	Machinery in the Mining Industry VR and AR in online gaming
Access Technologies	Improved network deficiency	Economic loss as a result of replacing current equipment	patchy use cases
New Spectrum –mmWaves THz VLC	Increased bandwidth accessibility	Low penetrating strength	erratic use cases
Internet-of-Everything (IoE)	Low latency advanced data charge	Low liveliness efficiency	erratic use cases

CHALLENGES

To properly implement 6G communication systems, a number of technological issues must be resolved. A few potential issues are mentioned briefly below.

THz has a fast dissemination rate and is absorbed by the atmosphere, resulting in high connection speeds. The THz bands, however, present a significant hurdle for information transfer over longer distances due to maximum broadcast loss and air absorption factors (Tang, 2019). We could use a system which is able and coordinate design for THz network technologies. We need the transmitter to be able to operate at a different intensity, and we need to make sure that all dataset rates are used. The poor gain and directivity of the various THz band antennas issue with THz communicating. Concerns regarding the safety and health of THz band transmission

Resource Management Complication in 3D Networking

3D computer science has been expanded in the vertical direction. As a result, a broader definition has been introduced. Furthermore, numerous attackers may capture valid data, causing overall system performance to suffer (Wang & Hepsaydir, 2014). As a response, for vehicular networks, congestion control, and numerous requests, new micromanagement and technology to ensure are necessary. Development will necessitate an innovative service structure.

Heterogeneous Hardware Resource Limits

6G will include a large number of assorted communications network, including frequency bands, network topology, delivery of services, and so on. Furthermore, the hardware configurations and transportable terminals are vastly different. Massive MIMO is expanded from 5G to 6G, which will likely necessitate a more sophisticated design. It will also make the design of the communication protocol and process more difficult (Wu & Zhang, 2019).

Because there is a scarcity of licenced spectrum and there are worries about disruption, it is important to manage 6G spectrum, which includes frequency band methods and novel spectrum management methodologies. Effective wavelength management is necessary for optimal resource use and QoS maximisation. Research teams on 6G must address challenges including wireless channels and bandwidth allocation in wireless links where broadcasting on the same wavelength are synchronised. Investigators must also consider how common parallelizing techniques, such as diffraction, can be utilised to balance out the noise.

Device Capabilities

The 6G includes number of new functionalities. Smartphones, for example, should be capable of handling the new functionalities. Supporting Tbps speed, XR,AI, and sensing with communication features utilising different networks is particularly difficult. Some 6G capabilities may not be supported by 5G devices, and the increased capability in 6G devices. The number of devices that will be available for 5G is estimated to be in the billions (Yaacoub & Alouini, 2019). The interoperability of 5G devices with 6G networks is a major problem as communication network transitions from 5G to 6G. This interoperability makes it easy for end-users to utilise while also saving money. As a result, based on technological compatibility with 5G, 6G should prioritise integrating communication services systems, computational efficiency improvement, and so on.

Planning of Economic Prospect

For the introduction of 6G connectivity, the economic outlook is equally critical. A new 6G deployment will cost a lot of money in terms of network transportation. Moreover, cost is diminished by converting 5G to 6G and properly planned (Yang & Li, 2019). To make the 6G network expense, the possibilities for infrastructural, communication, and bandwidth allocation must be thoroughly examined. The combination involving expected 6G allowing developments, workloads, controller design, and corporate strategy, as well as the peaceful transfer of power, is important to construct the expense 6G infrastructures (Yastrebova & Koucheryavy, 2018).

Autonomous Wireless Systems

Innovative gadgets such as driverless cars, surveillance aircraft, and AI-based Industry 4.0 will be fully enabled by the 6G network (See Figure 2). To develop high wireless communication systems, we need the merging of many diverse comment thread such independent computing, compatibility techniques, system of systems, deep learning, independent internet, machineries of systems, and diverse transceivers (Zhang & Leung, 2016). As a result, the entire system growth gets complicated and hard. Designing a fully independent network for a driverless car, for example, will be considerably more difficult since 6G engineers will have to create totally automated self-driving vehicles that outperform human-driven automobiles.

Capacity should, in an ideal world, grow exponentially with TP density. However, problems like as interference and mobility may significantly impede densification gains. Furthermore, practical installation and maintenance elements like as backhaul and energy supplies are critical to UDN's viability. The difficulties of UDN are discussed in this section (Zhang & Nallanathan., 2019). The 4G solutions that go with them are also explored. While these technologies may not be enough, they can serve as a point of departure and inspire for the advancement of technology.

Figure 2. Interference in HetNet

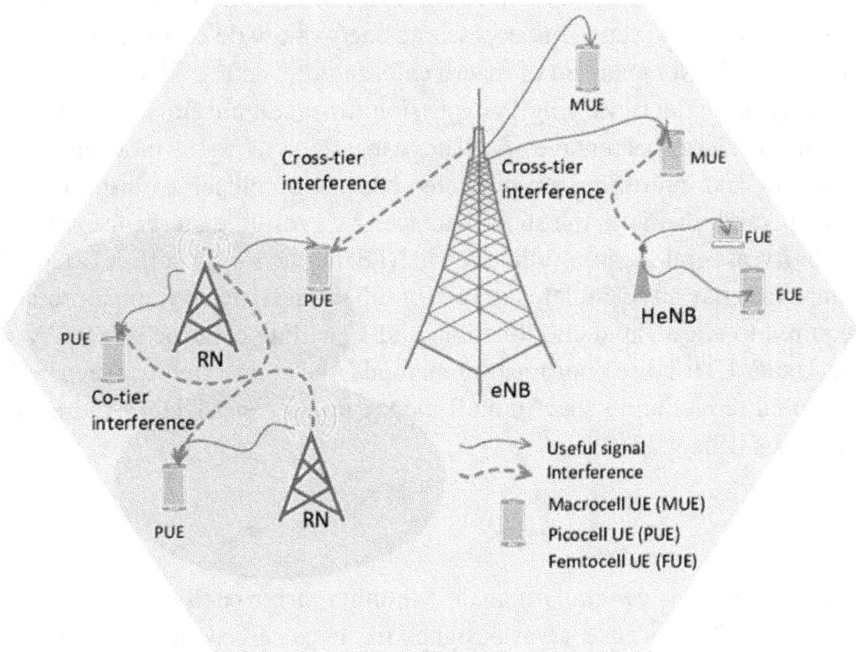

Inference

Disturbance is a major problem with LTE since nearby cells share the same frequencies supply, which has been a concern for decades of mobile networks. Static and semi-static techniques are presented in Release 8 to deal with massive MIMO disturbance. Cell planning and partial frequency reuse are examples of static methods (PFR). In terms of semi-static techniques, eNBs communicate for disturbance coordination through the X2 interface with a period range of 10 to 200 milliseconds (Zhang, 2019). In the downlink, RNTP is used to notify neighboringeNBs of the high-power bands. In the uplink, OI denotes the observed method can cause, whereas HII denotes the bands where high-power uplink communication may be possible.

Release 10 of LTE HetNet uses cell range expansion (CRE) to offload macro information in the same coverage area as macro cells. The entanglement issue in HetNet is more difficult than in a homogeneity macro system due to overlapping reception, and the problem is considerably more acute close to the edge of micro cell coverage. In Releases 10, time domain eICIC can mute specific macro cell sub-frames to avoid problems with the small cell's general and controlling signal. feICIC(further improved ICIC) and CoMP (Coordinated Multipoint Process) are added on the recipient side in Release 11. The interference produced by shared transmitter would be substantial if more adjacent cells were present. As a result, Release 12 uses dynamic cell on/off to turn off idle cells, reducing interference and system power usage (Zhang, 2019). However, studies demonstrate that the speed with which a cell may be activated or deactivated has a significant performance impact.

The situation of interfering in UDN is considerably more difficult than in LTE HetNet because of the dense cells. More adjacent cells are expected to produce considerable influence as ISD shrinks. In the UDN, one cell generally has two or more dominating carried out in this study, but in the macro network, only one dominant interferer is seen. This necessitates network-wide cooperation with far more cells engaged. UDN will be a HetNet made up of macro cells and tiny cells of diverse kinds. Because of the increasing cell density and widely varying transmission time, cell coverage is more likely to overlap, resulting in poor rates in a large percentage of cell edge locations. With the rapid entry and exit of users, large variations in flow and interfering are predicted because a cell serves just a few users as well as the pace would just be high owing to the short distance. As a result, disturbance coordination requires greater time granularity, as semi-static coordination in LTE may no longer suffice (Zhang & Song, 2020). Because the X2 interface has a typical latency of 20 milliseconds, an additional protocol communication system is required to allow rapid coordination, and signalling over the air (SoTA) appears to be a promising option. Under LTE noise coordination methods, the connection between cell efficiency and density. The substantial reduction in spectrum efficiency implies that UDN requires new interference coordination techniques (Zhao, 2019).

Mobility

In mobile networks, mobility is generally managed through turnover (See Figure 3). Handover is not expected to happen very often because it was designed for mega cells with broad coverage in LTE. As a result, a convoluted approach with significant signalling overhead is used. Furthermore, because of the hard failover method, the contact to the source point is released even before connection to the destinations cell is established. The load - balancing mechanism of TCP may be triggered by the interruption in transmission, resulting in data rate variations. Even if the link quality improves, the new link cannot achieve its peak capacity immediately. By retaining a link to the macro cell when travelling between small cells, the dual connectivity (DC) suggested in Release 12 can mitigate such transmitting interruption to some amount (Zong, 2019). Nevertheless, near cell boundaries, the rate of macro cell connection is often low, allowing only control signals and rudimentary connectivity to be preserved. During the handover phase, huge fluctuations in throughputs would emerge.

In mobile networks, mobility is generally managed through turnover. Handoff is not expected to happen very often because it was designed for mega cells with broad coverage in LTE. As a result, a convoluted approach with significant signalling overhead is used. Furthermore, because of the hard changeover method, the link to the source node is released even before connect to the destinations cell is established. The congestion management mechanism of TCP may be triggered by the interruption in transmission, resulting in data rate variations (Zuo, 2019). At cell boundaries, the rate of macro cell connection is typically low, so only control signals and basic connectivity may be kept. During the handover phase, huge fluctuations in user throughput would occur. Because UDN cells have such limited coverage, handover is likely to occur considerably more regularly. It is self-evident that an elegant solution is required to address the mobility issue and provide a more uniform user experience. The proposed approach should be clear to UE in the ideal scenario, reducing signalling time and rate variations.

Figure 3. Mobility in HetNet

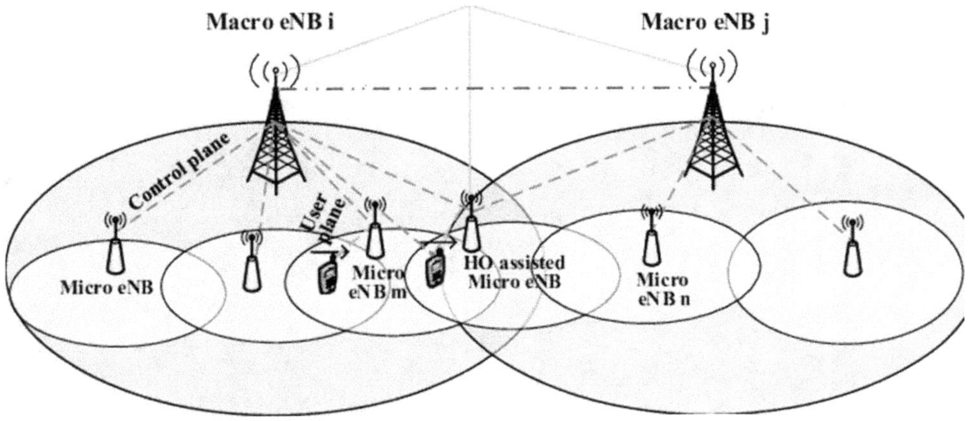

Backhaul

Because there are so many tiny cells, the cost of backhaul equipment, placement, and maintenance must be low enough because UDN to be commercially viable. Because it's difficult to find a good site placement for every cell, UDN is likely to have a less controlled rollout. Because UDN small cells may be mounted on streetlight towers and billboards, the backhaul must be adaptable. The functionality of CoMP declines significantly when the backhaul delay reaches 10ms owing to channel ageing, according to a study. For rapid coordination, operations like collision avoidance and handover require a minimal backhaul latency. Finally, UDN requires a cost-effective and adaptable backhaul with short distances.

Wired logistics managers, such as fibre and cable, are not only expensive to build and maintain, but they also lack the flexibility to address coverage gaps. Because tiny cell traffic is typically highly bursty, cable backhauls intended for peak output do not appear to be cost effective. Microwave backhaul has a great capacity but is prone to blockage owing to the necessity for line-of-sight (LOS), severely restricting its applications. Furthermore, if functioning in a licenced band, it entails additional hardware expenses, as well as bandwidth rental fees, which would be included in the. Wlan backhaul using unlicensed airwaves, such as 802.11, can be inexpensive. Unfortunately, due to uncontrolled interference in unlicensed channels and its central argument access strategy, mobile communications quality of such backhaul may not be ensured.

A promising possibility is self-backhaul (sBL) of series / parallel operation, in which the backhaul and access link (AL) (See Figure 4) share the same bandwidth supplies. The same equipment may be used for both selfbackhaul and access link AL, lowering costs. Numerous small units can be aggregated on a backhaul provision node to handle capacity variations of single cells, and link throughput can be managed by combining accessibility and backend link efficiency. The concept of sBL has already been incorporated in LTE, and Releasing 10 relay may be considered a form of sBL. MBSFN subframes are used for backhaul in Released 10, while R-PDCCH is used to get downstream control from of the source TP (dTP). Release 10 relay, on the other hand, has a number of difficulties with efficiency, flexibility, and capacity. Backhaul data could employ 13 Signals at the start of the MBSFN sub-frame, due to ineffective backhaul downlink transmission. Release 10 relay will struggle to respond to highly dynamic UDN traffic situations due to a lack of resource provisioning flexibility. Because no enhancements to backhaul connection capacity have been performed, backhaul availability can constitute a bottleneck when accessible link quality is excellent.

Figure 4. Backhaul in HetNet

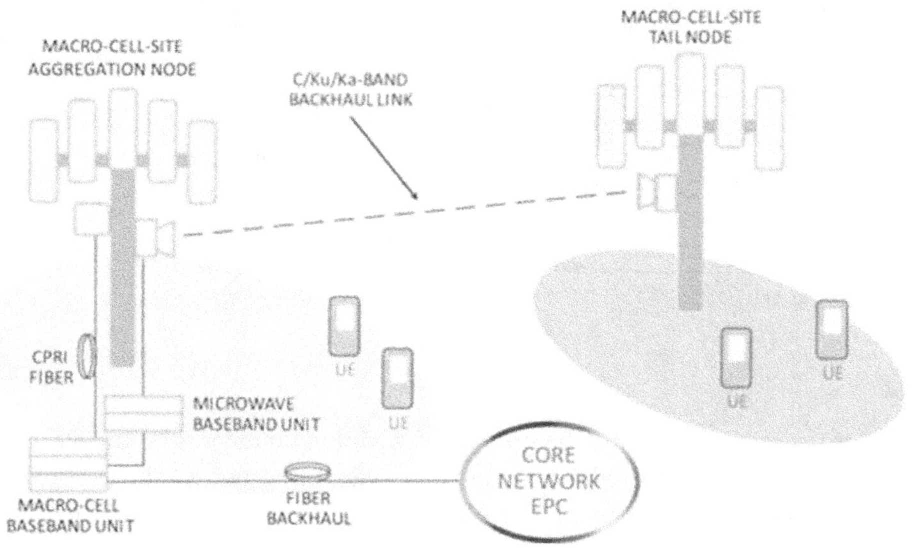

Energy Consumption

In 2007, the digital cellular sector produced 86 million tonnes of CO_2, equivalent for 0.2 percent of the worldwide carbon footprint, according to statistics. And with a conservative estimate, mobile telephony carbon emissions are anticipated to quadruple by 2020, accounting for more than a third of the UK's yearly emissions. Because UDN entails a lot more tiny cells, there's a great deal of concern about energy usage if UDN's energy efficiency stays the same as 4G. High energy usage not only harms the environment, but it also puts a strain on operators' energy costs. As a result, green communication has been a popular study issue in recent, and one of 5G's main goals is to improve it (Zong, 2019).

CONCLUSION

Each young generation of telecommunication brings with it new and exciting capabilities. 5G digital communication is available worldwide in 2020 with number of interesting features. 5G, subsequently, were incapable to fulfill the growing demand for wireless connection by 2030. As an outcome, 6G technology is required. 6G innovation is still an upcoming phases and closely investigated. The options and techniques for accomplishing the goal of 6Relevant techniques are elaborated in this article. In this post, we covered the many tools and applications that will be employed in 6G communication. We also spoke about potential barriers and research directions options for attaining our goals.

ACKNOWLEDGMENT

The authors greatly acknowledge the anonymous reviewers for their valuable suggestions and comments in improving the article in an efficient manner.

REFERENCES

Alsharif, M. H., & Nordin, R. (2017). Evolution towards fifth generation (5G) wireless networks: Current trends and challenges in the deployment of millimetre wave, massive MIMO, and small cells. *Telecommunication Systems, 64*(4), 617–637. doi:10.100711235-016-0195-x

Chiaraviglio, L., Cacciapuoti, A. S., Martino, G. D., Fiore, M., Montesano, M., Trucchi, D., & Melazzi, N. B. (2018). Planning 5G networks under EMF constraints: State of the art and vision. *IEEE Access: Practical Innovations, Open Solutions, 6*, 51021–51037. doi:10.1109/ACCESS.2018.2868347

Ding & Poor. (2019). Simple semi-grant-free transmission strategies assisted by non-orthogonal multiple access. *IEEE Transactions on Communications, 67*(6), 4464-4478.

Giordani & Zorzi. (2019). Toward 6G networks: Use cases and technologies. *IEEE Communications Magazine, 58*(3), 55-61.

Gu. (2018). Network topology reconfiguration for FSO-based fronthaul/backhaul in 5G+ wireless networks. *IEEE Access, 6*, 69426-69437.

Han. (2018). Licensed-assisted access for LTE in unlicensed spectrum: A MAC protocol design. *IEEE Journal on Selected Areas in Communications, 34*(10), 2550-2561.

Huq & Busari. (2019). Terahertz-enabled wireless system for beyond-5G ultra-fast networks: A brief survey. *IEEE Network, 33*(4), 89-95.

Jaber & Tukmano. (2019). 5G backhaul challenges and emerging research directions: A survey. *IEEE Access, 4*, 1743-1766.

Letaief & Zhang. (2019). The roadmap to 6G: AI empowered wireless networks. *IEEE Communications Magazine, 57*(8), 84-90,

Li & Avestimehr. (2017). A scalable framework for wireless distributed computing. *IEEE/ACM Transactions on Networking, 25*(5), 2643-2654.

Li & Wu. (2014). Wireless backhaul of dense small cell networks with high dimension MIMO. *IEEE Globecom Workshops (GC Wkshps)*, 1254-1259.

Liu & Vucetic. (2018). Backscatter multiplicative multiple-access systems: Fundamental limits and practical design. *IEEE Transactions on Wireless Communications, 17*(9), 5713-5728.

Liu. (2017). Polar Codes for Block Fading Channels. *IEEE Wireless Communications and Networking Conference Workshops (WCNCW)*, 1-6.

Mao. (2020). AI-based joint optimization of QoS and security for 6G energy harvesting Internet of Things. *IEEE Internet of Things Journal, 7*(8), 7032-7042.

Mozaffari & Debbah. (2018). Beyond 5G with UAVs: Foundations of a 3D wireless cellular network. *IEEE Transactions on Wireless Communications, 18*(1), 357-372.

Nawaz, S. J., Sharma, S. K., Wyne, S., Patwary, M. N., & Asaduzzaman, M. (2019). Quantum machine learning for 6G communication networks: State-of-the-art and vision for the future. *IEEE Access: Practical Innovations, Open Solutions*, *7*, 46317–46350. doi:10.1109/ACCESS.2019.2909490

Nguyen & Pirttikangas. (2020). Privacy-aware blockchain innovation for 6G: Challenges and opportunities. *6G Wireless Summit (6G SUMMIT)* (pp. 1-5),

Pan, L., Yi, J., Yin, C., Yu, J., & Li, X. (2019). Joint 3D UAV placement and resource allocation in software-defined cellular networks with wireless backhaul. *IEEE Access: Practical Innovations, Open Solutions*, *7*, 104279–104293. doi:10.1109/ACCESS.2019.2927521

Popovski. (2018). 5G wireless network slicing for eMBB, URLLC, and mMTC: A communication-theoretic view. *IEEE Access, 6*, 55765-79.

Qi & Li. (2016). Energy-efficient cooperative network coding with joint relay scheduling and power allocation. *IEEE Transactions on Communications, 64*(11), 4506-19.

Rappaport, T. S., Xing, Y., Kanhere, O., Ju, S., Madanayake, A., Mandal, S., & Trichopoulos, G. C. (2008). Wireless communications and applications above 100 GHz: Opportunities and challenges for 6G and beyond. *IEEE Access: Practical Innovations, Open Solutions*, *7*, 78729–78757. doi:10.1109/ACCESS.2019.2921522

Saad, C., Bennis, M., & Chen, M. (2019). A vision of 6G wireless systems: Applications, trends, technologies, and open research problems. *IEEE Network*, *34*(3), 134–142. doi:10.1109/MNET.001.1900287

Sergey, A., Vitaly, P., Mischa, D., & Halim, Y. (2019). Future of ultra-dense networks beyond 5G: Harnessing heterogeneous moving cells. *IEEE Communications Magazine*, *57*(6), 86–92. doi:10.1109/MCOM.2019.1800056

Shanzhi, C., Ying-Chang, L., Shaohui, S., Shaoli, K., Wenchi, C., & Mugen, P. (2020). Vision, requirements, and technology trend of 6G: How to tackle the challenges of system coverage, capacity, user data-rate and movement speed. *IEEE Wireless Communications*, *27*(2), 218–228. doi:10.1109/MWC.001.1900333

Strinati. (2019). 6G: The next frontier: From holographic messaging to artificial intelligence using subterahertz and visible light communication. *IEEE Vehicular Technology Magazine*, *14*(3), 42-50.

Su & Chih-Lin. (2019) Energy and spectral efficient frequency reuse of ultra-dense networks. *IEEE Transactions on Wireless Communications*, *15*(8), 5384-98.

Tang. (2019). Future intelligent and secure vehicular network toward 6G: Machine-learning approaches. *Proceedings of the IEEE*, *108*(2), 292-307.

Wang & Hepsaydir. (2014). Cellular architecture and key technologies for 5G wireless communication networks. *IEEE Communications Magazine*, *52*(2), 122-130.

Wu & Zhang. (2019). Intelligent reflecting surface enhanced wireless network via joint active and passive beam-forming. *IEEE Transactions on Wireless Communications*, *18*(11), 5394-09.

Yaacoub & Alouini. (2019). *A key 6G challenge and opportunity--connecting the remaining 4 billions: a survey on rural connectivity*. arXiv preprint arXiv:1906.14.

Yang & Li. (2019). 6G wireless communications: Vision and potential techniques. *IEEE Network*, *33*(4), 70-75.

Yastrebova & Koucheryavy. (2018). Future Networks 2030: Architecture & Requirements. *Proceedings of 10th International Congress on Ultra-Modern Telecommunications and Control Systems and Workshops (ICUMT)*, 1–8.

Yueyue, D. (2019). Blockchain and deep reinforcement learning empowered intelligent 5G beyond. *IEEE Network*, *33*(3), 10–17. doi:10.1109/MNET.2019.1800376

Zhang & Leung. (2016). Fronthauling for 5G LTE-U ultra-dense cloud small cell networks. *IEEE Wireless Communications*, *23*(6), 48-53.

Zhang & Nallanathan. (2019). Artificial intelligence-based resource allocation in ultradense networks: Applying event-triggered Q-learning algorithms. *IEEE Vehicular Technology Magazine*, *14*(4), 56-63.

Zhang & Song. (2020). HetMEC: Heterogeneous multi-layer mobile edge computing in the 6 G era. *IEEE Transactions on Vehicular Technology*, *69*(4), 4388-00.

Zhang. (2019). 6G Visions: Mobile ultra-broadband, super internet-of-things, and artificial intelligence. *China Communications*, *16*(8), 1-14.

Zhang. (2019). 6G wireless networks: Vision, requirements, architecture, and key technologies. *IEEE Vehicular Technology Magazine*, *14*(3), 28-41.

Zhao. (2019). *A survey of intelligent reflecting surfaces (IRSs): Towards 6G wireless communication networks*. arXiv preprint arXiv:1907.04789.

Zong. (2019). 6G technologies: Key drivers, core requirements, system architectures, and enabling technologies. *IEEE Vehicular Technology Magazine*, *14*(3), 18-27.

Zuo. (2019). Improving the tolerance of stochastic LDPC decoders to overclocking-induced timing errors: A tutorial and a design example. *IEEE Access*, *4*, 1607-1629.

Chapter 3
6G-Based Undersea Communication

Harshitha . B.
BMS College of Engineering, India

Madhusudhan . K. N.
https://orcid.org/0000-0002-6863-1103
BMS College of Engineering, India

Anantha Sunil Maligi
BMS College of Engineering, India

Lalitha S.
BMS College of Engineering, India

Supreetha M. A.
BMS College of Engineering, India

ABSTRACT

Evolution in mobile network technology has seen a transition from 1G to 5G. Successful launching of 5G in 61 countries of the world has led to advance research on the upcoming 6th generation or 6G networks. Exploiting the several advantages of 6G throws light on its application in undersea communication. The chapter briefs the conventional communication methods and discusses the benefits of 6G in this field. UVLC (underwater visible light communication) and many other advanced acoustic methods like BBICCF (basic bio-inspired camouflage communication frame). Applications and upcoming challenges are highlighted.

DOI: 10.4018/978-1-6684-3804-6.ch003

INTRODUCTION

Wireless communication has proven to be an extraordinary fulfilment tale. No other technology to date—not FM radio, now not the washing machine, neither tv nor the personal computer, and not even the internet— has controlled to draw billions of customers in such short span of time. Nearly every ten years since the introduction of the principal analog communications system in the 1980s, a new generation of cellular communication system has been delivered.

Evolution in mobile network technology has seen a transition from 1G to 5G. Successful launching of 5G in 61 countries of the world has led to advance research on the upcoming next 6th generation or 6G networks. Exploiting the several advantages of 6G, throws light on its application in the undersea communication. This article briefs the conventional communication methods and discuss the benefits of 6G in this field. UVLC (Underwater Visible Light communication) and many other advanced acoustic methods like BBICF (Basic Bio-Inspired Camouflage Communication Frame), applications and upcoming challenges are highlighted.

With the quick improvement of brilliant terminals and arising new applications, wireless data traffic has definitely expanded, and current cell networks can't totally coordinate with the swiftly mounting dedicated requirements. To cope with future challenges, 6th generation (6G) mobile network is relied on to get a high level of tailored norm enabling unique range and energy-efficient transmission plans.

The 6th generation (6G) mobile network may be relied on to get a high level of tailored norm enabling unique range and energy-efficient transmission plans.

To satisfy vision of interfacing everything around the world, 6G will be answerable for the outrageous correspondence necessities that can be seldom upheld by 5G. 5G and beyond, in contrast to "connected persons," is expected to function with human-to-machine or even machine-to-machine communication for "connected society" rather than "connected individuals."

An important design tactic for 6G would be socio-ethical design considering the legal, psychological and economic requirements this may lead the development of new systems and applications at the ease of almost every user, The technological advances are much faster than the understanding of their implications on the personal, community and societal level.in spite of this each generation of mobile communication is loaded with a number of applications. with the advancement of 6G, another method of reasoning ought to be embraced from the start that incorporates social mindfulness and obligation on an individual, community, and societal level.

This article also focusses on research challenges in 6G such as 6G novel networking technologies for example, Nanonetworking ; N-IoT which is based on molecular communication, Bionetworking ; B-IoT, optical networking, 3D networking. With the quick improvement of brilliant terminals and arising new applications, wireless data traffic has definitely expanded, and current cell networks can't entirely coordinate with swiftly growing particular necessities. To cope with challenges ahead, 6th generation (6G) mobile network is trusted relied on to project high dedicated norm of innovative variety and energy-productive transmission tactics.

A large portion of earth is enclosed with water in the form of oceans and sea, almost 70% of earth's surface is covered with water,. Ocean is an important part of human life as it regulates the climate and weather pattern. Sea water has significant impact on the world climate with oceans acting as the large heat reservoir. A larger portion of the underwater areas still remain unexplored. Since ocean plays a vital role in human's life, it is a dire necessity to explore the huge ocean volume. Underwater survey has led to significant exploration of ocean over the last decade. The primary aim of these surveys is environmental monitoring and structural scrutiny of underwater amenities namely drilling devices and pipelines and safeguarding marine infrastructures against intruders. The primary technology to perform these tasks is mainly relies on ability to wirelessly exchange information among submerged devices.

Marine and undersea communication has been developing lately with novel ideas. The chapter aims at briefing the conventional acoustic, optical communication methods and their possible challenges faced. It further proceeds with the advancements in marine communication with the assistance of 6G network.

The 6th generation network or 6G aims at achieving Internet of Everything (IOE) and merge computation, navigation and sensing with communications. (W.Saad et al, 2020) discusses the applications and challenges of 6G. Various driving applications like multisensory XR application, CRAS, wireless brain computer interaction, holographic telecommunications are mentioned. The notable characteristics of 6G communication are as follows

- Developing a ubiquitous and integrated network to achieve device to device communication around the world.
- Higher frequency, wider bandwidth and a peak rate of 1Tb/s to suffice the ever-increasing demand from users.
- An intelligent network combining artificial intelligence technologies.
- In-built security assured network by real time dynamic analysis.

The maritime application of 6G is tremendous. Various research identifying the same has been proposed. A heterogeneous network using LEO satellites, unmanned aerial and marine vehicles to establish an automated deep-sea network has been introduced in many researches undertaken under 6G communication. An overall architecture of 6G highlighting the unmanned intelligent network integrating aerial, marine, terrestrial and satellite communication can be viewed in *Figure 1*.

UNDER-SEA COMMUNICATION

Marine and undersea communication has been developing lately with novel ideas. The chapter aims at briefing the conventional acoustic, optical communication methods and their possible challenges faced. It further proceeds with the advancements in marine communication with the assistance of the 6G network.

Conventional Methods

The possible modes of underwater communication are

- Acoustic or sound wave communication
- RF - Electromagnetic wave communication
- Optical or Light communication

Figure 1. 6G architecture

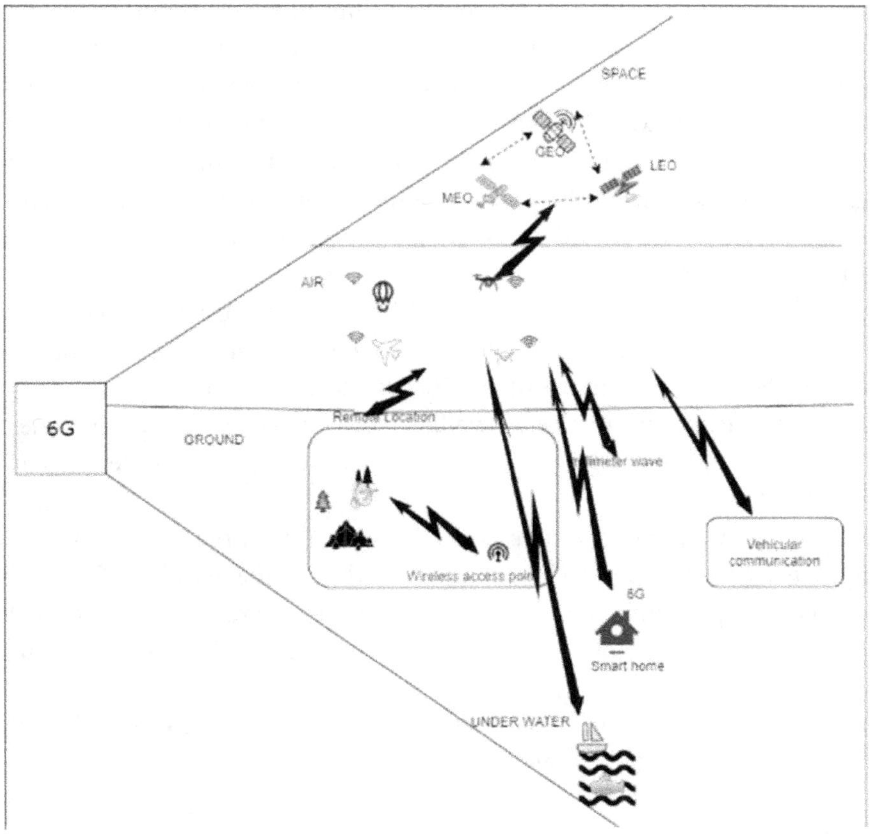

Presently, underwater acoustic networking is a popular communication domain. Major reasons for its success are as follows

- Large range of propagation (few kilometers)
- No Line of Sight (LOS) restriction.

However, the drawback of this conventional method is with speed of propagation which is very slow in comparison to speed of light. According to (Seongwon Han et al, 2014) it supports a limited data rate of 20 kb/s. Other disadvantages include narrow bandwidth, low carrier frequency, large propagation delay, and inadequate security.

Electromagnetic waves used in terrestrial communication are not a good option in undersea communication as waves get absorbed leading to poor signal strength. However they can provide a better performance at the air-water interface.

Optical communication provides a higher data rate and operates with higher efficiency. But the short distance communication is the major disadvantage of this communication.

Underwater wireless communication is a rapidly emerging arena. Underwater optics has been established to enable the high-bandwidth applications that these technologies offer. Including its complementary technology - underwater acoustics – it generated a great deal of interest. Due to its order-of-magnitude larger bandwidth, underwater wireless optical communication (UWOC) has spurred interest in a variety of underwater activities. There are numerous factors that contribute to the complexity of the ocean water ecosystem. Despite the fact that underwater optical wireless communication is still a work in progress (UOWC) though systems are capable of providing high-data-rate communications with great security, their performance is limited by the maximum distance that can be covered, due to scattering and attenuation effects. Underwater wireless communication (UWC) is a rapidly emerging turf resulting in emergence of number of technologies. Underwater optics has been established to enable the high-bandwidth applications that these technologies offer. Including its complementary technology - underwater acoustics – it generated a great deal of interest.

Over decades, humanity endeavoured to create viable UWC infrastructures in order to explore these undiscovered places. Numerous significant and unique challenges come with obtaining a reliable UWC, establishing reliable UWC poses a number of significant and unique challenges.

The first fundamental difficulty is that land-dwelling radio frequency (RF) technologies are inconceivable in underwater environment owing to radio waves' extreme dilution in water, which is exacerbated in conductive ocean water.

Underwater domain's severe energy constraints are second significant obstacle for wireless communication. Requirement for UWC is nodes need to be active for the duration of mission ranging from couple of days to years despite limited battery storage capacity. As recharging or substituting batteries in nodes is not cost effective more over a complicated job, energy-efficient communication is a dire necessity for UWC.

An amalgamation of the possible methods to culminate a more advanced technology that can encompass air, water and satellite communication is a substantial aim of 6G technology.

Novel Methods

VLC is an unlicensed high-speed communication technology that operates in the 400–800 THz frequency range. Different colour emitting LEDs and LDs (Laser Diodes) are utilised to achieve high transmission rates and appreciable speed. The UVLC system uses two kinds of LEDs to form a group for improved assessment. Theoretical foundation for UVLC is based on the fact that blue-green light had negligible attenuation in 450 to 550 nm range. *Figure 2* depicts the UVLC structure consisting of Automated Underwater Vehicles (AUV) sensor and optical fiber networks coordinating and communicating via VLC links. The advantages of UVLC over conventional methods include economical, high transmission rates, strong anti-interference capability, and high safety. LD based UVLCs offer higher transmission range and data rate. The preferred techniques to overcome challenges like fading involve identifying an optimum error detection and correction methods. (Hongyan Jiang & Ning He, 2020) discusses LDPC (Low-Density Parity Check) coding as a good technique and CAP (Carrierless amplitude and phase) modulation. CAP demonstrates a wider advantage for optical communication with spectral efficiency matching OFDM (Orthogonal Frequency Division Multiplexing).

Underwater Optical Wireless Communication (UOWC) offers extraordinary data rate at little power consumption. UOWC employs optical beams to travel and communicate underwater. Inherent optical properties of water like absorption coefficient, scattering coefficient, attenuation coefficient are used in determining the link budget of UOWC. Photomultiplier tubes (PMT) used in pure sea water can enlarge the field of view (FOV) of a receiver. By retro-reflecting modulated light back to the interrogating source, modulating retro-reflectors (MRR) are also utilised to lessen pointing and tracking needs. The

Figure 2. UVLC network

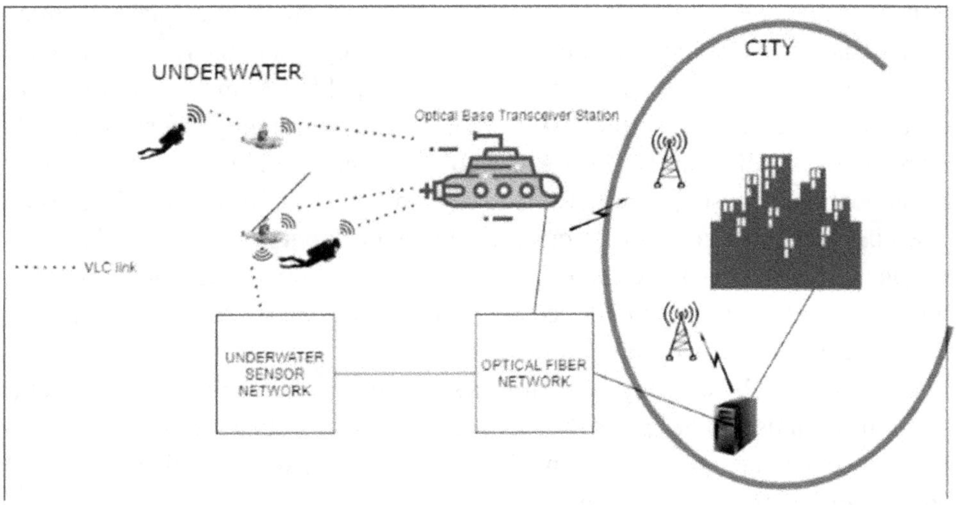

information data is modulated in an optical channel and transmitted to a longer distance. The transmitter employs projection optics and beam steering devices to focus and steer the optical beam towards the receiver's position. Information traverses underwater with parameters depending on the characteristics of water. An optical to electrical conversion is preceded with signal processing. *Figure 3* depicts the UOWC framework. No Line Of Sight (NLOS) configuration is also proposed which overcomes the hindrances of Line Of Sight (LOS) communication. The principle of total internal reflection is used to achieve the same. The light beam is projected with an angle greater than critical angle. This mode is energy efficient and cost effective in comparison to its counterparts.

Figure 3. Underwater optical wireless communication

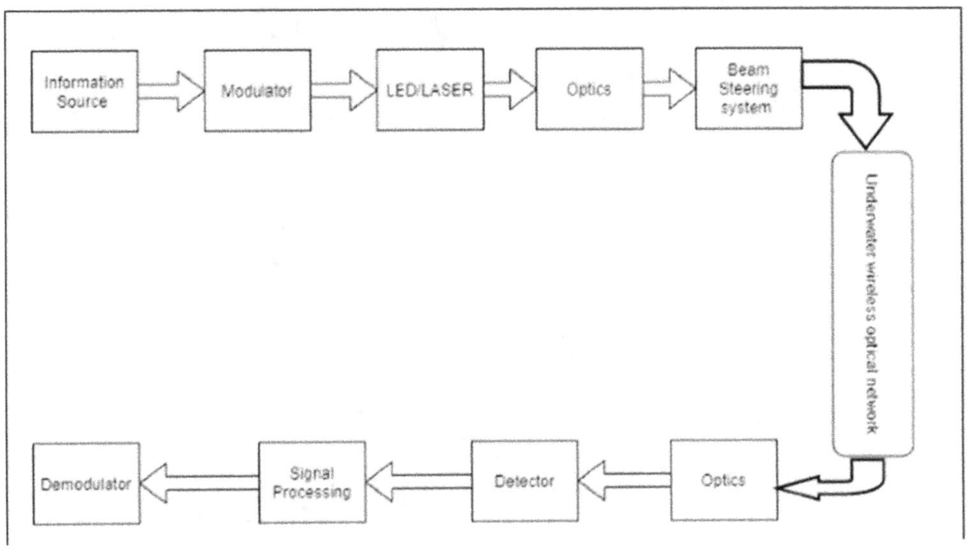

Underwater wireless Sensor network (UWSN) builds an efficient network for real time communication. The network's nodes which are connected by an autonomously adaptable wireless link that can automatically adapt to changing environmental conditions by changing system parameters. Sensor nodes at the bottom of ocean are interconnected via gateways and follow multi-hop paths. Sensor nodes are deployed underwater and perform the task of sharing and exchanging information with the other nodes. There are nodes installed on the surface as well to enhance communication efficiency. All three modes of communication (acoustic, optical, RF) can be used. The major functions of these nodes include monitoring, tracking and actuating applications. MIMO (Multi Input Multi Output) has an appreciable transmission range and enhances the channel capacity. RF- Electromagnetic mode of communication has greater advantage at the water-air interface over other modes and is tolerant to turbidity. Figure 4 delineates the different types of nodes in a layered form.

Optical and acoustic hybrid technology attempts at compensating the drawbacks noticed in both the communication when used independently. Figure 5 depicts an Autonomous underwater vehicle, an AUV (Autonomous underwater vehicle) is armed with both optical and acoustic modems. This vehicle establishes optical communication between the nodes and also can switch to acoustic mode to ensure a successful communication between far away nodes. Acoustic communication is highly appreciated for its long range communication, while this is the bottleneck in optical communication. No Line of Sight (NLOS) advantage of acoustic is used to transmit signal to remote, unreachable nodes. The co-ordination of optical and acoustic via an AUV has magnificent potential to overcome the mutual disadvantages. *Figure 5* depicts an abstract view of hybrid technology. It demonstrates an AUV communicating among all nodes using the hybrid mode.

Figure 4. Layered underwater wireless sensor network

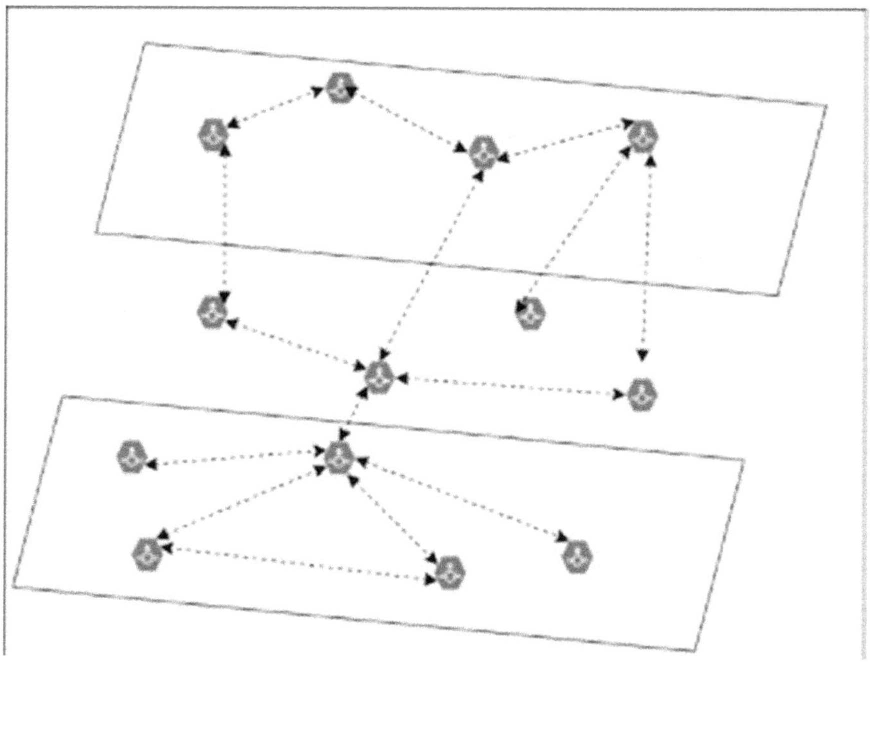

Figure 5. Hybrid model

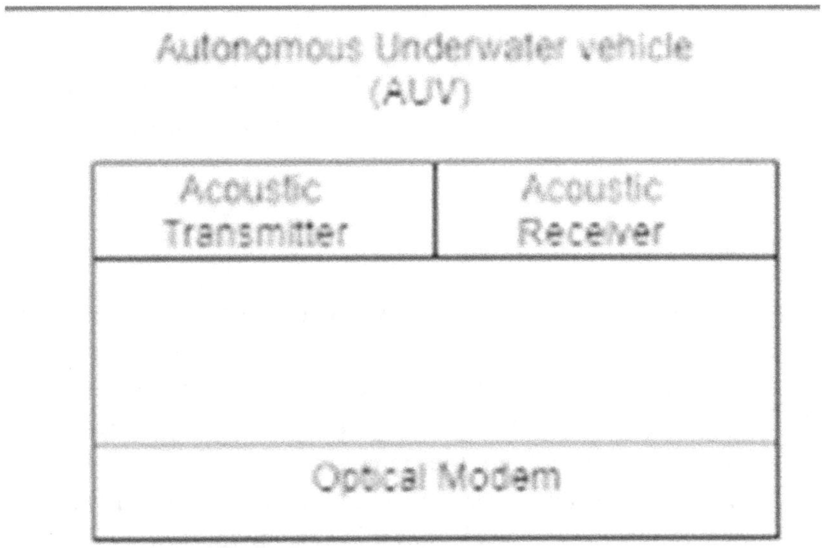

Basic Bio-Inspired Camouflage Communication (BBICCF) described in (J. Jiajia et al, 2020) broaches possible use of camouflaged killer whale sound as a mode of covert communication in military specific applications. The research aims to camouflage the information with the whistles and clicks of killer whales and dolphins. The AUV platforms created underwater mimics and recognises the sound of marine animals. Military underwater platforms (MUPs) are created for the purpose. Whistles, pulsed calls and clicks are few among many different kinds of sounds whales can generate from its vocals. Centralised, distributed and multi-hop communication can be established using platforms that mimic the whales and their sounds as pulse waves. The structure of BBICCF frame is as described in figure 6. The frame consists of destination address, source address, Synchronisation Code, Group number indicator, Group hop indicator and information Code. Destination and Source addresses are long duration pulses which resemble the whistles of whales. All other information is generated as the click sounds. The procedure involves calculating the pulse duration and identifying if the information favours a whistle or click. Based on the deduction, whistles and clicks are decoded as address and data respectively.

APPLICATIONS

Few of the notable applications and of achieving 6G based undersea communication is listed below

- The relatively least unexplored communication domain opens the door for many **military based covert communication**. AUV, gliders used in various military applications like anti-submarines, reconnaissance and surveillance.
- Deep ocean **marine animals' behaviour** can be observed and contemplated. This is necessary information to understand our marine ecosystem.

- The all-time demand **Internet** reaches every door today with the help of optic fibre cable web created undersea.
- Assisting the navigation in sea by identifying dangerous rocks and submerged wrecks.
- Identifying the characteristics of the ocean is a perk that comes with other advantages.
- UWSN method can also identify seismic activity and help in disaster management.

CHALLENGES

The expected launching of 6G networks is by 2030. Many advancements in research fields are yet to come. The suggested novel ideas present some practical inhibitions which are to be addressed.

(N. Chi et al, 2020; M. Katz & I. Ahmed, 2017) discusses the various challenges encountered in the VLC technique. Presence of marine animals in the communication path underwater can lead to LOS (Line of Sight) issues leading to attenuation of the signal strength. Bandwidth limitation of LEDs used and differing water attenuation coefficient are few other hindrances. Underwater optical turbulence, Scattering, absorption are summarises the challenges in the technique proposed.

Geographical location, complex physio-chemical characteristics of seas and ocean greatly affect the wireless optical beam transmission in (UOWC). Scattering is another parameter which can affect the direction of the beam. In pure water seas, presence of salts like KCl, $CaCl_2$ results in absorption and hence dip in blue-green region of the visible spectrum. Any unpredictable turbulence on the water surface like wind and rains can pose a major challenge to NLOS communication. Signal dispersion is the result of such disturbances. Presence of phytoplankton and detritus in water increases the scattering of light.

Establishing communication underwater is not a cakewalk. Characteristics of water are not as synchronising as that of air-based communication. One foremost challenge that needs to be addressed for installing UWSN network is security. Authorisation, confidentiality, authentication are some of the security issues. Sensor failure is yet another disadvantage

BBICCF method mimics the voice of marine animals. However, presence of many killer whales and dolphins in a region asks researchers to go in detail on voice of each individual animal. Speed is also a major concern as it proposes for a sound wave inside the water. The errors in the extraction of messages from signals need to be addressed yet.

CONCLUSION

The paper highlights the advanced possible 6G based under-water communications. All the novel hybrid techniques and their advantage over conventional technique has been the focus of the paper. The summarised methods discussed sheds light on the smart world that can be achieved by integrating all modes of communication via 6G. The marine and underwater communication is one among the unexplored domains which can be exploited to utilise for various applications. Both military and civilian can avail undue benefits from the following innovations made in the field of networking and communication.

REFERENCES

Bergano, N. S. (2012). Undersea fber-optic cables make the web worldwide. *2012 17th Opto-Electronics and Communications Conference.*

Chi, N., Zhou, Y., Wei, Y., & Hu, F. (2020). Visible Light Communication in 6G: Advances, Challenges, and Prospects. *IEEE Vehicular Technology Magazine, 15*(4), 93–102. doi:10.1109/MVT.2020.3017153

Elamassie, M., & Uysal, M. (2020). Vertical Underwater Visible Light Communication Links: Channel Modeling and Performance Analysis. In IEEE Transactions on Wireless Communications (Vol. 19). IEEE.

Gkikopouli, A., Nikolakopoulos, G., & Manesis, S. (2012). A survey on Underwater Wireless Sensor Networks and applications. *2012 20th Mediterranean Conference on Control & Automation (MED).*

Han, S., Noh, Y., Liang, R., Chen, R., Cheng, Y.-J., & Gerla, M. (2014). Evaluation of underwater optical-acoustic hybrid network. *China Communications, 11*(5), 49–59. doi:10.1109/CC.2014.6880460

Jiajia, J., Xianquan, W., Fajie, D., Xiao, F., Chunyue, L., & Zhongbo, S. (2020). A basic bio-inspired camouflage communication frame design and applications for secure underwater communication among military underwater platforms. *IEEE Access: Practical Innovations, Open Solutions, 8*, 24927–24940. doi:10.1109/ACCESS.2020.2970746

Jiang, H., Qiu, H., He, N., Zhao, Z., Popoola, W., Ahmad, Z., & Rajbhandari, S. (2020). LDPC-coded CAP with spatial diversity for UVLC systems over generalized-gamma fading channel. *Sensors (Basel), 20*(12), 3378. doi:10.339020123378 PMID:32549316

Jouhari, M., Ibrahimi, K., Tembine, H., & Ben-Othman, J. (2019). Underwater wireless sensor networks: A survey on enabling technologies, localization protocols, and internet of underwater things. *IEEE Access: Practical Innovations, Open Solutions, 7*, 96879–96899. doi:10.1109/ACCESS.2019.2928876

Katz, M., & Ahmed, I. (2020). Opportunities and Challenges for Visible Light Communications in 6G. *2020 2nd 6G Wireless Summit (6G SUMMIT).*

Kaushal, H., & Kaddoum, G. (2016). Underwater Optical Wireless Communication,&quot. *IEEE Access: Practical Innovations, Open Solutions, 4*, 1518–1547. doi:10.1109/ACCESS.2016.2552538

Li, W., Su, Z., Li, R., Zhang, K., & Wang, Y. (2020). Blockchain-based data security for artificial intelligence applications in 6G networks. *IEEE Network, 34*(6), 31–37. doi:10.1109/MNET.021.1900629

Saad, W., Bennis, M., & Chen, M. (2020). A vision of 6G wireless systems: Applications, trends, technologies, and open research problems. *IEEE Network, 34*(3), 134–142. doi:10.1109/MNET.001.1900287

Chapter 4
AI-Based Wireless Communication

Sanjana T,
BMS College of Engineering, India

Lalitha S.
BMS College of Engineering, India

Surendra H. H.
BMS College of Engineering, India

Madhusudhan K. N.
https://orcid.org/0000-0002-6863-1103
BMS College of Engineering, India

ABSTRACT

Artificial intelligence (AI) is one of the key enablers among quantum technology, smart meta-surfaces, dense antenna arrays, and mobile edge communication in 6G. The level of maturity achieved in the field of AI and development of computationally efficient hardware architectures with reduced costs have powered up the use of AI in different layers of wireless communication. Based on the learning, reasoning, and decision-making capability of AI, performance of wireless communication can be optimized. In addition, a whole new range of smart applications such as augmented reality (AR), virtual reality (VR), unmanned aerial vehicle (UAV), extended reality (XR) and holography, and autonomous driving, which demands high precision and low latency, can easily be accomplished by integrating AI into wireless communication. This chapter covers the role of AI in different layers, utilization of deep unfolding in physical layer, AI in mobile edge computing, explainable AI, federated learning, and AI for energy-efficient communication. The chapter concludes with research challenges and opportunities.

DOI: 10.4018/978-1-6684-3804-6.ch004

INTRODUCTION

The advent of technologically powerful applications such as Autonomous driving systems in Automobile, Internet of Everything, Industrial IoT and connected robotics in industries, Holographic Communications, multi-sensory networking, gaming in Entertainment sector, Augmented reality, virtual reality in Entertainment and Education sector, Wireless brain and computer interactions, heterogeneous networks operating in smart environments, remote surgery in healthcare sector have been the driving forces for 6G communication.

The different features offered by 6G include data rate of 1Tbps, low latency upto 1ms, maximum spectral efficiency, high mobility support, and AI empowered. It supports machine to machine and machine to human interactions, Internet to Everything dense networks (Jagannath, 2021), smart environments and cell-free massive MIMO. In order to provide very high bandwidth, communications are pushed to THz and mm-wave frequencies. The key features of 6G include Computation Oriented Communications (Edge intelligence), Contextually Agile enhanced Mobile Broadband communications (situational awareness) and Event Defined ultra-reliable low latency communications (K. B. Letaief, 2019).

Edge intelligence will play a key role in Intelligent IoT as it adds trust, resilience, monitoring and detection, reliability and guarantees efficiency and value-added services in the network. Depending on the computing resources and storage at the edge devices there are different cases that could be realized. For example, training the model can be either done at the edge or cloud, and also its inference can be either done at the edge or device itself. This would definitely affect the quality of service provided to the end users. AI can be realized for the edge services or on the edge. AI for edge services includes managing energy efficiency and optimization based on location. AI on the edge includes data intelligence, computing, satisfying real-time requirements (Peltonen, 2020). The AI algorithms in 6G networks are generally applicable at the edge, or on different layers of wireless communication or in mobile applications or in self-configuring networks. The various network entities that would be needed to realize AI applications include caching, computing, wireless power transfer and communication. AI would assist in design and optimization of networks, network monitoring and management, radio resource management and security (Md Arifur Rahman, 2019). Artificial intelligence (AI) plays an important role as it not only contributes in design and optimization of core network, but also provides exciting services and applications for the end devices. Strategy Analytic have predicted that by 2023, 80% of the smart phones will be AI enabled (Strategy Analytics, 2021).

The highly dynamic nature of environment and huge volume and variety of data generated motivates the use of AI from network edge to the core. AI makes the 6G architecture intelligent in terms of flexibility, adaptivity, computational, fast learning, reasoning, smart decision making and agile. It is called Intelligent Radio (IR) as the operating system which exists between hardware and transceiver algorithms, is capable of estimating the abilities of hardware and based on which configures the algorithms (K. B. Letaief, 2019).

Machine Learning (ML) and Deep Learning (DL) which are part of AI have emerged to be applied to different layers of 6G communication such as physical layer, MAC layer, network layer and application layer. For example, they are used for network optimization, channel estimation, equalization, encoding, decoding, spectrum allocation (Naser, 2021). Apart from these AI and ML algorithms are so powerful and scalable as they use large volumes data as well computing capabilities to adaptively learn and perform predictions so as to realize smart, efficient and self-aware networks. There will be lot of data traffic and high energy consumption by devices operating in heterogenous and ultra-dense networks.

This motivates the use of data-drive AI and ML techniques to sense the variations in data traffic and accordingly allocate resources satisfying user demand, smartly coordinate between different network entities and also improve energy efficiency (You, X., Wang, 2021).

BACKGROUND

There are several significant contributions and ideas proposed in the field of AI in 6G networks. AI enabled 6G includes contribution in radio interface, traffic control, network orchestration, network security and optimization. As a part of air interface, AI is used for channel estimation, decoding, detection, modulation recognition and optimization. Traffic prediction, classification, caching and routing is done intelligently using AI techniques. Managing the network faults, QoE, QoS and resource management can be effective by incorporating AI. Additional capabilities such as mobility management, energy efficiency and ensuring security by detecting intrusion and abnormalities in network traffic are offered by AI enabled 6G network (Shunliang Zhang et al., 2020). Privacy and security has been a major concern even in 5G. This can be overcome in 6G using ML algorithms for detection and prediction of traffic, intrusion and anomalies and implementation of security policies. Data privacy can be eliminated by implementing distributed AI and federated learning (B. McMahan, et al., 2017).

Data driven type of DL has an advantage to learn channel model and new channel states without any mathematical model for the channel (H. Huang et al.,2020). Further due to parallel processing nature of DL, it is faster and energy efficient compared to conventional approaches (T.O. Shea, J. Hoydis et al., 2017). Deep reinforcement learning (DRL) has been popularly used for efficient resource utilization and the overall end to end optimization of the air interface is possible by the use of Autoencoders. ML algorithms are light weight algorithms which can be used for energy saving (A. Zappone et al., 2018). Some of the supervised learning techniques that can be used for classification and regression type of problems in the air interface and network layer are KNN, decision tree, random forest, SVM and Naïve Bayes. In order to categorize the unknown data, unsupervised learning techniques such as k-means, Self-organizing map, Restricted Boltzmann machine and Hidden Markov model can be used. In practical scenarios, where there is a mixture of labelled and unlabelled data, semi-supervised learning techniques would prove to be useful. Both model based and model free RL is widely used for spectrum sharing, handover and user scheduling. However, this won't be practical for situations with large action space. There are few typical DL approaches that are suitable for prediction, decision making, and address sequential problems and capable of handling large data set are generative adversarial network (GAN), DRL, RNN, CNN, deep belief network (DBN) and multilayer perceptron (MLP). Both ML and DL techniques are very much suitable for highly dynamic environments.

Channel estimation is the most important part of air interface. The accuracy of channel estimation is crucial for reduced error rates. DL (W.S. C. Wen et al., 2018), DNN, CNN (W.U. David Neumann et al., 2017) methods are extensively reported in literature for both channel estimation and detection. Further, modulation recognition, classification, traffic prediction and classification can be effectively performed using LSTM and CNN (C.C. C. Huang, 2017). End to end optimization of the 6G communication system can also be achieved through AI like combination of DNN and GAN (B.F.J. H. Ye G. Y. Li, 2018) and Autoencoders.

DL, RL, DRL techniques are proposed to effective solutions for resource management and routing in network layer. DL plays important role in caching of traffic. Estimation of performance indicators of network such as QoS, QoE is possible using Neural network models. Self-organizing map (SOM) performs well in detecting faults and DNN in fault prediction. Apart from these, mobility management and energy efficiency are also important concerns which can be efficiently managed by NN, RL and ML algorithms. Some of the open research challenges in adopting AI in 6G are developing high quality data set, real-time and intelligent network security and optimization (Shunliang Zhang, 2020).

Some of the Key Performance Indicators (KPIs) that are used to evaluate the performance of 6G networks include QoS, Quality of Experience (QoE), spectrum efficiency, energy efficiency, latency, reliability, mobility and localization. Most of the recent applications are computationally intensive and data-driven, they demand low delay such as autonomous vehicle, AR, VR etc. In such scenarios, AI would be the ideal choice to simplify computation, perform data caching and intelligently communicate. Integration of AI in the 6G architecture is easily supported as processing capability is improved to satisfy high data rates. The applications of 6G where AI plays importance is shown in Figure 1.

Figure 1. Applications of AI in 6G

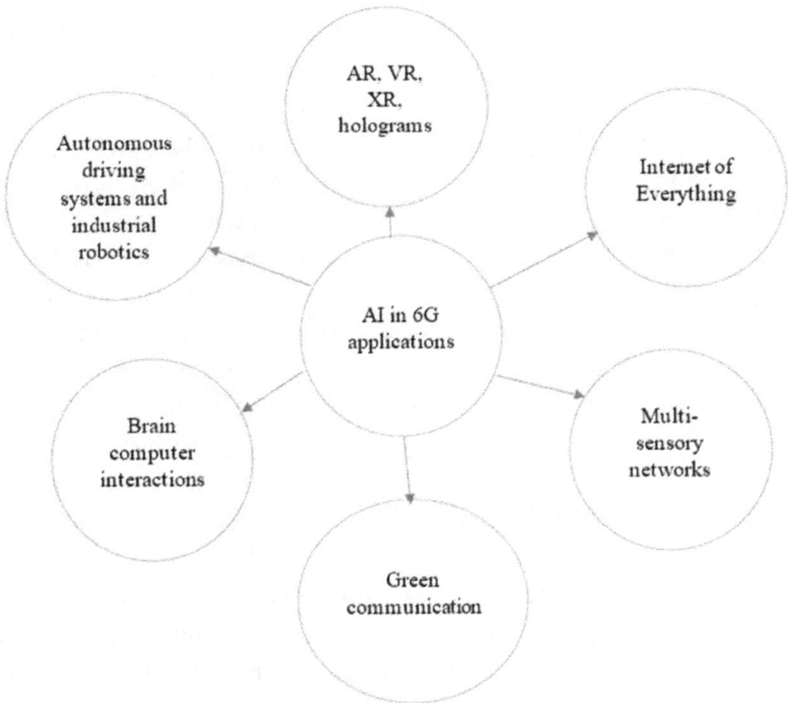

An architecture of 6G network enabled by AI would consist of sensing layer, data analytics layer, control layer and application layer augmented with intelligence. There are two ways of viewing AI enabled 6G network, either in terms of its use at device, edge and cloud level or physical, MAC layer, data link layer, network layer and application layer. AI which includes Machine Learning (ML) and Deep Learning (DL) can be used to implement tasks which were conventionally not possible or difficult to achieve because of tedious computations. They could be simply used for solving problems involving classification, regression, dimensionality reduction, clustering, pattern recognition and decision making.

They give deep insights about the data collected from the network with the help of which accuracy and other performance metrics could be improved. But the AI based algorithms should satisfy requirements such as being computationally efficient, hardware and energy efficient, optimized and robust to dynamic radio environment so as to be used in the next generation networks.

MAIN FOCUS OF THE CHAPTER[1]

In this chapter, explicit interest is given to the key distinguishing factors of AI in 6G networks compared to all the previous generation networks that is envisaged to revolutionize the future. They include Deep unfolding in the physical layer, AI enabled Mobile Edge computing, Explainable AI, Federated learning (Distributed AI) and AI for energy-efficient communication. The importance of data driven, model driven approaches and end-to-end communication system modelling with their advantages and disadvantages will be discussed. Since 6G mostly supports mission critical, safe and tactile applications, importance is given to reducing latency, traffic as well improving energy efficiency and security. 6G will be open to changes at the edge as well core computing operations from AI. Design of efficient lightweight and model driven algorithms are encouraged at the edge. Deep unfolding can replace all iterative and computationally intensive tasks especially present at the physical layer. Based on the requirements at the edge such as training data size, privacy level and hardware constraints, Federated learning can be enabled at the edge and device levels. This would solve problems like intrusion detection and mobility prediction in wireless communication. Some of the hardware efficient lightweight algorithms will be discussed for localization, sensing, tracking and security purposes. The important task would be combining all these scattered capabilities of AI to meet the needs as well as provide best experience to the users in diversified scenarios. The chapter ends with future research opportunities in integrating AI in 6G networks.

Present your perspective on the issues, controversies, problems, etc., as they relate to theme and arguments supporting your position. Compare and contrast with what has been, or is currently being done as it relates to the chapter's specific topic and the main theme of the book.

AI IN DIFFERENT LAYERS OF 6G NETWORK

AI is an important part of 6G network architecture. It brings in changes and adds intelligence in the core network. There are two ways in which AI could be used for wireless communication. AI could be either used to realize the entire end-to-end system which has the capabilities of self-learning and optimization or it can be used to individually realize certain signal processing functions. AI empowered network architecture will have the ability to self-learn, self-organize, optimize, self-protect, perform reasoning and take decisions wisely (M. J. Piran, 2019).

In this section, the role of AI in different layers of 6G network is briefly described. As a part of physical layer which is also called the sensing layer, AI would provide appreciable solutions to dynamic spectrum sensing, interference detection, monitoring of ever-changing radio environment, resource allocation, security, beam forming, channel estimation and equalization, link adaptation, error correction, coding, synchronization, positioning, signal detection etc. Classification of data whether it is important or not for transmission to the upper layers and fusion of data from different sources is required for improving energy efficiency and meaningful interpretation of data respectively. These two tasks are very challenging, yet can be implemented using various ML and DL techniques. A wide range of spectrum including sub 6GHz, mm wave, Tera Hertz, optical frequency bands are explored in 6G, hence the dynamic use of frequency and unknown scenarios can be predicted by applying AI and ML algorithms. The entire physical layer can be replaced using Autoencoders or model driven approaches can be used for various signal processing modules. Open radio access network (O-RAN) is inclined to incorporate intelligence in the core operations. Some of the commonly used ML and DL algorithms used in physical layer includes CNN, RNN, SVM and KNN for spectrum sensing and resource allocation, CNN and Bayesian learning for data fusion, ANN to detect anomalies in data over network.

Data link and analytics layer performs data mining, dimensionality reduction, resource management, mobility management, congestion control, privacy and security and asymmetric traffic management. Principal Component Analysis (PCA) and isometric mapping techniques are popularly used for dimensionality reduction and filtering of data which in turn reduces computation time and storage. Markov decision process (MDP), RL, game theory, Q-learning and Deep Reinforcement Learning (DRL). AI techniques are primarily used in Decision making concerned with handover, beamforming in mm wave and THz etc.

Control layer functions are routing management, power control, optimization of network parameters, network slicing, virtualization and resource management. Integrating AI with the network layer enables self-healing, organization, configuration, handover optimization, fault recovery, QoS and energy optimization. AI aids the control layer in choosing appropriate actions by its learning and optimization capabilities. DL has majorly contributed to Software Defined Network (SDN) and Network Function Virtualization (NFV). Load balancing, security and routing can be automated by AI techniques.

AI-enabled application layer offers intelligent and automated service provisioning. Application layer is designed to be more client centric. Applications such as intelligent transport system and vehicular networks, smart city, Intelligent IoT, smart health, UAV controls and many more are supported. AI provides real-time and low latency support especially for applications such as guiding UAV, autonomous driving, remote surgery and industrial control. The performance metrics which includes Quality of Service, spectrum utilization, storage and energy efficiency, Quality of experience (QoE) and Quality of Trust (QoT) can be enhanced by incorporating ML and DL algorithms in different layers of the network. The role of AI in different layers of wireless communication is as shown in Figure 2.

AI for Physical Layer

ML and DL algorithms are used for channel encoding and decoding, sensing and localization, channel estimation and equalization, signal detection and classification (Md Arifur Rahman, 2019). Optimization of the physical layer is extremely important in order enhance the capabilities of the higher layers. (Zhang et al., 2019) proposed the use of deep neural networks (DNNs), supervised and unsupervised learning to optimize different functions of the physical layer and enhance its security. Hong et al. (Hong, T, 2019) proposed ML based antenna design for physical layer communications. AI is employed in Radio Access Network RAN design. artificial neural networks (NNs) can be used to jointly optimize several functions of the physical layer. The implementation of AI based solutions is a real challenge in practical scenarios. AI and DL are used for adaptive modulation, coding and signal processing. Convolutional neural networks (CNNs) are used to improve the accuracy of channel estimation and reduce overhead (P. Dong et al., 2019). As Neural Networks are suitable for parallel computations, they are best candidates for low latency applications (A. Zappone, 2019). Most of the computationally intensive tasks can be

Figure 2. Role of AI in different layers of wireless communication

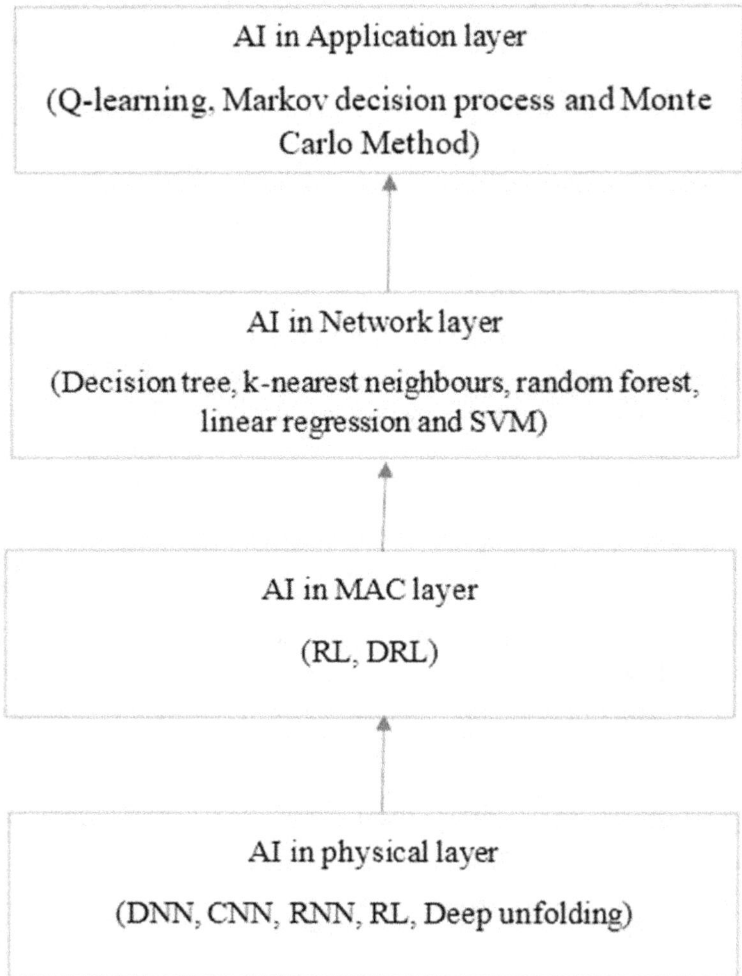

simplified by using AI models. Amalgamation of human intelligence and AI is preferred for the physical layer design (S. Han et al, 2020). The air interface can be optimized in real time based on the information obtained from the application layer. Minimum mean square error (MMSE) estimation is outperformed by supervised learning methods. Bidirectional recurrent neural network (RNN) is suitable for detection without the need for channel state information (CSI) (N. Farsad, 2018). Supervised learning approaches such as k-means clustering, isometric mapping, principle component analysis (PCA) are used for channel estimation, decoding, equalization, beamforming etc. Modulation, precoding, beam switching and interference cancellation can be optimized using unsupervised learning. Reinforcement learning (RL) can be used for beamforming, selection of modulation techniques, energy harvesting and monitoring channel. Data driven DL/DNN approach is well suited for channel estimation, decoding and signal detection in OFDM (Ye H, 2018) and MIMO system. Decoding using DL is efficient and minimizes bit error rate. Model driven DL approaches are less complex compared to data driven approaches as domain knowledge is involved. Bayesian optimal estimators can be used to realize many physical layer functions. Massive MIMO detection (Tan X S, 2020, He H T, 2020) and polar decoding (Xu W H, 2017) are the amin applications of model driven DL techniques. Model driven approaches are eagerly used in open Radio Access network (RAN) of the 6G architecture. Deep unfolding is popularly used for signal processing tasks in the physical layer to reduce complexity, enhance efficiency with limited resources.

An extended version of model driven approach is interactive learning design paradigm (ILDP) which does adaptive learning by using domain knowledge and extracting information in real-time from dynamic environments. It is mainly used along with DRL for beam tracking without the knowledge of channel model (Zhang J J, 2019). Because of its smart sensing capabilities in dynamic environments, it is preferred to be used in the physical layer. (You X H, 2019) proposes the iterative Bayesian solver with deep learning for the implementation of the entire end-to-end system. DNN are generally used to reduce computational complexity and helps in planning and coordination of caching and computing resources (Zhang Z M, 2019). The use of AI in the physical layer also brings in security concerns. But with the combination of ML and big data analytics, authentication and fault detection can be achieved (Shakiba-Herfeh M, 2020). Data analytics plays a very important role along with AI to improve 6G system performance. There are analytics concerned with description of network performance, diagnosis of faults and root cause of anomalies, prediction of network behaviour and prescriptive which uses the prediction results to optimize decisions on resource allocation and edge computing etc (K. B. Letaief, 2019).

AI for MAC Layer

Dynamic resource scheduling, multiple access, adaptive power control, fault detection, network planning and optimization, interference management and user clustering (Md Arifur Rahman, 2019). Reinforcement learning, supervised learning and transfer learning techniques are employed in the MAC layer for radio resource management and control, mobility management, prediction of data traffic, smart protocol selection. AI is used for scheduling and control operations. Radio resource management and control is customized for specific user based on the decisions taken by DL techniques used as a part of AI based scheduler and controller. Deep Reinforcement Learning (DRL) which outperforms conventional Reinforcement learning (RL) is popularly used for radio resource management and resource allocation. But the actions and the decisions taken by DRL cannot be understood by the radio engineers (Guo, Weisi, 2020). Thus, it needs to be explainable for the system to be transparent. ML algorithms can be used for managing congestion in Internet of Everything networks.

AI for Higher Layers

Dynamic network orchestration, Dynamic network slicing, Measurement and monitoring, Security enhancement, Handover and mobility management (Md Arifur Rahman, 2019). In the network layer, supervised learning approaches can be used for the purpose of caching, classification of data traffic, fault detection, and throughput optimization. Whereas routing, traffic management, mobility management, resource allocation, network slicing and handover management are addressed by unsupervised learning approaches such as decision tree, k-nearest neighbours, random forest, linear regression and SVM. In the application layer, Reinforcement learning approaches such as Q-learning, Markov decision process and Monte Carlo Method are applied for rate allocation, error prediction and caching. Packet scheduling, prediction and classification of data traffic, security of network layer can be achieved using RL.

DEEP UNFOLDING IN PHYSICAL LAYER

Deep unfolding is mainly popular in the physical layer as it is used to replace the computationally intensive iterative signal processing tasks. It uses a combination of DL/ML and domain knowledge. It can be used at the edge or in 6G radio. Deep unfolding mainly takes care of the computational resources and memory requirements. These features aid to reduce latency and improve reliability and throughput. Deep unfolding is a model driven approach which is used to perform authentication, data traffic prediction and control, channel estimation, coding, security and error correction (Jagannath, 2021).

Some of the important and computationally complex signal processing tasks include estimation of channel, interference, channel encoding, decoding and signal detection. These operations are majorly iterative in nature. Some tasks involve matrix inversion, singular value decomposition and Eigen value decomposition which takes more time to converge. One of the optimal signal detection schemes is Maximum likelihood (Mld) detection. This approach is highly complex. There are other alternative approaches such as Approximate Message Passing (AMP), minimum mean squared error (MMSE) and Zero-Forcing which are suboptimal and less complex. However, MLd detection can be effectively replaced by deep unfolding algorithm.

Deep unfolding is basically a low complexity neural network architecture that combines the best of DL and domain knowledge. Iterative algorithms are unfolded into different layers of the NN. For example in (M. Un, 019), Alternating Direction Method of Multipliers (ADMM) was proposed to unfold algorithm for MIMO detection. Maxlog maximum a posteriori (MAP) and DNN are combined to address error correction and MIMO detection in (Y. He, 2019).

AI ENABLED MOBILE EDGE COMPUTING

Edge intelligence supports those applications of 6G which require low latency, optimization of data traffic with improved security and privacy. Edge intelligence is the main factor that distinguished 6G from 5G. It basically means to move some processing and storage from cloud to edge i.e. near to the network nodes or end users or data providers. This is also called distributed AI as the intelligence is moved from cloud to the edge. There are several challenges that need to be addressed while implementing AI at the edge. They are hardware resource constraints, memory constraints, handling heterogenous data and energy limitations. In order to address this training and inferencing of AI is carried out at different levels of the architecture namely cloud, edge and devices. The key performance indicators to evaluate AI on edge are reliability, efficiency, cost and security. Some of the AI techniques such as generalized adversarial networks, swarm intelligence, game theory and genetic algorithms are used to address inconsistency due to heterogenous data at the edge. Light weight AI techniques are suitable for edge computing (Peltonen, 2020). Data integrity and authentication is utmost importance in edge intelligence. Federated learning is a preferred technique used at the edge distributed training of data, local model is trained using that and model parameters are sent on to the cloud. AI at the edge is used in many applications like autonomous driving, healthcare, IoT, sensing, AR, VR, robotics, vehicular networks etc.

Distributed intelligence is used in internet of everything. Challenges here include distribution of dataset among the edge devices, computation and storage limitations (I. Tomkos, 2020). However, the advantage is that it improves the security and privacy of the network. Blockchain and ML are also used at the edge for the purpose of authentication and prediction of malicious attacks in the network respectively (E. Calvanese Strinati et al, 2019, G. Gui, 2020).

Edge computing in 5G included software defined network and network function virtualization (Han, G, 2017). But in 6G it is visualized to support data processing and AI in order to reduce bandwidth and latency by moving resources closer to the end users. This reduces the response time. The benefit of using intelligent edge computing are proximity, low latency, support for location aware services and standalone operation using local resources. Open research challenges of intelligent edge computing in 6G includes its deployment in holographic communication, time engineered applications and multi-sense networks. Operations performed at the edge includes data acquisition, pre-processing, storage and computations. If these are effectively performed with the help of AI/ML algorithms then it is called Edge intelligence. Data needs to be aggregated from multiple devices connected to the edge, these are used for prediction, analysis and training the AI model. But since the results obtained from AI/ML algorithms are extremely dependent on data, prime importance would be to protect the data from attacks. Some of the possible data breaching attacks are data poisoning (affecting training phase), data evasion (affecting testing phase), denial of service (DoS) attacks, API-based attacks (P. Porambage, 2021). Adversarial training (F. Tramèr, 2017) and Defensive distillation (M. Soll, 2019) are possible solutions for data poisoning and data evasion attacks. Security and privacy of user information are closely related. One of the possible solutions to overcome the privacy attacks is Federated learning. Intelligence network management architecture called Zerotouch network and Service Management is proposed in (ZSM, 2021) which employs AI to provide end-to-end service and network management. When it is said that AI/ML is also used for security, then few questions arise regarding the accountability, visibility and trust. Both data and model are expected to be secure in training and testing phases. NN, DL and supervised learning algorithms are used for protecting the privacy of user data, location and communication among devices in IoT.

Generative Adversarial Nets (GANs) is proposed in (Y. Xiao, 2020) to be used at the edge, as it has the ability to synthesize and automatically learn. In general, Resource-efficient, Data-efficient, Scalable and Decomposable, Distributed and Personalized AI are required to be designed so that they operate with limited resources at the edge. DRL, federated and transfer learning used limited resources and data and still achieves low communication overhead. The best of the abilities of supervised and unsupervised learning is used for designing data efficient AI. Scalable and decomposable AI is required for managing parallel processing and heterogenous hardware and software. Federated learning comes under distributed AI. In situations where machine makes wrong decisions and there is need for human intelligence to intervene and take decisions, then personalized AI is required. All the social and economic impacts of AI in 6G need to evaluated using new performance metrics.

Self-supervised learning, generative adversarial networks (GANs) and automatic ML (AutoML) are some of the effective ML techniques used to enhance self-learning capabilities at the edge. AutoML automatically searches an appropriate ML approach which is suitable. But it is most important that the practical implementation of these algorithms must be simple and effective.

Self-supervised GAN with multiple generators is proposed in (N.-T. Tran et al., 2019) which synthetic data (with varied distribution of data) is generated from multiple generators, based on which it automatically learns the features and builds ML model to classify data. In order to reduce the computational complexity, transfer learning is used. Edge intelligence is always limited by the small coverage area of the edge servers and limited data set. Popular AI techniques used for edge intelligence include RL, transfer learning, semi-supervised learning and autoML. These are capable to adapt to the changes in the network.

EXPLAINABLE AI IN 6G

In case of time and safety critical applications such as remote surgery and autonomous driving, there is a need to understand the decisions of AI/ML models in order to build trust. Hence the AI algorithms must be designed and developed to be trustworthy and explainable to humans. Since AI is the core of 6G, it necessitates to build confidence of humans in its decisions or predictions. There is a lack of transparency in case of data, bias and decisions taken by DNN. DRL cannot explain its actions and the choice of bias. There is a need to develop AI algorithms to quantify and explain its decisions to the human experts with evidence and reasoning. There is a need to improve algorithm design by knowing which data is important for taking decisions, which features are essential and what are the values of hidden bias. The best way to build trust is to choose correct data features, explain the logic and convince that the prediction is correct and agreeable to human expert knowledge and experience.

Different ways of explaining includes visualization techniques (output highlighting the features based on weights in NN), Hypothesis Testing (formulated based on input and output decisions) and Didactic Statements (in natural language to explain the features and functions used to make decisions, explaining relevance of each layer in NN) (Guo, Weisi., 2020). In order to explain the process of learning and decision there needs to be good human to machine interface. Among different AI algorithms, decision trees and Bayesian learning have good explainability in terms estimation of parameters and confidence level in output decision. Whereas algorithms like DRL have zero explainability in terms of bias.

There are many factors that influence change in output of AI techniques. They are different sets of data, data with noise, changes in model parameters. RL and non-linear classification techniques have less explainability. There is always a chance that machine decisions and perception of human to the same situation is different, hence study of human psychology to be used for AI models is a recent topic of research.

There are better ways of designing the model to improve understandability. There are several local and global model reduction techniques that reduces its complexity. Examples that can reduce complexity and improve explainability includes pruning of NN, removing sparsity in data, reducing parameters, replacing fully connected layers by convolutional layers and quantization of data (W. Wen, 2016).

FEDERATED LEARNING IN 6G

Distributed training at the device level is called Federated learning. The training data and model training is carried out at the device. This upholds the security and privacy of the data because of its distributed nature rather than centralized. However, there needs to be frequent communications with the edge devices or the cloud to update the parameters of the local model running on the device. Global model is shared among distributed devices. Main challenge in federated learning would be able handle heterogeneity in hardware devices. One of the promising solutions for this challenge is transfer learning which transfers the design of one architecture to another (K. B. Letaief, 2019).

Federated learning is a decentralized approach where a global model is shared among a set of devices, and these devices individually trains the model using its local dataset. There are three steps involved in federated learning. First step is initialization phase where a device registers to its nearest cloud based on need/request and gets connected wired or wireless. It joins in training shared global model. Out of the registered devices, cloud chooses a group of registered devices to participate in training. Second step is the training phase where the selected devices perform training and send model updates to the cloud. Third step is the aggregation phase where updates from all the selected devices are used to build a new global model. This global model will be shared by cloud to again a selected number of devices in the next iteration. This process continued until the model converges or satisfies some criteria (Y. Liu, 2020). Evaluation of model parameters obtained from selected devices is done at the server and this is used to update the global model.

There are lot of challenges involved in Federated learning such as unreliable and slow devices used for training, prone to security attacks like poisoning attacks and data breaches, heterogeneity among devices, global model is huge for a single device, high cost because of multiple devices used for training and high communication overhead. Communication overhead can be reduced by reducing number of gradients and communication. Federated learning model must be secure and robust to different types of security attacks. Homomorphic encryption (HE) is also helpful in overcoming data breach attacks.

Communication efficiency between the device and cloud can be improved by either aggregating the model updates asynchronously or by reducing the number of iterations used for training the model using Stochastic gradient descent (SGD) algorithms (Y. Shi, 2020). Robust Aggregation Algorithm, detection mechanism and Reliable Reputation Management are used to perform effective aggregation, detect malicious devices and evaluate reliability based on metric. Privacy in case of federated learning can be improved by employing Differential privacy (DP) techniques (M. Abadi, 2016), pruning DNN, reduce number of gradients, use of parallelization and distillation for effective training and sharing of weights.

Wireless Federated Learning (WFL) came to importance in 6G as there was increased computing capabilities at end users and also there were privacy concerns. Privacy was maintained in WFL as the devices did not share the raw data with the cloud instead shared the model parameters. As the amount of data sent from the device to the cloud reduced, the communication cost and latency also reduced. Both model training and decisions are taken at the device level. But the WFL has to deal with heterogeneity with respect to resources, devices and data. Model shared by the cloud is trained in a distributed manner (Bouzinis, Pavlos S, 2021). There are several applications in which WFL plays an important role in smart grids, autonomous vehicles, Augmented Reality etc. There exists trade-off between model performance and speed, latency and number of training sessions. Future directions of research with regard to WFL would be asynchronous communication which would allow participants to join the process of training at any time. WFL will be widely used in Internet of Everything network.

FL is data driven learning in which group of devices collaboratively learn the behaviour of the network. It avoids the sharing of raw data to the cloud which is the case of centralized approaches. This will reduce the latency and waiting time. It also reduces communication overhead as large volumes of data need not be sent to the cloud as in case of centralized systems. Increased amount of computing power, resources and good communication capabilities of devices used in IoT networks have motivated the use of FL (Naser, 2021).

FL can be used effectively implement Visible light communication (VLC). GANs can be used to operate on local data sets which represent different conditions of the network when FL is used in VLC system (A. T. Z. Kasgari, 2020). Distributed GAN (DGAN) was proposed in (Q. Zhang, 2021) which is more generalized and is used for channel modelling and estimation. Basically, FL is a good technique which will ensure physical layer security. One of the interesting areas of research is optimizing FL in VLC with Reconfigurable intelligent surfaces.

FL can also be used to aid base stations in allocating resources. Interference and resource management, control of the network and clustering can be optimized by the prediction results obtained from the FL model. There are two types of FL namely federated reinforcement learning (FRL) and federated supervised learning (FSL). In case of FRL each device learns from the other devices which work in different environments. Whereas, in case of FSL, each device contributes its learning to the base station which aggregates it and updates the global model. Some of the signal processing operations performed at the lower layers of the network such as signal detection, radio resource management, beamforming, channel estimation and network behavior predictions have motivated the use of FL.

AI TOWARDS ENERGY-EFFICIENT COMMUNICATIONS

A combination of conventional and state-of-art ML and DL techniques can reduce the complexity, improve accuracy and also reduce latency. This will effectively reduce the energy consumption and also will be suitable for u ltra-reliable low latency applications of 6G. some of the conventional AI techniques include linear regression, Support Vector Machine, and K-means clustering which are suitable for reducing energy consumption. Conventional AI techniques also include Heuristic Algorithms such as Particle Swarm Optimization (PSO) and Genetic algorithm that gives better results with low latency. Some of the state-of-art AI techniques include DRL, Transfer Learning and Federated Learning. DRL improves energy efficiency by optimizing resource allocation, control of power allocation and computation management at the base station. Transfer learning uses knowledge acquired from the system to reduce computations, training data and hence the energy consumption. In Federated Learning, since instead of raw data only model parameters are sent to the cloud after training, the amount of communication overhead is less (Mao, 2021).

CONCLUSION, CHALLENGES AND RESEARCH OPPORTUNITIES

In this section, some of the challenges in 6G applications and challenges with respect to design of future AI techniques are discussed.

6G is expected to support multisensory extended reality application which gives completely immersive experience based on the human sensory inputs. The main being visual and haptic senses. There is a need to satisfy QoE and QoS requirements of the end users. Such applications expect the allocation of maximum resources and minimum delay. This is mainly used in the Education and Entertainment sector. The problems associated with its implementation using 5G included compression of video and images and latency issues. These issues can be overcome in 6G. But there is always a challenge with respect to security and privacy attacks. AI techniques need to be developed to address this issue.

Autonomous Systems in Industry and Automotive

Since these applications expect no human intervention, they demand for complete control over computation, resource and communications. UAV networks suffer from WiFi based attacks, spoofing, eavesdropping and DoS attack. Privacy and security is also a concern in case of autonomous driving system with respect to theft and hackers. There are several ML and DL techniques proposed in literature to address the issues of privacy and security, but still there is scope for improvement. Energy consumption is also an important research topic in this area.

Brain Computer Interactions (BCI)

Extended reality and BCI can be merged to help the disabled people. Wireless communication is supported by 6G. Security and privacy concerns about the sensitive information transmitted need to be protected by novel ML and DL techniques (Minghao Wang, 2020). Biases need to be set based on human response and mental state. There is a scope to develop explainable intelligence framework for the same (Guo, Weisi., 2020). Digital twin could be developed to improve the performance. Defensive mechanisms need to be developed to protect against ML/DL attacks.

Tactile Internet to Support AR VR Applications and Holographic Communication

This involves interaction among human and machine. It involves interactive experiences with very low latency. It is used to give remote support, industrial maintenance and surgery using robotics etc. Algorithms to support super low latency is required. Creating digital holograms require data rates above 1Tbps which is supported by 6G (M. J. Piran, 2019).

Challenges Related to FL

Due to dynamic environments, some devices may quit FL because of poor channel conditions, this may be led to problems in convergence. Other concerns in FL will be to be improve privacy protecting algorithms, operate with limited bandwidth and collaborate it with emerging technologies (Yang, 2021). Fairness with respect to allocation of resources in FL is required to be maintained and decisions taken by the devices should be explainable (Y. Liu, 2020). Performance of FL can be enhanced by incorporating asynchronous system, incentives for participation and personalized services.

Interoperability between Heterogenous Infrastructure

AI techniques designed for 6G applications should be capable of handling and processing heterogenous data. There are heterogenous devices operating in the network with different hardware and software provisions. In case of FL, global model must be able to run on heterogenous devices.

Industry and Healthcare

Tele-surgery and robotics for control system require precision and accuracy. There is no provision for errors. ML and DL algorithms employed in these applications are not only expected to have low latency but also be able to efficiently use resources and reduce human risk because of inaccuracies (Al-Ansi, 2021).

Green Communication in 6G

AI is expected to reduce computations, communication overhead, optimize resource management, RF energy harvesting in space, air and ground integrated networks. RL and DRL techniques can optimize caching and computational offloading in heterogenous networks. Harvesting efficiencies can be improved using AI techniques. There should be a balance between performance of ML/DL algorithms and energy efficiency. Thus there is lot of scope to develop light weight AI algorithms with low energy consumption without compromising the performance (Mao, 2021).

ACKNOWLEDGMENT

Authors would like to acknowledge B.M.S. College of Engineering for supporting and providing materials and resources to carry out the research. This research received no specific grant from any funding agency in the public, commercial, or not-for-profit sectors.

REFERENCES

Abadi, M., Chu, A., Goodfellow, I., McMahan, H. B., Mironov, I., Talwar, K., & Zhang, L. (2016). Deep learning with differential privacy. *Proceedings of the 2016 ACM SIGSAC Conference on Computer and Communications Security,* 308–318.

Al-Ansi, A., Al-Ansi, A. M., Muthanna, A., Elgendy, I. A., & Koucheryavy, A. (2021). Survey on Intelligence Edge Computing in 6G: Characteristics, Challenges, Potential Use Cases, and Market Drivers. *Future Internet*, *13*(5), 118. doi:10.3390/fi13050118

Analytics, S. (n.d.). *Smartphones: Global Artificial Intelligence Technologies Forecast: 2010 to 2023.* Retrieved September 26, 2021, https://www.strategyanalytics.com/access-services/devices/mobile-phones/emerging-device-technologies/market-data/report-detail/smartphones-global-artificial-intelligence-technologies-forecast-2010-to-2023

Bouzinis, P. S., Diamantoulakis, P. D., & Karagiannidis, G. K. (2021) *Wireless Federated Learning (WFL) for 6G Networks — Part I: Research Challenges and Future Trends*. arXiv:2105.00842.

Calvanese Strinati, E., Barbarossa, S., Gonzalez-Jimenez, J. L., Ktenas, D., Cassiau, N., Maret, L., & Dehos, C. (2019). 6G: The Next Frontier: From Holographic Messaging to Artificial Intelligence Using Subterahertz and Visible Light Communication. *IEEE Vehicular Technology Magazine*, *14*(3), 42–50. doi:10.1109/MVT.2019.2921162

David Neumann, W. U., & Wiese, T. (2017), Deep channel estimation. *21th International ITG Workshop on Smart Antennas*, 1–6.

Dong, P., Zhang, H., Li, G. Y., Gaspar, I. S., & NaderiAlizadeh, N. (2019). Deep CNN-Based Channel Estimation for mmWave Massive MIMO Systems. *IEEE Journal of Selected Topics in Signal Processing*, *13*(5), 989–1000. doi:10.1109/JSTSP.2019.2925975

Farsad, N., & Goldsmith, A. (2018). Neural network detection of data sequences in communication systems. *IEEE Transactions on Signal Processing*, *66*(21), 5663–5678. doi:10.1109/TSP.2018.2868322

Gui, G., Liu, M., Tang, F., Kato, N., & Adachi, F. (2020). 6G: Opening New Horizons for Integration of Comfort, Security, and Intelligence. *IEEE Wireless Communications*, *27*(5), 126–132. doi:10.1109/MWC.001.1900516

Guo, W. (2020). Explainable Artificial Intelligence for 6G: Improving Trust between Human and Machine. *IEEE Communications Magazine*, *58*(6), 39–45. doi:10.1109/MCOM.001.2000050

Han, G., Guizani, M., Bi, Y., Luan, T. H., Ota, K., Zhou, H., Guibene, W., & Rayes, A. (2017). Software-defined vehicular networks: Architecture, algorithms, and applications: Part 1. *IEEE Commun.*, *55*(7), 78–79. doi:10.1109/MCOM.2017.7981529

Han, S., Xie, T., I, C.-L., Chai, L., Liu, Z., Yuan, Y., & Cui, C. (2020). Artificial-Intelligence-Enabled Air Interface for 6G. *Solutions, Challenges, and Standardization Impacts IEEE Communications Magazine*, *58*(10), 73–79. doi:10.1109/MCOM.001.2000218

He, H. T., Wen, C. K., Jin, S., & Li, G. Y. (2020). Model-driven deep learning for MIMO detection. *IEEE Transactions on Signal Processing*, *68*, 1702–1715. doi:10.1109/TSP.2020.2976585

He, Y., Zhang, J., Wen, C., & Jin, S. (2019). Turbonet: A model-driven DNN decoder based on max-log-map algorithm for turbo code. *Proc. Of IEEE VTS Asia Pacific Wireless Comms. Symp. (APWCS).* 10.1109/VTS-APWCS.2019.8851650

Hong, T., Liu, C., & Kadoch, M. (2019). *Machine learning based antenna design for physical layer security in ambient backscatter communications. Wireless Communications and Mobile Computing. How AI will Enable Network Intelligentizing?* Vision for Future Communications Summit, Lisbon, Portugal. Retrieved September 25, 2021, https://www.is-wireless.com/wp-content/uploads/2019/12/Network-Intelligentizing-presentation-on-NetWorld2020-summit.pdf

Huang, C. C. C., & Li, Q. (2017). A study of deep learning networks on mobile traffic forecasting. *IEEE 28th Annual International Symposium on Personal, Indoor, and Mobile Radio Communications (PIMRC),* 1–6. 10.1109/PIMRC.2017.8292737

Huang, H., Guo, S., Gui, G., Yang, Z., Zhang, J., Sari, H., & Adachi, F. (2020, February). Deep Learning for Physical-Layer 5G Wireless Techniques: Opportunities, Challenges and Solutions. *IEEE Wireless Communications*, *27*(1), 214–222. doi:10.1109/MWC.2019.1900027

Jagannath, A. J., & Melodia, J. (2021). Redefining Wireless Communication for 6G: Signal Processing Meets Deep Learning with Deep Unfolding. doi:10.36227/techrxiv.16434855.v2

Kasgari, Saad, Mozaffari, & Poor. (2020). Experienced deep reinforcement learning with generative adversarial networks (GANs) for model-free ultra reliable low latency communication. *IEEE Trans. Commun.*

Letaief, K. B., Chen, W., Shi, Y., Zhang, J., & Zhang, Y. A. (2019). The Roadmap to 6G. *AI Empowered Wireless Networks IEEE Communications Magazine*, *57*(8), 84–90. doi:10.1109/MCOM.2019.1900271

Liu, Y., Yuan, X., Xiong, Z., Kang, J., Wang, X., & Niyato, D. (2020). Federated learning for 6G communications: Challenges, methods, and future directions. *China Communications*, *17*(9), 105–118. doi:10.23919/JCC.2020.09.009

Mao, B., Tang, F., Yuichi, K., & Kato, N. (2021). *AI based Service Management for 6G Green Communications.* arXiv preprint arXiv:2101.01588.

McMahan, B. (2017). Communication-efficient learning of deep networks from decentralized data. *Int'l. Conf. Artificial Intell. Stat.*, *54*, 1273–1282.

Naser, S., & Bariah, L. (2021). Towards Federated Learning-Enabled Visible Light Communication in 6G Systems. doi:10.36227/techrxiv.14632551.v2

Peltonen, E., Bennis, M., Capobianco, M., Debbah, M., Ding, A., Gil-Castiñeira, F., Jurmu, M., Karvonen, T., Kelanti, M., Kliks, A., Leppänen, T., Lovén, L., Mikkonen, T., Rao, A., Samarakoon, S., Seppänen, K., Sroka, P., Tarkoma, S., & Yang, T. (2020). *6G White Paper on Edge Intelligence* (White paper) (6G Research Visions, No. 8). University of Oulu. http://urn.fi/urn:isbn:978952622677

Peltonen, E., Bennis, M., Capobianco, M., Debbah, M., Ding, A., Gil-Castiñeira, F., . . . Yang, T. (2020). *6G white paper on edge intelligence*. arXiv preprint arXiv:2004.14850.

Piran, M. J., & Suh, D. Y. (2019). Learning-Driven Wireless Communications, towards 6G. *2019 International Conference on Computing, Electronics & Communications Engineering (iCCECE),* 219-224. 10.1109/iCCECE46942.2019.8941882

Porambage, P., Gür, G., Osorio, D. P. M., Liyanage, M., Gurtov, A., & Ylianttila, M. (2021). The Roadmap to 6G Security and Privacy. *IEEE Open Journal of the Communications Society*, *2*, 1094–1122. doi:10.1109/OJCOMS.2021.3078081

Rahman. (2019). Network Intelligentizing for Future 6G Wireless Networks. *Digital Communications and Networks, 6*(3), 281-291. doi:10.1016/j.dcan.2020.07.003

Shakiba-Herfeh, M., Chorti, A., & Poor, H. V. (2020). *Physical layer security: authentication, integrity and confidentiality*. https://arxiv.org/pdf/2001.07153.pdf

Shea, T. O., & Hoydis, J. (2017). An introduction to deep learning for the physical layer. *IEEE Transactions on Cognitive Communications and Networking*, *3*(4), 563–575. doi:10.1109/TCCN.2017.2758370

Shi, Y., Yang, K., Jiang, T., Zhang, J., & Letaief, K. B. (2020). *Communication-efficient edge ai: Algorithms and systems*. arXiv preprint arXiv:2002.09668.

Soll, M., Hinz, T., Magg, S., & Wermter, S. (2019). Evaluating defensive distillation for defending text processing neural networks against adversarial examples. In I. V. Tetko, V. Kþurková, P. Karpov, & F. Theis (Eds.), Artificial Neural Networks and Machine Learning—ICANN Image Processing (pp. 685–696). Academic Press.

Tan, X. S., Xu, W. H., Sun, K., Xu, Y., Be'ery, Y., You, X., & Zhang, C. (2020). Improving massive MIMO message passing detectors with deep neural network. *IEEE Transactions on Vehicular Technology*, *69*(2), 1267–1280. doi:10.1109/TVT.2019.2960763

Tomkos, Klonidis, Pikasis, & Theodoridis. (2020). Toward the 6G Network Era: Opportunities and Challenges. *IT Professional, 22*(1), 34-38. . doi:10.1109/MITP.2019.2963491

Tramèr, F., Kurakin, A., Papernot, N., Goodfellow, I., Boneh, D., & McDaniel, P. (2017). Ensemble adversarial training: Attacks and defenses. *Proc. Int. Conf. Learn. Represent.* Available: https://arxiv.org/abs/1705.07204

Tran, N.-T. (2019). Self-Supervised GAN: Analysis And Improvement with Multi-Class Minimax Game. NeurIPS, Vancouver, Canada.

Un, M., Shao, M., Ma, W., & Ching, P. C. (2019). Deep mimo detection using ADMM unfolding. *Proc. of IEEE Data Science Workshop.*

Wen, Wu, Wang, Chen, & Li. (2016). Learning structured sparsity in deep neural networks. *NIPS*, 2082–2090.

Wen, W. S. C., & Jin, S. (2018). Deep learning for massive mimo csi feedback. *IEEE Wireless Communications Letters*, *7*(5), 748–751. doi:10.1109/LWC.2018.2818160

Xiao, Y., Shi, G., Li, Y., Saad, W., & Poor, H. V. (2020). Toward Self-Learning Edge Intelligence in 6G. *IEEE Communications Magazine*, *58*(12), 34–40. doi:10.1109/MCOM.001.2000388

Xu, W. H., Wu, Z. Z., & Ueng, Y. L. (2017). Improved polar decoder based on deep learning. *Proceedings of IEEE International Workshop on Signal Processing Systems*, 1–6. 10.1109/SiPS.2017.8109997

Yang, Z., Chen, M., Wong, K. K., Poor, H. V., & Cui, S. (2021). *Federated learning for 6G: Applications, challenges, and opportunities*. arXiv preprint arXiv:2101.01338.

Ye, B. F. J. H., Li, G. Y., & Sivanesan, K. (2018). Channel agnostic end-to-end learning based communication systems with conditional gan. *IEEE Globecom Workshops*, *16*, 1–5. doi:10.1109/GLOCOMW.2018.8644250

Ye, H., Li, G. Y., & Juang, B. H. (2018). Power of deep learning for channel estimation and signal detection in OFDM systems. *IEEE Wireless Communications Letters*, *7*(1), 114–117. doi:10.1109/LWC.2017.2757490

You, X., Wang, C. X., Huang, J., Gao, X., Zhang, Z., Wang, M., Huang, Y., Zhang, C., Jiang, Y., Wang, J., Zhu, M., Sheng, B., Wang, D., Pan, Z., Zhu, P., Yang, Y., Liu, Z., Zhang, P., Tao, X., ... Liang, Y.-C. (2021). Towards 6G wireless communication networks: Vision, enabling technologies, and new paradigm shifts. *Science China. Information Sciences*, *64*(1), 110301. doi:10.100711432-020-2955-6

You, X. H., Zhang, C., Tan, X. S., Jin, S., & Wu, H. (2019). AI for 5G: Research directions and paradigms. *Science China. Information Sciences*, *62*(2), 021301. doi:10.100711432-018-9596-5

Zappone, A., Debbah, M., & Altman, Z. (2018). Online energy efficient power control in wireless networks by deep neural networks. *IEEE 19th Int. Workshop Signal Process. Adv. Wireless Commun.*, 1–5. 10.1109/SPAWC.2018.8445857

Zappone, M. D. R., & Debbah, M. (2019). Wireless Networks Design in the Era of Deep Learning: Model-Based, AI-Based, or Both? *IEEE Transactions on Communications*, *67*(5), 7331–7376. doi:10.1109/TCOMM.2019.2924010

Zero-Touch Network and Service Management (ZSM). (2020). *Landscape*. ETSI Standard GR ZSM 004.

Zhang, J. J., Huang, Y. M., & Wang, J. H. (2019) Intelligent beam training for millimeter-wave communications via deep reinforcement learning. *Proceedings of IEEE Global Communications Conference*, 1–7. 10.1109/GLOBECOM38437.2019.9014113

Zhang, Q., Ferdowsi, A., Saad, W., & Bennis, M. (2021) *Distributed conditional generative adversarial networks (GANs) for data-driven millimeter wave communications in UAV networks arXiv preprint.* arXiv:2102.01751.

Zhang, S., & Zhu, D. (2020). Towards artificial intelligence enabled 6G: State of the art, challenges, and opportunities. *Computer Networks, 183.* doi:10.1016/j.comnet.2020.107556

Zhang, Z., Xiao, Y., Ma, Z., Xiao, M., Ding, Z., Lei, X., Karagiannidis, G. K., & Fan, P. (2019). 6G Wireless Networks: Vision, Requirements, Architecture, and Key Technologies. *IEEE Vehicular Technology Magazine, 14*(3), 28–41. doi:10.1109/MVT.2019.2921208

Zhang, Z. M., Hua, M., Li, C. G., Huang, Y., & Yang, L. (2019). Placement delivery array design via attention-based sequence-to-sequence model with deep neural network. *IEEE Wireless Communications Letters, 8*(2), 372–375. doi:10.1109/LWC.2018.2873334

KEY TERMS AND DEFINITIONS

AI in Mobile Edge Computing: It is a unique feature of 6G where most of the processing is moved from cloud to edge devices because of their enhanced capabilities.

Challenges: Several key enablers of 6G such as AI, quantum technology, new frequency bands, smart meta-surfaces, dense antenna arrays and mobile edge communication have benefits as well challenges associated. Challenges in terms of design and implementation leads to new research opportunities.

Deep Reinforcement Learning: It is a combination of DL and RL. DRL is effective in several applications.

Deep Unfolding: It is a technique of using Neural Networks to implement complex iterative signal processing approaches in a simplified way especially at the physical layer.

Distributed AI: It simplifies complex tasks by utilizing spatial distribution of computing resources. It is very much helpful to process large data sets using distributed nodes.

Explainable AI: It is essential to develop trust in the system when decisions are taken by AI techniques. The techniques should be transparent in terms of operations performed and reasoning for the decisions taken.

Federated Learning: It is decentralized approach employed at the edge nodes which learns global model collaboratively using local data sets. It is basically a distributed training technique.

Physical Layer: It is most important to have great capabilities at the lower layers in order to support the upper layers to have an overall efficient communication system. Physical layer mostly involves signal processing tasks suitable for effective transmission.

Chapter 5
AI-Empowered 6G and Next Generation Networks

Narasimha Reddy K.
VISTAS, India

Monica K. M.
VISTAS, India

Sridevi S.
VISTAS, India

Bindu G.
VISTAS, India

ABSTRACT

As a future leading technology, sixth generation (6G) networks should be capable of dynamic allocation of resources, process signals, and change in traffic flow. AI/ML can bring about such solutions to manage such tasks. AI-empowered techniques can be used in optimizing the performance of the network efficiently together with mobile edge computing. AI/ML techniques provide the possibility of producing automatic-learning models for optimized network for 6G wireless networks, which grants operators/providers the access for optimizing parameters of network and automatic network adjustment. In this chapter, the authors explore the various applications of AI/ML in sixth-generation and next generation networks and provide detailed explanation on how AI/ML may be implemented in 6G network effectively. Moreover, they provide possible issues while implementing AI-empowered 6G networks.

INTRODUCTION

Wireless communication systems are the factors which play a vital role in modern society for health, commercial, business as well as entertainment purposes. These technologies have been evolving and as of now, industrial consultants predict the development of Sixth-generation (6G) network. 6G, the superseder of fifth-generation (5G) network, is expected to use higher frequencies than 5G and is anticipated to use lower latency of below 1 ms of end-to-end lag, elevated mobility of up to 1000 km, large frequency bands of around 1THz – 3THz and it supports data rates of 1 terabyte (TB) per second. It enables users to get better QoE and QoS, along with enormous improvement in network performance. Moreover, the combination of selection of frequency and sub-millimeter bring forth relative electromagnetic absorption rates which can lead to improvement of wireless sensing technology to a higher level.

DOI: 10.4018/978-1-6684-3804-6.ch005

One of the primary elements of 6G network will be the utilization of AI and ML techniques. ML is one of the types of AI which learns and trains itself using historical data through algorithms and makes a decision as well as predicts the new input without being explicitly programmed. ML algorithms can be splitted into three types: 1) Supervised Learning, 2) Unsupervised Learning and 3) Reinforcement Learning. This chapter discusses the applications of AI/ML in 6G network along with its implementation of its architecture.

The 6G and next generation networks are expected to be the required phases of society that may need specific essential qualities (Curtis Watson, 2021) such as

- Flexibility – The recent need of the next generation network has epidemically increased with unmanned autonomous vehicles (UAVs) and internet of things (IoT). So, it should be flexible to be able to merge with the developing technologies.
- Sustainability – Consumption of energy, lifetime of its battery, and its impacts on the environment should be taken into consideration for the prolongation of uninterruptible communication network.
- Intellect – AI-enabled smart network, mobile edge technology and network sensing will boost the network and takes it to new value paradigms.
- Trust – The network should meet the privacy and security demands.

The features of the 6G network are given in Figure 1. The requirements and the use of AI/ML enabled 6G network is introduced in the introduction section. Background Section provides the literature survey of the related work. AI/ML enabled 6G Section provides the details on the AI/ML empowered 6G network. Applications of AI enabled 6G Section describes the various applications of 6G combined with AI/Ml technologies. While issues that may occur in implementing 6G network are presented in issues section, some potential solutions are discussed in the solutions and recommendations section. Finally, the chapter is arrived at the conclusion section.

BACKGROUND

In (K. B. Letaief, 2019), Khaled B Letaief, et. al. discussed about potential technologies for 6G network which can enable various mobile AI applications. Moreover, several AI based methodologies for optimization and network design were presented along with the key trends which will be factors for the evolution of 6G.

In (Razvan-Andrei stoica, 2019), Razvan-Andrei Stoica, et. al. proposed an AI based paradigm shift for wireless communication which can be combined with technologies like (NOMA) and full duplex radio. Finally, they provided a method called randomized incoherent tight frames which can lead to optimized maximum likelihood detection.

In (Karan Sheth, Keyur Patel, 2020), Karan Sheth, et. al. presented various future applications which can be integrated AI enabled 6G like drone communication, preservation of security and privacy, object localization etc. Finally, they discussed on the use cases which shows how different AI techniques can be adopted in the intelligent transport system and provided solutions for issues like low communication overhead cost.

In (Z. Zhang, 2019), Zhengquan Zhang, et. al. presented the vision 6G wireless network and usage scenarios which can satisfy emerging services. Moreover, the basic needs for multi-terabyte/ps as well as smart 6G network were provided along with an autonomous network architecture integrating air, space and ground networks to get unlimited wireless connectivity.

In (T. Bin Ahammed, 2020), Tareq Bin Ahammed, et. al. presented the subdivisions of Artificial Intelligence which can be fused with 6G communication technology. It mainly focuses on combination of aspects like machine learning, federated learning and deep learning to prepare the 6G network with fully deployable in an AI eco system.

In (H. Yang, 2020), Helin Yang, et. al. proposed an intelligent architecture for AI enabled 6G network which can comprehend automatic network adjustment, and intelligent service provisioning. The proposed architecture consists of 4 layers, data mining layer, intelligent sensing layer, analytics layer and intelligent control layer. They finally discuss on the ways to implement effective AI techniques in 6G for optimizing network performance along with AI-empowered mobile edge computing and smart spectrum management.

AI/ML – Enabled 6G

The wireless networks which are still in developing stage massively depend on mathematical models which provides the structure of the communication system. These models usually do not define the systems accurately. Furthermore, for a few building blocks of the wireless networks, there's currently no suitable mathematical model which, in result, became a challenging task for modelling such blocks (Ali, 2020). Alternatively, it also requires huge mathematical solutions for optimizing the wireless networks. These solutions do not meet the requirements which are to be set by 6G applications. Therefore, Machine Learning will play a vital part in 6G wireless network since it is more suitable for modelling of systems which cannot be defined by the mathematical equation. It is anticipated that ML tools can be used instead of algorithms such as brute-force and heuristic algorithm for optimizing localized tasks. Additionally, it may also enable real-time analysis along with automated no touch operation in 6G network. The comparison of various wireless communication generations are given in Figure 2.

Machine Learning

Machine learning, the backbone of 6G wireless network, is a subset of AI which allows software to predict the outcomes without being explicitly programmed by training itself through historical data. It has the capability to imitate an intelligent human behavior. Machine learning can be used to predict the text, translate languages, detects images to diagnose medical conditions and even powers the autonomous vehicles for self-driving (Sara Brown, 2021). Machine learning models are able to learn particular features of the provided system which cannot be defined by a mathematical model. ML models can perform tasks such as classification and regression and arithmetic calculations effectively which allow modeling for accessibility, mobility and manage the network communications of 6G data. Furthermore, it improves the performance of the network management to maintain the current Key Performance Indicator (KPI) between pre-defined thresholds. The Figure 3 shows the categories of ML models.

Supervised Learning

Supervised learning is one of the ML methods which uses labeled data sets for training the model. After the input data is provided to the model, cross-validation process is carried out to fit the model appropriately by adjusting its weight. Supervised Learning is mainly used to classifying the data or predicting the outcomes accurately. The commonly used techniques in supervised learning are Naïve Bayes, Linear Regression, Logistic Regression, Support Vector Machine (SVM), and Decision Tree. The working of the Supervised Learning is given in Figure 4.

Unsupervised Learning

Unsupervised Learning method takes the unlabeled data set and learns the functions in the data to represent the hidden data structures and patterns without any human intervention. The ability of Unsupervised Learning to detect the similarities and differences in the information makes it an ideal solution for data analysis, image recognition and customer segmentation. The commonly used techniques in Unsupervised Learning are K-Nearest neighbor (KNN), Principal Component Analysis (PCA), Apriori algorithm, Hierarchical Clustering and Neural Networks. The working of the Unsupervised Learning is given in Figure 5.

Reinforcement Learning

Reinforcement Learning works in accordance with merit based approach. The model rewards the action if it is a desired behavior and punishes the action if it is an undesired behavior and takes the actions by learning through trial and error. This method achieves an optimal solution by seeking long-term and maximum reward. The commonly used Reinforcement Learning techniques are State-action-reward-state-action, Markov Decision Process (MDP), Deep Q-Networks and Q-Learning, (J. Kaur, 2021). The workflow of the Reinforcement Learning is given in Figure 6.

APPLICATIONS OF AI-ENABLED 6G

There can be various applications for AI – enabled 6G network.

Smart Factories

The current industries are under a transformational phase which requires exchange of data via automation processes in manufacturing technologies. This phase is being referred to as Industry 4.0. With exponential growth in smart devices, requiring ultra-low latency and high in data rate, the need for the optimized protocols along with fog computing architecture is essential. The smart industry includes robotics and distributed automated systems. This goal of smart factory can be made possible by using AI technology and ML algorithms. The interconnection sensors, microchips, actuators, controller and robots can automate the processes such as asset monitoring for checking the materials and products condition and digital twins for simulating the outcomes throughout the lifecycle of an asset. These inspection systems not only automate the process but will also self-improve by using previous data and analytics to make adjustments for correcting the issues (Ryan Szporer, 2020). The use of technologies such as Autonomous Mobile Robots (AMR) and collaborative robots can be able to maintain the whole lifecycles of the products automated devices with zero-touch.

Smart Product

The driving factors of the Industry 4.0 are big data, Internet of Things (IoT) and cloud computing. Industry 4.0 products are considered smart as they combine with microchips and sensors. The existing manufacture system must be fused with industry 4.0 architecture (IoT+WSN+CPS) which will enable communication between products and humas. Devices such as smart wearables and automated smart vehicles are the some of the output of the Smart products which can make communication between humans and machines easier. To fulfil the complicate requirements of the products, attributes such as self-optimization and high level automation will be enabled.

Smart Cities

Around 6.5 billion people are anticipated to be part of the world's cities in 2050 (Basir, 2019). This sudden rise in population may result in Big Data as well as demand for services. To build the future smart city, many areas are needed to be smart health centers, smart agriculture, smart institutions, smart home, smart office and smart transportation. A lot of research has been done in the field of smart cities. For example, (Bibri, 2017) provided a method by which the current limitations can be knocked over by the smart cities.

Intelligent Healthcare

To avail an affordable high quality healthcare services, there should be a better understanding on diseases and health of the people. The public health sectors are today confronted with a massive evolution increasingly focused on the integration of the assorted health pathways defined not only by a conceptual view but also from a methodological view through overhaul of services and the dissemination of IoT and sensors (L. Mucchi, 2020). Monitoring the state of health of people for prevention, well-being and treatment of complicated chronic diseases demands integration of data from clinical tests and the data from real-world. Artificial intelligence and machine learning algorithms based advanced data analysis systems allow versatile interpretation of large amount of the acquired data. The 6G technology plays a crucial role in enabling interconnection of IoNT and IoBNT to the IoT, which provides a meaning for the realization of "all-things connected" vision.

ISSUES

6G relies on the AI/ML to achieve fully automated networks. In spite of the significance of AI/ML in 6G system, there are several concerns for privacy, security and ethics. In addition, artificial intelligence can be a tool for launching smart attacks to steal the user data (Siriwardhana, 2021). This section provides security, privacy and ethical issues in AI/ML in 6G network.

Security Issues

6G enables connected intelligence through functions compatible with AI, mainly with machine learning systems that are prone to security threats. The ML models can be affected by the poisoning attacks which will make the system to learn inaccurately by influencing the learning phase of an ML model. Few of the poisoning attacks are data manipulation, logic corruption and data injection. ML models can even be attacked based on API such as model inversion, model extraction and membership inference.

Privacy Issues

Privacy can be easily compromised as AI/ML is required to analyze data at a large scale in combination with the automation and speed of the future computers. An enormous amount of data is required by 6G system which will be collected through millions of devices while the users cannot comprehend how these data are handled by external systems. For instance, the proposed intelligent authentication system (H. Fang, 2019) which requires physical attributes may require user's private data.

Ethical Issues

AI/ML based fully automated 6G network requires less human intervention in network operations. Machines learn in a different way than humans and do not consider ethics. These systems only act in accordance with the way they learn and train. They don't have the capability to behave in contrast to the logic in exceptional cases like humans.

SOLUTIONS AND RECOMMENDATIONS

For security, countermeasures such as moving target defense and adversarial machine learning may produce flexible AI/ML systems while robust learning and input validation can be used as defense mechanisms against poisoning attacks. Data privacy can be preserved using Edge-based federated learning which will maintain the data using a physical control. Meanwhile, homomorphic encryption can be used to perform mathematical operations without requiring any need for decrypting the data and provides a technical control over privacy. AI/Ml enabled 6G systems can be more useful if ethics are considered at an early stage. To address an ownership or data ethics in the context of AI/ML enabled 6G measures such as laws, regulations and guidelines can be used to balance the benefits and risks (K. B. Letaief, 2019, E. Bird, 2020).

CONCLUSION

AI/ML is a crucial factor for 6G and next generation networks and providing security is an important concern for realizing the 6G vision. Although 6G provides stimulating anticipation, high frequencies affect the health conditions of many. To establish infallible cyber security, AI/ML empowered 6G network equips intelligent and robust solutions. This chapter provides an outline to comprehend various applications, challenges and risks involved in implementing 6G and NextGen Networks using AI/ML techniques. Moreover, it also provides potential solutions to overcome various security and privacy issues.

REFERENCES

Ali, S., Saad, W., Rajatheva, N., Chang, K., Steinbach, D., Sliwa, B., Wietfeld, C., Mei, K., Shiri, H., Zepernick, H.-J., Chu, T. M. C., Ahmad, I., Huusko, J., Suutala, J., Bhadauria, S., Bhatia, V., Mitra, R., Amuru, S., Abbas, R., & Malik, H. (2020). *6G White Paper on Machine Learning in Wireless Communication Networks*. Academic Press.

Basir, R., Qaisar, S., Ali, M., Aldwairi, M., Ashraf, M., Mahmood, A., & Gidlund, M. (2019). Fog Computing Enabling Industrial Internet of Things: State-of-the-Art and Research Challenges. *Sensors (Basel)*, *19*, 4807. doi:10.339019214807

Bibri, S. E., & Krogstie, J. (2017). Smart sustainable cities of the future: An extensive interdisciplinary literature review. *Sustainable Cities and Society*, *31*, 183–212.

Bin Ahammed & Patgiri. (2020). 6G and AI: The Emergence of Future Forefront Technology. *Advanced Communication Technologies and Signal Processing (ACTS),* 1-6. . doi:10.1109/ACTS49415.2020.9350396

Bird, E., Fox-Skelly, J., Jenner, N., Larbey, R., Weitkamp, E., & Winfield, A. (2020). *The Ethics of Artificial Intelligence: Issues and Initiatives*. European Parliamentary Research Service. Available: https://www.europarl.europa.eu/RegData/etudes/STUD/2020/634452/EPRS_STU(2020)634452_EN.pdf

Brown, S. (2021). *Machine learning, explained.* https://mitsloan.mit.edu/ideas-made-to-matter/machine-learning-explained

Fang, H., Wang, X., & Tomasin, X. (2019). Machine Learning for Intelligent Authentication in 5G and Beyond Wireless Networks. *IEEE Wireless Communications*, *26*(5), 55–61.

Kaur, J., Khan, M. A., Iftikhar, M., Imran, M., & Emad Ul Haq, Q. (2021). Machine Learning Techniques for 5G and Beyond. *IEEE Access: Practical Innovations, Open Solutions*, *9*, 23472–23488. doi:10.1109/ACCESS.2021.3051557

Letaief, K. B., Chen, W., Shi, Y., Zhang, J., & Zhang, Y. A. (2019). The Roadmap to 6G: AI Empowered Wireless Networks. *IEEE Communications Magazine*, *57*(8), 84–90. doi:10.1109/MCOM.2019.1900271

Letaief, K. B., Chen, W., Shi, Y., Zhang, J., & Zhang, Y.-J. A. (2019). The Roadmap to 6G: AI Empowered Wireless Networks. *IEEE Communications Magazine*, *57*(8), 84–90.

Mucchi. (2020). How 6G Technology Can Change the Future Wireless Healthcare. *2nd 6G Wireless Summit (6G SUMMIT),* 1-6. doi: . doi:10.1109/6GSUMMIT49458.2020.9083916

Sheth, K., Patel, K., Shah, H., Tanwar, S., Gupta, R., & Kumar, N. (2020). A Taxonomy of AI Techniques for 6G Communication Networks. *Computer Communications*, *161*, 279–303. doi:10.1016/j.comcom.2020.07.035

Siriwardhana, Y., Porambage, P., Liyanage, M., & Ylianttila, M. (2021). *AI and 6G Security: Opportunities and Challenges*. 10.1109 / EuCNC / 6GSummit 51104.2021.9482503

Stoica & Freitas de Abreu. (2019). *6G: the Wireless Communications Network for Collaborative and AI Applications.* arXiv. arXiv:1904.03413v1.

Szporer, R. (2020). *The Climb to the Top of the Smart Factory Data Pyramid.* https://www.6gworld.com/exclusives/the-climb-to-the-top-of-the-smart-factory-data-pyramid

Watson, C., & Woods, K. (2021). TW: 6G and Artificial Intelligence & Machine Learning. The MITRE Corporation.

Yang, H., Alphones, A., Xiong, Z., Niyato, D., Zhao, J., & Wu, K. (2020). Artificial-Intelligence-Enabled Intelligent 6G Networks. *IEEE Network, 34*(6), 272–280. doi:10.1109/MNET.011.2000195

Zhang, Z., Xiao, Y., Ma, Z., Xiao, M., Ding, Z., Lei, X., Karagiannidis, G. K., & Fan, P. (2019). 6G Wireless Networks: Vision, Requirements, Architecture, and Key Technologies. *IEEE Vehicular Technology Magazine, 14*(3), 28–41. doi:10.1109/MVT.2019.2921208

KEY TERMS AND DEFINITIONS

Actuators: A device which produces a motion by converting energy and signals into the system.

Cross-Validation: A procedure for evaluating ML models.

Electromagnetic Absorption Rate: A measure of the energy which is absorbed per unit mass when a body is exposed to radio frequency electromagnetic field.

Optimization: The process of making something perfect or effective as possible.

Poisoning Attack: The practice of manipulating the training data of the system itself.

Protocol: A pre-defined set of rules for transferring data between electronic devices.

Unmanned Aerial Vehicle (UAV): A drone without a pilot on board.

APPENDIX

Figure 1.

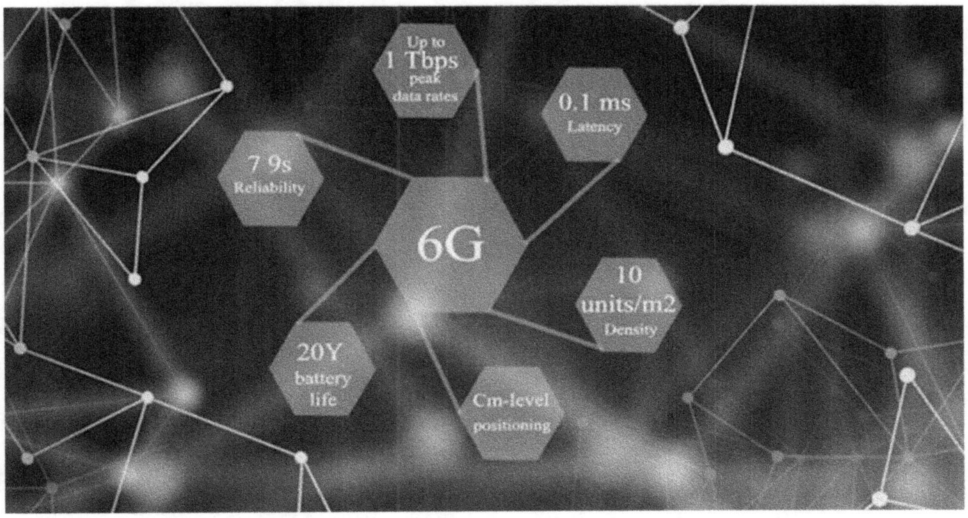

Figure 2.

Features	1G	2G	3G	4G	5G	6G
Time Span	1980-1990	1990-2000	2000-2010	2010-2020	2020-2030	2030-2040
Highlight	Mobility	Digitization	Internet	Real-Time	Heavy Data Rates	Privacy and Security
Core Network	PSTN	PSTN	Packet N/W	Internet	IoT	IoE
Architecture	Voice	Text	Picture	Video	3D VR/AR	Tactile
Maximum Frequency	~900 MHz	~1900 MHz	~2100 MHz	~5 GHz	~100 GHz	~10 THz
Max. Data Rate	2 kbps	150 kbps	2 Mbps	1 Gbps	35 Gbps	100 Gbps

Figure 3.

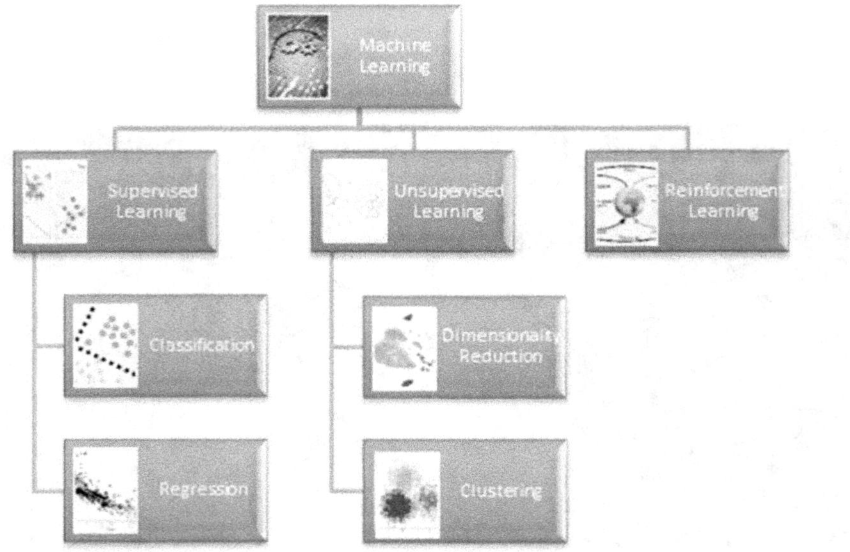

Figure 4.

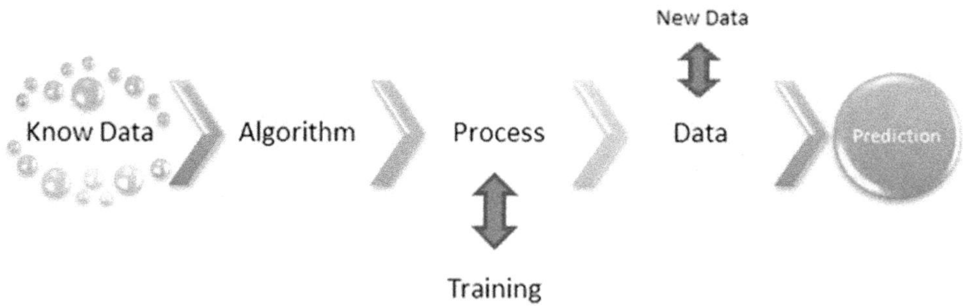

Figure 5.

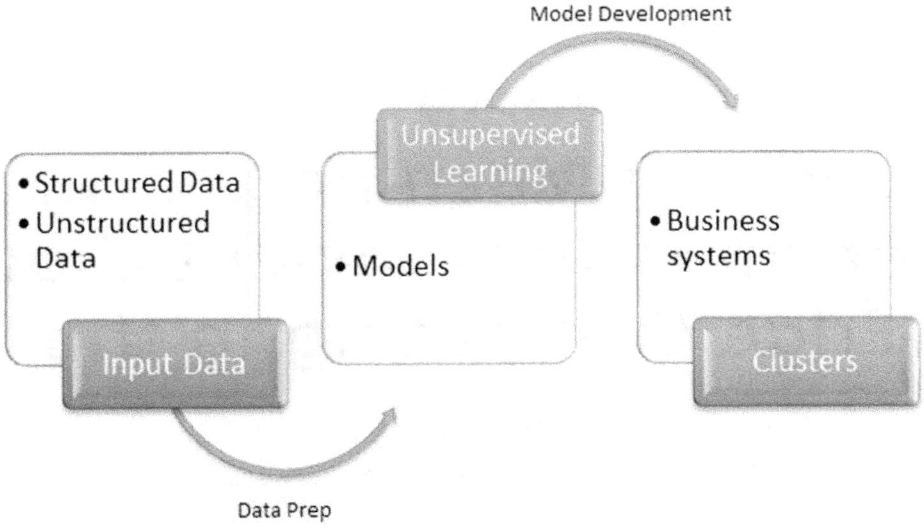

Figure 6.

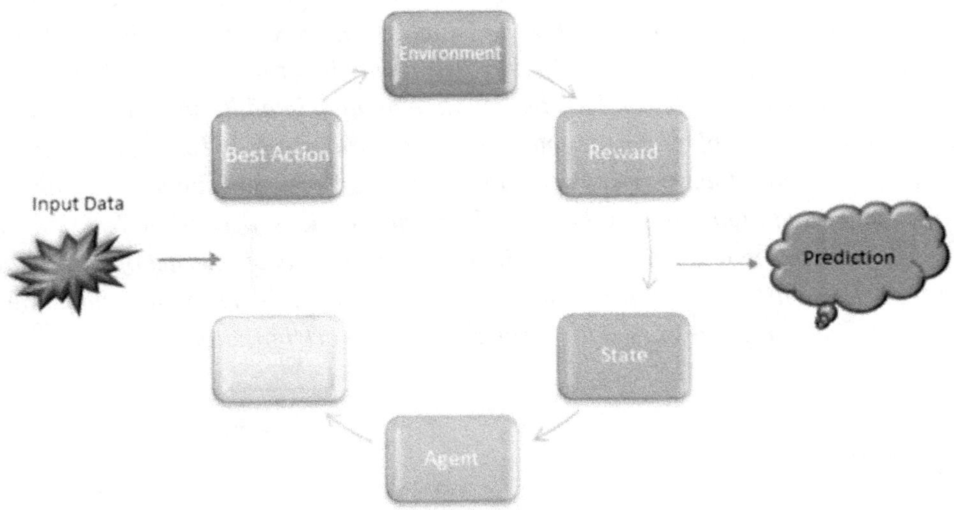

Chapter 6
Analysis of Machine Learning Algorithms for Efficient Cloud and Edge Computing in the IoT

Ramya R.
SRM Institute of Science and Technology, India

Ramamoorthy S.
SRM Institute of Science and Technology, India

ABSTRACT

The internet of things (IoT) technology, which connects internet-connected devices, continues to extend the current internet by allowing communication and interactions across the physical and cyber worlds. The IoT generates big data that is characterized by its velocity in terms of time and place dependency, with a diversity of diverse modalities and fluctuating data quality, in addition to increased volume. The key to designing smart IoT applications is intelligent data processing and analysis. This chapter first describes the details of how cloud services and edge technology work and support the internet of things with many challenges and limitations in the overall internet services. Second, it describes the support of different machine learning algorithms (MLA) in the different fields of internet of things applications. Finally, there is a description of the future research scopes and open issues in the field of the internet of things with machine learning algorithms for further research work initiation.

INTRODUCTION

A plenty of infamy in recent years has gained by the Internet of Things components. It is a collection of different hardware devices, and they are operated by various software programming, those compounds are involved to connect real world environment with the internet(Merenda,2020). There is a dramatic increment in the count IoT device because of extraordinary development in IoT environment. The year wise increment of devices from the year 2016 to 2021 with respect to the Exabyte per month according

DOI: 10.4018/978-1-6684-3804-6.ch006

to all the devices as M2M(Machine to Machine), Smartphones, Non-smartphones, TVs, PCs, Tablets and others are comparatively twice.

Usually the IoT devices have restricted computing efficiency, limited storage and will generate huge volume of data. A real world environment of homes, vehicles and industries has many sensors with low efficiency and connected with IoT. Here there is a requirement of cloud computing concept to provide efficient computing and large storage services to the connected devices in the internet(Statista, 2020). The processing data in computing framework is an important part of IoT, those are named as edge and cloud computing(M. Aazam, 2015). The edge computing is a technology to support cloud computing services in the IoT for end user devices(Joakar 2016).

Cloud computing is a technology to provide services to the end user devices through internet[5]. The cloud computing provides services in various types, they are

- Infrastructure as a Service(IaaS): This type of cloud computing provides hardware and servers utilization to the end user or industrial users through internet.
- Platform as a Service(PaaS): In this type of cloud computing support to use operating systems and software environments to develop applications without installing it.
- Software as a Service(SaaS): Here there are many software applications provided as a service to the end users.

Unfortunately, there are many limitations and complexities to provide those services for different type of end user devices. Here there are different edge computing implemented with the support of Machine Learning (ML) algorithms for efficient and reliable services to the IoT devices(Liu, 2019).

This paper provides the detailed view of Internet of Things with different kind of challenges in the next section, then the different types of Machine learning algorithms which are supporting to the Internet of Things, and the research trends and open issues are discussed, finally there is a conclusion regarding this paper.

CLOUD AND EDGE COMPUTING IN THE INTERNET OF THINGS

The concept of Internet of Things(IoT) used to achieve a smarter environment and comfortable life with low cost, efficient in the sense of time and energy. As a result massive investments and numerous studies are increased in field IoT technology in recent years(C. Cortes, 1995) . The technology named as internet of things includes the different kind of service provider and service receivers. The cloud computing technology act as a service provider to provide different kind of internet services through internet. The end user devices act as service receivers to receive the requirements through the internet. In between the service provider and the end receiver there must be a intermediate layer named as intimidator or edge which includes routers, gateways, base station etc. The overall IoT environment can be categorized into different layer according to the efficiency of processing, those are shown in the following figure 1. with three layers(Statista, 2020). The name cloud refers the collection of large storage devices, high resource processors and virtual application platforms, these all are accesses and used by the end user devices through edge middle ware. The name edge refers with the collection of gateways, routers and base stations for providing support of communication between service provider to the end user devices. The name end devices refers the collection of all kind of devices used by the end users.

Figure 1. Illustration of edge computing

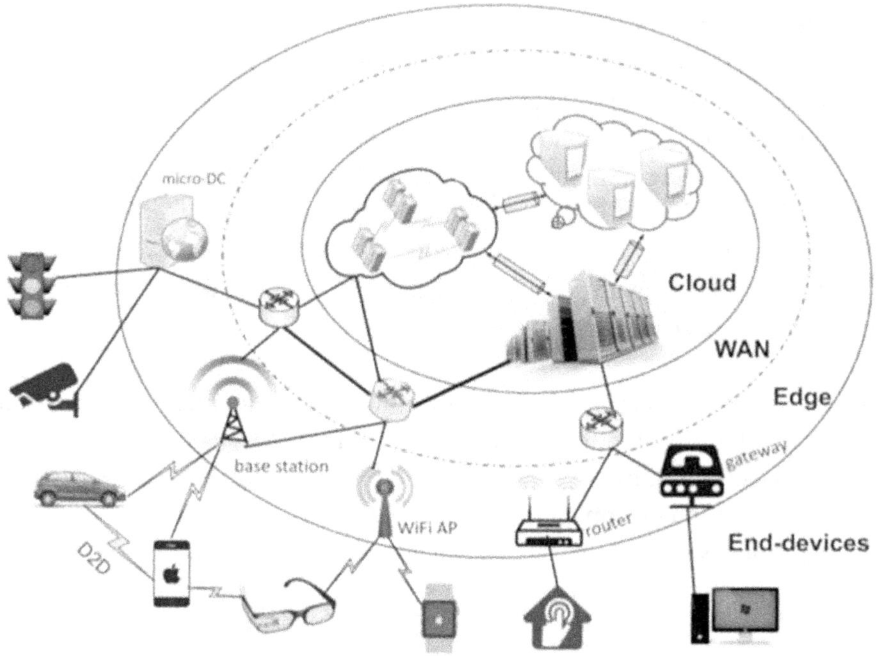

Figure 2. Layers of IoT devices

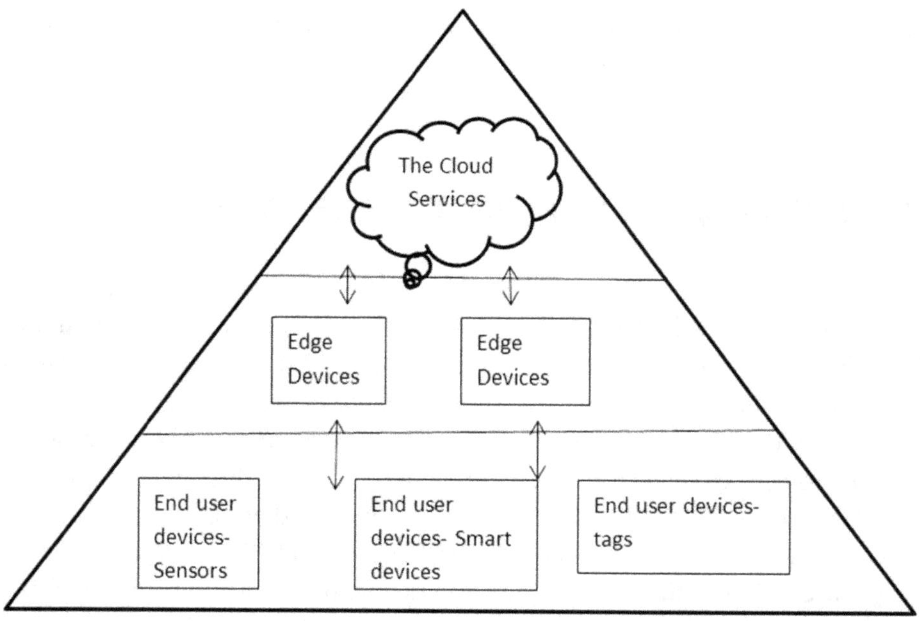

In the Figure 2. the top layer is cloud infrastructure to provide services to the bottom layer devices of IoT, which include large amount of storage, efficient processor as a server and network providers. The middle layer is edge layer to connect and compute efficiently between cloud infrastructure and end user devices. The bottom layer has collection of connected end user devices to get services from cloud infrastructure through edge gateways.

The end user devices in the IoT can receive efficient connection and communication services from the cloud infrastructure with the support of edge technology. As per the report of Google Trends(Merenda, 2020) the edge computing technology has become as increase of interest in the scientific community to study and research.

The comparison between the cloud and the edge computing has shown in the Table 1. According to the various aspects of computing support for the end users(Mahdavinejad, 2018). The latency is high in the cloud computing because the cloud providers are not near to the end users, but the edge provides low latency services because it is near to the end user devices. The data load is large and continuous in the cloud because it has many servers and storage, but the edge provides small and intermittent data load because it has limited devices to support data and computing processes. The service access is headquartered, on a main server in the cloud, but in the edge it is on the edge devices. Cost per device is very high for cloud, but that is low for the edge.

Table 1. The cloud and the edge comparison with different aspects

	cloud	Edge
Latency	High (eventual consistency)	Low(locality)
Dataload	Large and Continuous	Small and intermittent
Control	Centralized/ hierarchical	Distributed
Service Access	Heeadquartered,on a main server	At the edge / on edge device
Cost per device	$$$$$	$

Based on the various aspects and character sets makes the use cases selection of the cloud and the edge for various real time applications(Mahdavinejad, 2018). Through this analysis there might be the selection of supporting processors, even though there is a requirement of algorithms to handle the different processing models, those are become in the next section.

MACHINE LEARNING ALGORITHMS

The enhancement of the IoT framework the efficiency of end user processing can be achieved through efficient data handling methods. Here the efficient data handling means extract knowledge from the raw data of IoT system devices. The Machine Learning algorithms can be used as a tool to categorize and handle the raw data in efficient way for efficient IoT processing.

Machine Learning (ML) is a subset of Artificial intelligence (AI) in the field of computer science technology(D. Barber, 2012), which makes machine ability to learn themselves without human support. Machine learning algorithms are applied in different applications in the field of computing technology to make intelligent environments. The machine learning algorithms are classified according to the learning approach, they are unsupervised, supervised and reinforcement(K.P. Murphy, 2012). Under each type there are many algorithms available as per the application requirements.

- Supervised learning: The supervised learning is one of the types of Machine learning. The supervised learning gets input data and labelled data(that is required output data) to train the data set to generate algorithm to handle and process on the for future input data. There are many types under this supervised learning algorithm for various applications. The applications are Fraud detection, image classification, forecasting, prediction and process optimizations. The major divisions of supervised learning algorithms are Classification and Regression.
    - Classification: This type takes discrete values as input data set for classification. The input values are categorical. "Decision trees, Support vector Machine (SVM), Navie Bayes and K-Nearest Neighbors(KNN)" (Mahdavinejad, 2018) algorithms are come under the classification type of supervised learning.
    - Regression: This type takes continuous values as input data set for regression. This input values are numerical values. Linear regression, Logistic regression, polynomial regression and super vector regression algorithms are come under the regression type of supervised or labelled learning.
- Unsupervised learning: Unlike first type of learning the Unsupervised learning get only input data set without label data. The dimensionality reduction, clustering, pattern searching algorithms are come under this unsupervised learning of machine learning. The applications are segmentation, Big data visualisation, Structure discovery and feature elicitation.
- Semi supervised learning: This is one of the Machine learning algorithm with combination of supervised and unsupervised learning algorithms characteristics.
- Reinforcement learning: The reinforcement learning is a special type of machine learning methodology. In reinforcement method there is an intelligent agent to get input data and to do action on the environment. For each action the agent will get reward or punishment based on the action towards to reach goal or destination. Genetic algorithm, Q learning and Deep Q network algorithms are come under the reinforcement learning type of machine learning. The applications are Real time decision, game AI, learning tasks, Skill acquisition and Robot Nvigation.

Table 2. Shows the different types of Machine Learning algorithms for different types of applications in the IoT and Smart City with optimization metrics (Mahdavinejad, 2018)

Classification

K-Nearest Neighbors (K-NN)

In the given K data set there is a classification based on the with new objects which are not seen before in the data set named as K-nearest Neighbors(KNN) classification. The distance is the main factor for this classification based on this which makes classification those are near to the given datasets(Mahdavinejad, 2018). Through this algorithm the IOT devices can get the efficient path to complete the data transfer over the internet between the cloud services to edge devices.

Table 2. Machine learning algorithm types and applications on IoT and smart city use cases

Machine learning Algorithm	IoT, Smart City use cases	Type of data	Where data processed
Classification	Smart Traffic	Stream/Massive Data	Edge
Clustering	Smart Health	Stream/Massive Data	Edge/Cloud
Anomaly Detection	Smart Environment	Stream/Massive Data	Cloud
Support Vector Regression	Smart Weather Prediction	Stream Data	Edge
Density-Based Clustering,Naive Bayes	Smart Citizen	Stream Data	Cloud
Naive Bayes	Smart Agriculture	Stream Data	Edge/Cloud
K-Means	Smart Home	Massive/Historical Data	Cloud
K-Means	Smart Air Controlling	Massive/Historical Data	Cloud
Principal Component Analysis	Smart Public Place Monitoring	Historical Data	Cloud
K-Means	Smart Human Activity Control	Stream/Historical Data	Edge/Cloud

The K-NN algorithms can processes with large volume of dataset for data analysis and data mining with the support of data classifications. At the same time the K-NN also support to the distributed computing and storage management with minimal complexity in the bases of time. The main storage database also distributed to the end edge devices with support of K-NN methodology to manage stored database in distributed method.

Navie Bayes

Bayes theorem is base for this classification the classification is with the assumption of feature attributes named as 'naive'. Which classify unseen data point z=(z1,...zm), so it is named as group of probabilistic classifiers. In this very small number of data point were trained for the classifications(Mahdavinejad, 2018). This algorithm also makes the efficient processing between the service providers to the end user devices in the IoT through making probabilistic classifications on networks.

Support Vector Machine(SVM)

Classical SVM are binary classifiers that are non-probabilistic and try to locate the hyper-plane for the best segregation of two classes in the group of elements or nodes in the network. Then, based on which side of the hyper plane a fresh, unknown data point falls that data point is derived through the predicted label(Cortes, 1995). This type of algorithm can provide prediction for efficient processing of data in IoT for correct path selection in nearby network nodes.

Regression

Linear Regression

It is a supervised machine learning technique that utilize the statistical method to predict the value of output variable (y) with the use of input variable(x). It depicts the s(Cortes, 1995)equence relationship with one or more autonomous variables. The linear regression methodology used in data analysis and management to make efficient data transferring between service provider and end user devices. This type of algorithm also can provide prediction for efficient processing of data analysis in IoT environment.

Support Vector Regression

Which extends the already seen SVM method by solving regression problems, that is names as Support Vector Regression(SVR's) process(Cortes, 1995). There are a number of applications with the use of SVR for example nu-support vector regression. This algorithm also makes the efficient processing between the service providers to the end user devices in the IoT.

Combining Models

Classification and Regression Trees (CART)

A CART is a machine learning predictive technique. It illustrates how the values of a target variable can be predicted using other values. It's a decision tree with each fork separated into a predictor variable and a prediction for the target variable at the end of each node(Mahdavinejad, 2018). This algorithm also makes the efficient processing between the service providers to the end user devices in the IoT.

Random Forests

Rather than training a single set of network or tree, any number of network or trees is trained in random woods. The process of training one set of tree can be continued by the all other sub set of sub trees to do training, which bring the reduced work load to train and classify the sub sets (Mahdavinejad, 2018). This algorithm also makes the efficient processing between the service providers to the end user devices in the IoT along with the large number of connected devices for communications.

Bagging

also known as bootstrap aggregating, is an ensemble strategy for improving the accuracy and stability while reducing overfitting in machine learning algorithms. It can classify and aggregate according to the requirements in the network management processing of data transfer and data analysis. This algorithm also makes the efficient processing between the service providers to the end user devices in the IoT for accuracy and stability(Mahdavinejad, 2018).

Clustering

K-Means

The K-means clustering method divides an unlabeled data set into K clusters (groups), with data sets in almost the same cluster sharing certain commonalities. In the fundamental K-means technique, the distance between data sets is the measure of association. As a result, a set of K cluster centres are located with the k- means guidlines, indicated as fs1;...; skg, where the distances between data points are minimised. This algorithm can be used to cluster the available input data set devices to find the shortest distance between the IOT service providers to the end user devices.

Density-Based Spatial Clustering of Applications with Noise

A Density-Based Approach to Spatial Clustering of Applications with Noise (DBSCAN) clusters an unlabeled set of data based on the population of its data sets. Clusters of dense points are called clusters in this model, while outliers are data points in low-density regions.

Feature Extraction

Principal Component Analysis(PCA)

Principal Component Analysis is an unsupervised learning approach used in machine learning to reduce dimensionality. With the help of orthogonal transformation, it is a statistical technique that turns observations of correlated features into a set of linearly uncorrelated data. The Principal Components are the newly altered features. It's one of the most widely used programmes for experimental statistical computing. It's a method for extracting strong patterns from a dataset by lowering variances. Graphics rendering, movie recommendation systems, and optimising power allocation in various means of communication are some of the real-world uses of PCA.

Canonical Correlation Analysis

In two different techniques, Canonical Correlation Analysis may be used to represent the correlations among two datasets:

Design the two datasets in a linear extrapolation way, focusing on a relational system: As a consequence of data item x, data item y without designating any set of data as the dependent or explanatory variables, the focus is on the discovery of correlations between the two datasets.

Neural Network

Feed Forward Neural Network

A Feed Forward Neural Network is a subset of ANN in which nodes are connected in a circular pattern. A recurrent neural network is the polar opposite of a feed forward neural network, in which some paths are cycled. Because input is only processing in one way, the feed forward model is a simple form of neural network. Regardless of how many hidden nodes the data passes through, it always goes in one way and never reverses(Mahdavinejad, 2018).

Time Series and Sequential Data

All of the algorithms given in this article have worked with a set of data points that are dispersed and distributed uniformly. However, in many cases, the set of sample points is derived from time series measurements, such as the Dow Jones Industrial Average's daily closing value or acoustic qualities over time. An example of a non-i.i.d. is a non-i.i.d. In a German phrase, a character sequence is a set of data points in a context other than a time series. Dataset points are made of sequences of x; y pairings in these conditions, rather than being drawn in an i.i.d. fashion from a joint distribution px; y, and the series demonstrate considerable consecutive connections(Papageorgiou, 2014).

Anomaly Detection

The process of finding things or patterns in a data set that do not correspond to other things or an expected pattern is known as anomaly detection. Anomalies, novelties, exceptions, outliers, noise, surprises, or deviations are terms used to describe these unexpected patterns. This is used to make efficient IoT processing with good quality of services.

MACHINE LEARNING OPTIMIZATION FOR EFFICIENT IOT PROCESSING

General optimisation techniques are important for making many efficient service providing processes between cloud service providers to the end user devices. To make more efficiency in many applications there is addition of Machine learning and Artificial Intelligence with these general optimisation techniques as newly named as Intelligent Optimisation Algorithms. The intelligent optimisation algorithms are used in IOT processing to make efficient and quality processing between the cloud service providers to the end user devices. The important types of intelligent optimisation algorithms are listed in the Table 3.

Table 3. Types of intelligent optimisation algorithms

<table>
<tr><td rowspan="7">Intelligent Optimisation Algorithms</td><td rowspan="3">Gradient based Algorithms</td><td>Conjugate gradient algorithms</td></tr>
<tr><td>Sequential quadratic algorithms</td></tr>
<tr><td>Programming algorithms</td></tr>
<tr><td rowspan="2">Evolutionary Algorithms</td><td>Genetic Algorithms</td></tr>
<tr><td>Evolutionary programming</td></tr>
<tr><td colspan="2">PSO algorithms, EDAs, ant colony, and others</td></tr>
<tr><td colspan="2">Multi objective optimisation algorithms</td></tr>
</table>

- Gradient based Algorithm: This algorithm also known as Gradient decent algorithm. It is a first order logic algorithm to find optimised solutions. It has sub divisions such as conjugate gradient algorithms, sequential quadratic algorithms and Programming algorithms.
    - Conjugate gradient algorithms: It is used in linear applications to get numerical solutions. It is also used in the unconstrained optimisation problem for example energy minimization problems.
    - Sequential quadratic algorithms: It is used in the non-linear applications. It is also named as Sequential quadratic programming (SQP) algorithm. It is iterative method to solve optimisation problems.
    - Programming algorithms: It is gradient programming algorithms simply named as programming algorithms. It is used in simple network optimisation problem solving with the use of single iteration methods.

When handling unconstrained optimization issues, gradient descent (GD) optimization methods are widely utilised as black box optimizers. Using the gradient's basis functions parameters, each iteration of a gradient-based system seeks to achieve the minimizer/maximizer objective functions. The Gradient Descent (GD) has some other application oriented methods to do the optimization are "Batch Gradient Descent (BGD), Stochastic Gradient Descent (SGD) and Mini-batch GD". The machine learning method usually uses cost functions for finding optimised solutions, the GD method can use minimum number of cost functions for the network optimised solutions compared to the normal algorithms. GD is one of the most popular and common optimisation method to work for finding a local minimum.

§ Batch Gradient Descent (BGD): It revises the training set for getting best evaluations to detect error with the dataset. It can provide best optimal solution for network evaluation .

§ Stochastic Gradient Descent (SGD): By modifying the network model after each training step, SGD tries to discover the global minimum [9]. This method simply decreases error by estimating the gradient for a randomly selected batch rather than determining the gradient for the whole collection.

§ Mini-batch gradient descent: Mini-batch GD is a variant on the ideas of SGD and BGD. It splits the training dataset into smaller batches and performs an update in each of those batches. This strikes a compromise between the stability of SGD and the performance of batch GD.

- Evolutionary Algorithms: It is one of the old and efficient algorithms in many applications. It has many types from which here listed important types as Genetic algorithms and Evolutionary Programming.
    - Genetic algorithms: It is a search heuristic algorithm that is used in natural selection process. The important applications are image processing, parallelisation, scheduling and vehicle routing problems.
    - Evolutionary programming: It is a subset of evolutionary algorithm. It is used in biological reproduction and optimisation problems.
- PSO Algorithms: Particle Swarm optimization algorithms are used to improve the optimise solutions by iteratively processing with the given input data sets. This can be used in the applications of Energy storage optimisation and movie film processing. Metaheuristics like PSO, on the other hand, do not ensure that an optimal solution will ever be identified.

Resource scheduling is one of the most important issue in the cloud edge processing, but still there is no reasonable algorithm or model for efficient resource allocation on cloud for edge devices. The PSO can be referred as one of the machine learning based optimisation algorithm for resource allocation in the cloud for edge devices requirement management.

- EDAs: Estimation of Distribution algorithms are sampling and building explicit models to get the optimal solutions for given input data set models. It is used in solve the real world problems to get realistic solutions. EDAs are a subset of evolutionary algorithms. EDAs, like other evolutionary algorithms, may handle optimization problems expressed in a variety of formats, from matrices to LISP-style S statements, and the quality of potential solutions is frequently assessed using that or even more optimization methods.

The task scheduling is one of the important issues in the cloud service providing to the end user requirements. The task scheduling in the cloud services can be achieved with the use of Estimation of Distribution algorithms, the EDAs can provide some feasible solutions with base of sampling models and probability models of EDAs.

- Ant colony: Ant colony optimisation algorithms (ACO) find optimised path or solution from group of solutions through computationally. For example knapsack problem solving method is one of the best examples for the ant colony algorithms.

The ant colony optimisation algorithms can solve the problem of load balancing and distribution of workload between the nodes in the cloud services. The ant colony optimisation models can bring their self-sample result, from this they can develop a complete result required by the problem statement of load balancing in the cloud services.

- Multi objective optimisation algorithms: It is one of the multiple objective optimisation algorithms for multiple decision making in single demine problem. That is more than one objectives are utilised for problem formation to get optimised solution. The applications are in economic, finance, optimal control and design, radio power systems and electric power systems. Now a days all are become IoT devices to get the services through internet services.

There is no one solution that simultaneously optimises each objective for a significant multi-objective optimization problem. The goal tasks are said to be conflicting in this instance. If none of the objective functions can be enhanced in value without deteriorating a few of the other objective values, the solution is considered non dominated, Non optimum, Pareto effective, or non-inferior. There could be a (potentially endless) amount of Pareto optimal solutions without extra subjective preference information, all of which are equally valued desirable. Researchers approach multi-objective optimization issues from many perspectives, resulting in a variety of solution philosophies and goals when formulating and solving them. The goal could be to locate a representative group of Pareto optimal solutions, quantify the trade-offs in achieving several objectives, or find a single solution that satisfies a human decision maker's subjective preferences (DM).

With the use of all above different intelligent optimisation algorithms many different real time problems can be solved to get the best optimisation solutions. Specifically in the internet services to the end users can be achieved with the use of these intelligent optimisation algorithms to get optimised path to get internet services.

ARCHITECTURE FOR AN INTELLIGENT EDGE MANAGEMENT WITH AI/ML

The concept of Edge computing implementation is one of the important technology to support networking management for IoT devices. The distributed and heterogeneous network devices makes difficulties to the edge computing processing, to manage these difficulties in efficient way the artificial intelligent (AI) and Machine learning (ML) provides support to edge computing to work as intelligent edge cloud management that is in an architecture representation.

In Figure 3. IEC Management Functional Architecture shows the overall processing of Intelligent Edge Cloud (IEC) management functional architecture between global site and intelligent edge. There are four internal processing units in both sides. They are global model management, global data management, global intelligent operations and global monitoring in the global site processing with one of overall managed system together for connecting and communicating with outside processing. The local model management, local intelligent, local data management and local monitoring are the units of intelligent edge processing with one of overall management system to connect and communicate with the central management global site.

Figure 3. IEC management functional architecture

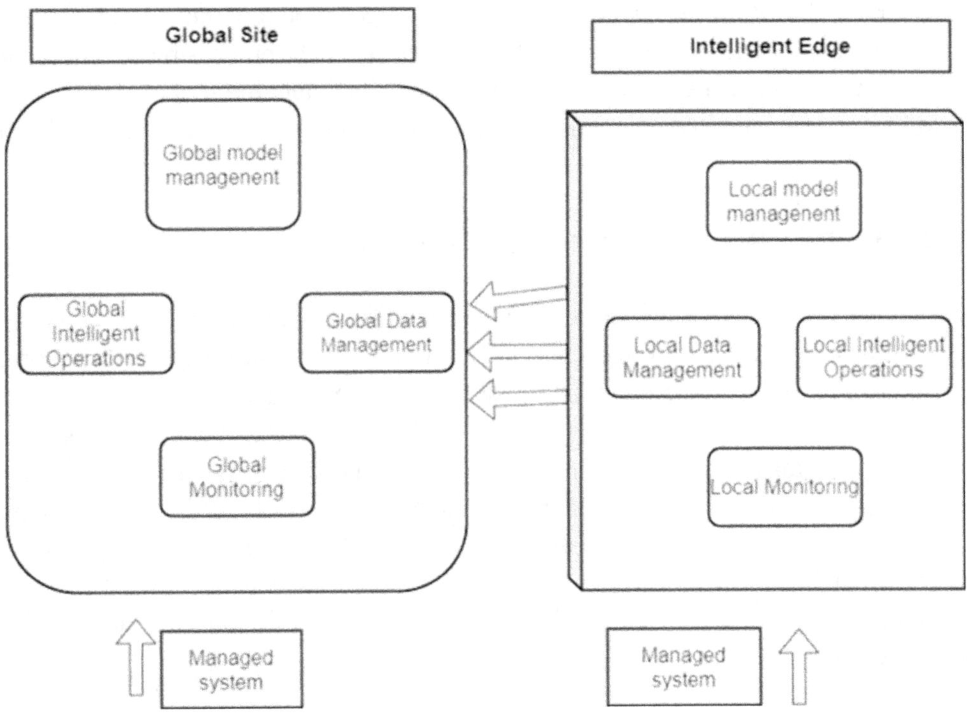

The common identified subsystems are:

- Intelligent Operations (IO)
- Monitoring (MON)
- Data Management (DM)
- Model Management (MM)

Global site has each subsystem as a global component and the intelligent edge has each subsystem as multiple local components with in each of edge sites. The IO subsystem assures the connection and communication between the local and global IO subsystem through data management (DM) subsystem. The global site can provide services like platform as a service (Paas) and Software as a service (Saas) to the local edge devices through the interfaces of intelligent management systems. AI/ML based intelligent management functions are used to create efficient interfaces for supporting global service providers. With the use of these AI/ML based intelligent management functions the result of services includes the benefits of fault detection, root cause analysis, fault prediction, dynamic reconfiguration and predictive workload analysis.

The future implementation of the intelligent edge is work together for providing efficient edge processing with multiple intelligent edges. Each and every edge intelligent agent can have the inner unit of all type of data processing.

WHY AND HOW TO RUN MACHINE LEARNING ALGORITHMS ON EDGE DEVICES

The machine learning algorithms can handle complex and large amount of database, but this processing requires some basics computational capabilities. So the large database processing's possible in the cloud based infrastructure and data centers at first. The invention of efficient internet of things (IoT) with low energy consuming devices can do the complex computations on edge devices for example robots in all kind of fields.

Machine Learning Frameworks for Edge Devices

There are some most popular machine learning (ML) frameworks with corresponding requirements of operating systems or software's in the edge devices. The edge device requirements and the corresponding machine learning frameworks are listed in the Table 4. Well-known Machine Learning frameworks for edge devices in IoT. The frameworks are pre trained models used in many applications such as object detection, image recognition and classification, natural language processing (NLP) and speech recognition.

Google develops a machine learning framework named as "TensorFlow Lite", it has capable APIs (Application Programming Interfaces) for different software languages such as Python, c++, Java, Objective-C and Swift. It must be designed for on-device operations and includes an interpreter optimised for on-device machine learning. To boost efficiency, customized models are translated to TensorFlow Lite form and its complexity is optimised.

Table 4. Well-known Machine Learning frameworks for edge devices in IoT

Framework name	Edge device requirements
TensorFlow Lite – Google	Android, iOS, Linux, microcontrollers (ARM Cortex-M, ESP32)
ML Kit for Firebase – Google	Android, iOS
PyTorch Mobile – Facebook	Android, iOS
Core ML 3 – Apple	iOS
Embedded Learning Library (ELL) – Microsoft	Raspberry Pi, Arduino, micro:bit
Apache MXNet – Apache Software Foundation (ASF)	Linux, Raspberry Pi, NVIDIA Jetson

Google develops one more machine learning framework named as "ML Kit for Firebase". It uses the existing Tensorflow Lite framework and some other APIs mainly to provide support for mobile plartforms such as bar-code scanning, facial detection and object detection.

Facebook develops a machine learning framework named as PyTorch Mobile. It is mainly for mobile platform models, currently it has trained models as experimental release and it is saved as torchscript models.

Apple develops a machine learning framework named as Core ML3. The most important thing in the Core ML3 is its support to the many machine learning models and in particular it supports to the deep neural network processing.

Microsoft develops a machine learning framework named as Embedded Learning Library (ELL). These machine learning frameworks are only for tiny, single board devices with interfaces fro c++ and Python. Edge devices can invoke these complied and deployed models for edge processing.

Apache Software Foundation (ASF) develops a machine learning framework named as "Apache MXNet". It has the ability to support different software languages such as c++,R, Scala, Python among others, but the most efficient training model provided by the Python based APIs.

Hardware Specifications of Edge Devices

The most important requirement of edge devices are in the real life use cases such as speech recognition, image processing, anomaly detection and natural language processing. These kinds of applications are achieved with good efficiency with the use of multiple layers processing of deep learning, the deep learning models are required in the overall machine learning algorithms framework.

The characteristics of deep learning methods demand huge parallel matrix multiplications, the best equipment for edge devices involves "application-specific integrated circuits (ASICs), field-programmable gate arrays (FPGAs), RISC-based CPUs, and embedded graphics processing units (GPUs)".

In the Table 5. Well-known edge devices with their corresponding hardware specifications, there are details about different edge devices with corresponding GPU, CPU, ML software support.

The edge device developed by Google named as Coral SoM is a completely integrated model for machine learning applications with Vivante GC7000Lite of GPU and Quad ARM Cortex A53 + Cortex-M4F of CPU specifications and edge Tensor processing unit (TPU). The TensorFlow Lite, and AutoML Vision Edge are the supporting software of machine learning framework. The Edge TPU used to speeds deep learning network execution and can execute 4 trillion operations (tera-operations) per second (TOPS).

Table 5. Well-known edge devices with their corresponding hardware specifications

Edge device	GPU	CPU	ML software support
Coral SoM – Google	Vivante GC7000Lite	Quad ARM Cortex-A53 + Cortex-M4F	TensorFlow Lite, AutoML Vision Edge
Intel NCS2	Movidius Myriad X VPU (not GPU)		TensorFlow, Caffe, OpenVINO toolkit
Raspberry Pi 4	VideoCore VC6	Quad ARM Cortex-A72	TensorFlow, TensorFlow Lite
NVIDIA Jetson TX2	NVIDIA Pascal	Dual Denver 2 64-bit + quad ARM A57	TensorFlow, Caffe
RISC-V GAP8			TensorFlow
ARM Ethos N-77	8 NPUs in cluster, 64 NPUs in mesh		TensorFlow, TensorFlow Lite, Caffe2, PyTorch, MXNet, ONNX
ECM3531 A – Eta Compute	ARM Cortex-M3 + NXP CoolFlux DSP		TensorFlow, Caffe

The edge device developed by The Intel Neural Compute Stick 2 (NCS2) has Movidius Myriad X Vision Processing Unit (VPU), It is a system-on-chip (SoC) device with a specialised Neural Compute Engine for speeding up deep-learning conclusions.

The edge device Raspberry Pi 4 with VideoCore VC6 of GPU and Quad ARM Cortex-A72 of CPU. The TensorFlow and TensorFlow Lite are the supporting software of machine learning framework. The Raspberry Pi 4 is a solitary computerised just on Broadcom BCM2711 SoC that runs all its edition of the Debian OS (Raspbian); ML algorithms may be enhanced when linked to its USB 3.0 port through the Coral USB.

The edge device NVIDIA Jetson TX2 with NVIDIA Pascal of GPU and Dual Denver 2 64-bit + quad ARM A57. The TensorFlow and Caffe are the supporting software of machine learning framework. The NVIDIA Jetson TX2 is an integrated SoC which is used to implement machine vision with artificial neural networks. The Jetson Xavier NX is also available from the firm.

The edge device developed by Greenwaves Technologies Greenwaves Technologies RISC-V GAP8. Greenwaves Technologies' ISC-V GAP8 is an ultra-low power, eight-core, RISC-V-based processor tailored to run audio and image recognizer. Before modeling can be implemented, they must be converted to TensorFLow using the Open Neural Network Exchange (ONNX) open standard.

The ARM Ethos N-77 is an inter Neural Processing Unit (NPU) that is one of the ARM Ethos, machine learning-focused technology. It provides up to four TOPs of capability and supports a wide range of Machine learning algorithms for image/speech/sound detection.

Eta Compute's ECM3531 is an ASIC based on the ARM Cortex-M3 design that can conduct deep learning models in microwatts. Deep neural networks may be operated on the DSP, which cuts power requirements even further.

APPLICATIONS OF MACHINE LEARNING BASED EDGE DEVICES

Edge based machine learning and already available software, hardware and trained models of machine learning are implemented and placed at network edge provides most benefits on many industries and many other real time applications.

Machine Learning Based Edge Devices in Healthcare

In healthcare industries there are many immediate decision making requirements and analysis arises on the data processing those are can be achieved by the use of edge devices with the support of machine learning models irrespective of connection with the data servers. There is no time for the data transfer from the data centers or database back end devices to make quick decision in emergency doesn't consider the connectivity at all.

Hospitalization is one field that might gain from edge-based ML, as actual data analysis and decision making are vital for natural systems which must keep important physiological parameters, such as blood glucose level or heart rate, within a certain range of outcomes.

As technology and machine learning approaches improve, increasingly complicated characteristics, such as neurological activity or heart rhythms, may be observed and evaluated by edge devices.

"Ambient intelligence" (AmI) is another field which may profit from edge-based information processing. Edge gadgets that are responsive and sensitive to the presence of humans are referred to as AmI. It has the potential to improve how people and surroundings interact with one another.

AmI is exemplified through daily activity monitoring for the elderly. The primary goal of the smart city environment for aged care is to identify abnormalities including a fall or a fire soon and take appropriate action by asking for help.

Mining, Oil, and Gas and Industrial Automation

The commercial advantage of edge-based ML is clear in the oil, gas, and mining industries, where personnel labour in remote locations with no internet. Devices on edge devices such robots may collect large amounts of data and correctly forecast factors like pressure across pumps or operating parameters which are outside of their regular value range.

Connectivity is also a problem in production, where predictive maintenance of machinery may save money and increase the life of industrial assets. Historically, factories shut down machinery at periodic intervals and perform complete examinations in accordance with the standards of the device makers. This strategy, however, is expensive and inefficient, because it does not take into account the unique working characteristics of each machine.

Instead, sensors embedded on all equipment within a warehouse facility can gather readings and apply machine learning to still photos, video, or sounds in order to predict the future events.

FUTURE IMPLEMENTATION OF 5G WITH EDGE INTELLIGENCE

The massive amount of data being moved first from Cloud facilitated Edge towards the decentralised Edge has prompted a drive for data processing at the edge. The challenges propelling Edge to clasp consistent environment include capacity, strength, intellectual power grid and operation along with information restrictions. The system gathers input from neighbouring bunches of expert systems and provides the relevance's that are operating beyond scattered edges, building lattice. The dilemma in latest invention of networks tune finer can enable the presumable architecture necessary to perform the non-symmetric operations in AI and ML components. As 5G advances ahead typical "MMTC for IoT, URLLC for IIoT, and eMBB"[11] to some relevant processing, these intelligent infrastructures must provide near-actual time facts of the order of sub milliseconds.

Figure 4. 5G Support architecture with cloud and edge technology

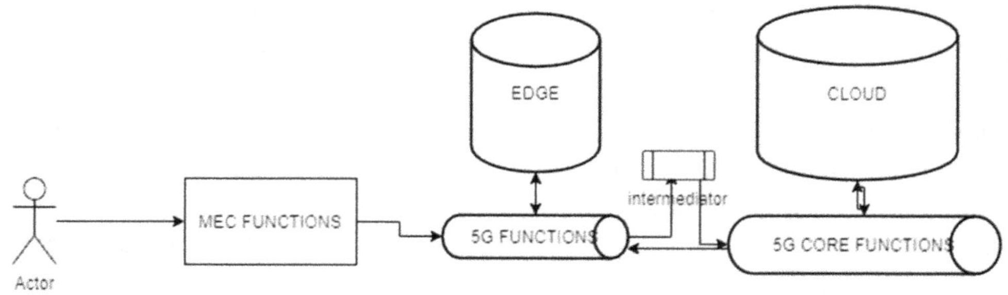

In Figure 4. 5Gsupport architecture with cloud and edge technology has the description about the 5G communication for end user named as 'actor', with the use of 'MEC Functions' that is Mobile edge communication functions from the end user devices to the edge supporting 5G functions. From the edge 5G functions to cloud supporting 5G core functions there may be some intermediate to support communication. The edge may work with intelligent for supporting overall 5G communication for the end users with efficiency.

The edge intelligence is one of the future research concepts for the updating of the 5G implementation in the distributed networks with efficient processing. The implementation of 5G communication can be supported by the concept of edge intelligence in many criteria such as efficient communication with low latency, low transfer time and cost.

RESEARCH TRENDS AND OPEN ISSUES

IoT applications are classified into numerous categories based on their attributes and characteristics. This makes many difficulties to provide good services to the end user devices as listed in following points. The security of group of information is fussy, because information gathering techniques can involve private or confidential working information, resulting in inescapable privacy concerns. Second, the large number of dataset and resources makes different kind of problems to handle resources and to provide services between cloud service provider and end user devices. There are many algorithms mentioned makes to choose and use difficulties in different applications of Internet of things for making efficient processing between the cloud service providers to the end user devices.

The machine learning based optimisation algorithms for efficient edge optimization with the support of machine learning trained model can bring many possibilities to try in research to find efficient solutions in end IoT devices. There are some benefits from the invention of architecture formation of intelligent edge management with AI/ML can provide some more possibilities to research in architectural point of view for making efficient edge devices with machine learning models. The details about why and how to run machine learning algorithms on the edge devices provides the details of all popular edge devices and supporting software's, that cam motivate to research on different software with corresponding hardware based implementations. Finally the future implementation of 5G with edge in intelligence can motivate the future research in communication efficiency processings.

CONCLUSION

This analysis describes about the Machine learning based Internet of Things importance with different cloud and edge application. There are large numbers of heterogeneous devices; they are connected in the Internet of Things for getting services like storing and computing with the use of cloud and edge computing technology through internet, which makes large number of data set collections and traffic in the transmission on the internet services. To the difficulties there are many algorithms become as solution for different applications those are described in this paper. The machine learning based optimisation algorithms are discussed for efficient edge optimization with the support of machine learning trained models. There are some benefits discussed from the invention of architecture formation of intelligent edge management with AI/ML. The discussion about why and how to run machine learning algorithms on the edge devices provides the details of all popular edge devices and supporting software's. Finally the discussion is about future implementation of 5G with edge in intelligence. The finding of this paper helps to choose right algorithm for any application in IoT and it will make to choose further research work on required field of machine learning algorithm.

REFERENCES

Aazam, M. E.-N., & Huh. (2015). Fog computing micro datacenter based dynamic resource estimation and pricing model for iot. *IEEE 29th International Conference on Advanced Information Networking and Applications*, 687–694.

Barber, D. (2012). *Bayesian Reasoning and Machine Learning*. Cambridge University Press.

Cortes, C., & Vapnik, V. (1995). Support-vector networks. *Machine Learning*, *20*(3), 273–297. doi:10.1007/BF00994018

Cover, T., & Hart, P. (1967). Nearest neighbor pattern classification. *IEEE Transactions on Information Theory*, *13*(1), 21–27. doi:10.1109/TIT.1967.1053964

Deng, Zhao, Fang, Yin, Dustdar, & Zomaya. (2020). Edge Intelligence: The Confluence of Edge Computing and Artificial Intelligence. *IEEE Internet of Things Journal, 7*(8).

Joakar. (2016). *A Methodology for Solving Problems with DataScience for Internet of Things*. Open Gardens.

Kiran, Kumar, & Kumar. (2020). *A Review of Machine Learning Algorithms on IoT Applications*. IEEE Xplore Part Number: CFP20V90-ART.

Liu, Y., Yang, C., Jiang, L., Xie, S., & Zhang, Y. (2019). Intelligent Edge Computing for IoT-Based Energy Management in Smart Cities. *IEEE Network*, *33*(2), 111–117. doi:10.1109/MNET.2019.1800254

Mahdavinejad, M. S., Rezvan, M., Barekatain, M., Adibi, P., Barnaghi, P., & Sheth, A. P. (2018). Machine learning for internet of things data analysis: A survey. *Digital Communications and Networks*, *4*(3), 161–175. doi:10.1016/j.dcan.2017.10.002

Merenda, M., Porcaro, C., & Iero, D. (2020). Edge Machine Learning for AI-Enabled IoT Devices: A Review. *Sensors (Basel)*, *20*(9), 2533. doi:10.339020092533 PMID:32365645

Murphy, K. P. (2012). *Machine Learning: A Probabilistic Perspective*. MIT press.

Pandey. (2017). *Machine Learning and IoT for Prediction and Detection of Stress*. Academic Press.

Papageorgiou, A., Zahn, M., & Kovacs, E. (2014). Efficient auto-configuration of energyrelated parameters in cloud-based iot platforms. *Cloud Networking (CloudNet), IEEE 3rd International Conference on*, 236–241.

Plastiras, Terzi, Kyrkou, & Theocharides. (2018). *Edge Intelligence: Challenges and Opportunities of Near-Sensor Machine Learning Applications*. IEEE.

Shanthamallu, Spanias, Tepedelenlioglu, & Stanley. (2020). *A Brief Survey of Machine Learning Methods and their Sensor and IoT Applications*. IEEE.

Statista. (2020). *IoT: Number of Connected Devices Worldwide 2012–2025*. Available online: https://www.statista. com/statistics/471264/iot-number-of-connected-devices-worldwide/

Su, Sperli, Moscato, Picariello, Esposito, & Choi. (2019). An Edge Intelligence Empowered Recommender System Enabling Cultural Heritage Applications. IEEE Transactions On Industrial Informatics, 15(7).

Sun, Liu. & Yue. (2019). *AI-Enhanced Offloading in Edge Computing: When Machine Learning Meets Industrial IoT*. IEEE Network.

Zantalis, F., Koulouras, G., Karabetsos, S., & Kandris, D. (2019). A Review of Machine Learning and IoT in Smart Transportation. *Future Internet*, *11*(4), 94. doi:10.3390/fi11040094

Zhou, C., & Li, Z., Luo, & Zhang. (2019). Edge Intelligence: Paving the Last Mile of Artificial Intelligence With Edge Computing. *Proceedings of the IEEE, 7*(8).

Chapter 7
Augmentation of Terahertz Communication in 6G and Its Dependency for Future State-of-the-Art Technology

Sivaramakrishnan S.
Dayananda Sagar University, India

Rathish C. R.
New Horizon College of Engineering, India

Lingasamy V.
Sterlite Technologies Limited, India

Premalatha S.
KSR Institute for Engineering and Technology, India

ABSTRACT

The communication technology has taken a long journey starting from telegraphy to high-definition video transmission and moving toward the high-density hologram. It is noticed that the data rate has raised linearly with reduced latency as the technology has progressed from first generation to the sixth generation. 5G technology has its limitation in meeting the challenge in creating the high-density hologram and future artificial intelligence application and that's one of the reasons for the focus in the 6G technology. The terahertz frequency that will be used in 6G communication will offer a wideband channel which can support multiple gigahertz channel. Terahertz communication in 6G can support data rate of terabits per second and the latency in 5G can be reduced to one tenth compared to the 6G. This chapter focuses on architecture, necessary hardware requirement, use cases, and deployment of machine learning/artificial intelligence algorithms for certain applications of interest.

DOI: 10.4018/978-1-6684-3804-6.ch007

INTRODUCTION

Once communication is related to human to human information exchange and this statement has taken a new course and in today's world, the communication is not limited to human but it is interconnecting things. The growth in the modern technology like Internet of Things (IOT), Machine Learning, Cloud and edge Computing and Virtual and augmented reality demands a high date rate resulting in an increased bandwidth along with very low latency. The predicted data generated by the number of device connected to the internet by 2025 will be closely 80 Zettabytes and the generated data will be stored in the cloud. Machine learning algorithm will be employed to process these data continuously and the end users will be having an access to the same based on the application of interest. To meet this demand there is certainly a need of high data rate with a bandwidth capable of handling various application of interest without any lag. The 6th generation communication can provide data rate of terabits per second and can provide gigahertz channel (Nayak et al., 2020). Thus to meet the demand of the future application it is mandate to move towards the 6G technology. The 5G technology has limitation in terms of latency for the application such a high density hologram which can be overcome in 6 G technology as the latency is reduced to one tenth in 6G compared to the 5G technology.

5G STANDARD

The 5th generation is the standard meant for the cellular technology and the telecom company started rolling it out from the year 2019. On comparison with the preceding 4th generation, the 5th generation offers maximum data rate of 10 gigabits per second, has less latency and offers better quality of service. Better capacity is achieved in 5G technology as it uses the millimetre wave and its offers higher throughput (Arias et al., 2021). Since the coverage area are limited in this technology more number of base stations have to be deployed. There was an issue in the 4G technology especially with high population density and this short fall was overcome in 5G technology as it could support millions of uses per square kilometre. To reduce the interference in the 5G technology Orthogonal frequency division multiplexing is used which is a technique used to modulate digital signal over the different channels.

REQUIREMENT OF 5G

The major difference between current generations and the 5G techniques includes

- Lower outage probability
- Better coverage
- Low battery consumption.
- More secure
- High Spectral efficiency.
- Many concurrent data transfer paths.
- Artificial sensors are used for communication which is not harmful to human health
- Ultra-densification
- Industrial IoT/Private networks

- Integration with existing network elements like WiFi etc.

WHAT IS BEYOND 5GENERATION?

The future overhaul of Nano-focus will be phenomenal as it gets together with Artificial Intelligence. One can handle his cleaver Robot using the application of his mobile device. The portable device can therefore can type the information that runs in our mind. There will be a circumstance where there won't be a need for any reach for correspondence. The Google hot examples have assessed the term 6G as the seventeenth most glanced through the word in the web records. 6G advancement hasn't been uncovered till now but anyway search questions like what is 6G convenient development, 6G flexible, 6G association, 6G wiki are getting more to know.

Obstacles in 5G Vs Advantage in 6G

The electrical ability of 5G has reached a point close to the limit and with progress in a immense Multiple Input Multiple results (MIMO), cellular density, and millimetre-wave transmission, for example, a progression of multiplex strategies acquired from 4G. The hindrances to convey 5G is addressed in figure 1. As the limits of Shannon are confined, it is incomprehensible that the ridiculous efficiency in 6G will foster a gigantic degree. Oppositely, 6G trades should further develop encryption standard, protection, and characterization with modern methodologies (Ziegler et al., 2020). In 5G associations, standard encryption computations set up on the principal Rivest-Shamir-Adleman (RSA) are as yet used to ensure the security and protection of transmissions. RSA crypto-spores are uncertain with regards to the strain of Big Data and man-made thinking advancements, impressively not exactly the security parts that weren't made in the 5G time frame. Table 1 addresses the examination somewhere in the range of 5G and 6G organizations (Yang et al., 2019).

Table 1. Comparison between 5G and 6G network factors

S.No	Major Factors	5G	6G
1	Traffic Density (Tb/s/Km^2)	10	>100
2	Mobility (Km/h)	350	>1000
3	Energy efficiency	1000x relative to 4G	>10x relative to 5G
4	Data Rate (Gbps)	10	>100

The ITU (2018) characterized three fundamental gatherings to relate the utilization situations. To begin with, Massive Machine Type Communications (mMTC), those needed to cover numerous gadgets, as IoT, Smart Agriculture, Smart Cities, and any brilliant observing, relies upon the volume of information, transmission capacity, and recurrence.

Second, Ultra Reliable Low Latency Communication (URLLC), focus mainly on the reliability of the 5G technology. It ensures one hundred percent reliability and also offers a very low latency of less than one millisecond for the transmission of the packets. The above mentioned futures of URLLC make it as an important feature of 5G communication coming into reality. URLCC is also an important technology in the Industrial 5.0 among the like of IoT, Machine Learning and quantum computing. Applications like TeleHealth, Smart Vehicles, Smart Grid, remote sensing, and modern mechanization depends on the URLLC.

Figure 1. Obstacles to deploy 5G

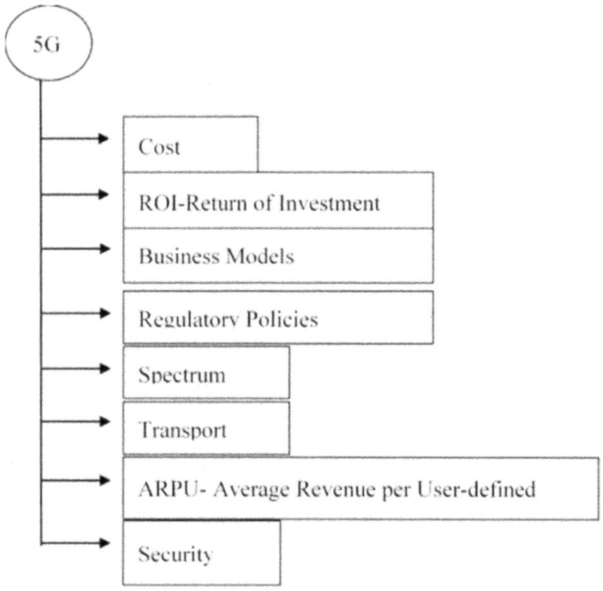

Third, Enhanced Mobile Broadband (eMBB) otherwise called as extreme mobile broadband is one of the important characteristic defining the 5G technology. The high data rate supported in this technology gives the user a great experience and also play a major role in the challenging scenarios such as providing seamless connectivity to thousands of fans in a stadium. eMMB works well for indoor as well as outdoor environment, were the density of users are high, also enables thousands of fans to upload the live content in their social platform without any latency. This technology can also connect more devices to the internet and thus place a major role in making the projects like smart city into a reality. It will also ensure to the growth of the cloud technology as all the application deployed in the cloud can be access even by the commuters without a lag. Thus the eMBB could support various use cases namely hotspot, broadband support to the commuters, smart office, smart cities, access to cloud, large gathering and many more.

6G is relied upon to help information paces of 1 Tera bits per second also it offers latency of less than one microsecond which is thousand times faster compared to the 5G technology (David et al., 2018). Communication path has to serve various clients as far as possible and 6G innovation is a pick innovation to satisfy the needs of the client. This level of cut off and torpidity will widen the display of 5G applications and develop the degree of abilities to help creative applications in far-off accessibility, discernment, distinguishing, and perception.

6G's higher frequencies will enable a great deal faster testing rates, just as giving by and large better throughput and higher data rates blend of sub-mm waves (e.g., frequencies short of what one millimeter) and repeat selectivity to conclude relative electromagnetic degrees of consistency may incite basic advances in remote identifying development (Akhtar et al.,2020).

Mobile edge processing (MEC) will be incorporated into all 6G organizations, while it should be added to existing 5G organizations. Edge and centre processing will be even more flawlessly incorporated as a component of a joined correspondences/calculation foundation structure when 6G organizations are conveyed. This methodology will give numerous likely benefits as 6G innovation becomes functional, including further developed admittance to Artificial Intelligence (AI) abilities (Yastrebova et al., 2018).

NECESSITY FOR 6G TECHNOLOGY

As all the devices is connected to internet, there is enormous amount of data collection happening in the cloud. Analysis of the data and subsequent statistical analysis will predict the future requirements of the customer or the requirement of the society (Sivaramakrishnan et.al, 2017). All the future technology like cloud computing, Artificial Intelligence, 3D networking, Data Science relay on the data generated and these technologies will satisfy the end user only when it is latency free (Sivaramakrishnan et al., 2021). The 6G technology with less than microsecond of latency and his high speed data rate in the range of terabits per second is the right match for the future technology to roll into reality. Also the application like 3D holographic projection can be met with the support of 6G seamlessly.

6G correspondences are relied upon to offer further developed types of assistance as far as inclusion, information rate and permit clients to associate each other all over the place. It is relied upon to embrace eccentric correspondence organizations to get to a few sorts of information and send them through regular worked on radio recurrence (RF) organizations, permitting new correspondence experience with virtual presence and association at anyplace. To unequivocally describe the possible features of 6G trades, it expects the potential application circumstances and hardships for 6G correspondences in this portion. It should be seen that as a speculative examination of 6G, it is relied upon to cover a tremendous extent of warmed subjects analyzed in late dispersed works.

The plan to meet the demand for the future is already laid and that plan can be executed successfully with the birth of the 6G technology and it will be more dominant by the year 2030. The Samsung white book on the 6G communication explain the path of travel from the 5G to the 6G and the glimpse of the timeline is represented in the figure 2 and 3 respectively (samsung, 2020).

The sixth generation of portable associations will facilitate a lot of currently different headways, including significant learning and colossal data examination. The introduction of 5G has arranged a ton of this gathering. The need to pass tense enlisting to ensure throughput and low dormancy for very strong, low-inactivity exchange plans is a critical driver of 6G. The need to help machine-to-machine correspondence in the internet of things is also a principal force.

Additionally, a strong relationship has been recognized among 6G and first-class execution figuring High Performance Computing (HPC).

Moving towards 6G, the technology related to the current trend will be updated regarding the following Application:

- Robotic communication will play a vital role in case of rescue operation and also will replace human beings were it will be difficult for the human to work especially in remote and difficult environmental conditions. Robotized expansion, arrangement, and improvement of new organization substances associated with existing Base stations by means of remote associations will be a key challenge and it will be supported by the 6G technology (Babukarthick et al., 2021).

- Computer vision and its application in the autonomous vehicle is challenging field of the future technology. The sensor collected thousands of images surrounding the vehicle and it has to be processed within a short span of time to make the decisions. Also the challenges faced in the communication between the vehicle even in the dense environment can be overcome with the 6G technology. Enhanced portability support for versatile organization elements considering the speed of transportation frameworks can be achieved (Elsayed et al., 2019).
- Enhanced assistance coherence for client gadgets served by the organization substances, which themselves might be moving and are associated with the cellular network through remote associations is also achieved (Vanitha et al, 2020).

Figure 2. Timeline of 5G and 6G communication

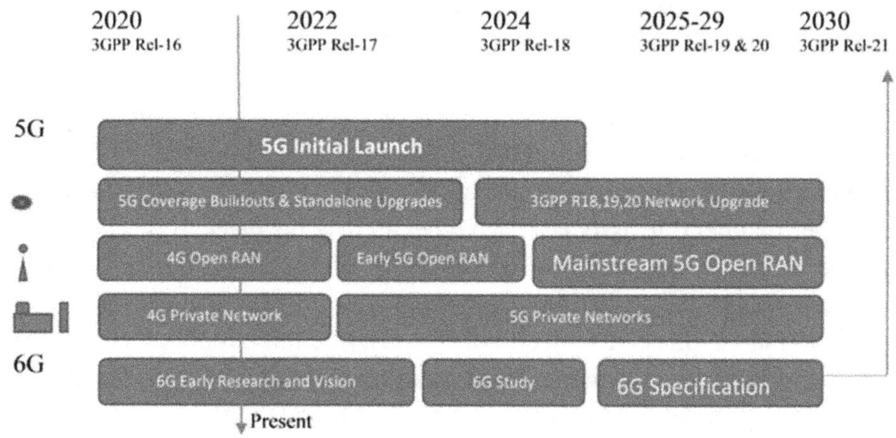

Figure 3. Evolution from 5G to 6G

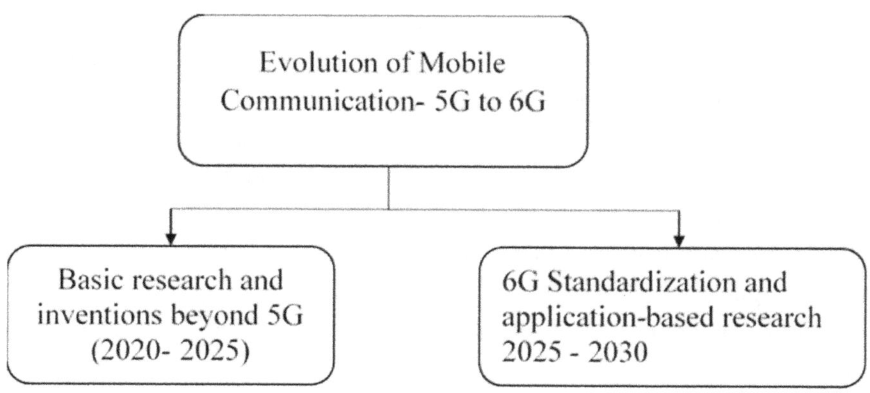

ARCHITECTURE FOR 6G TECHNOLOGY

The figure 4 shows the high level 6G Architecture. The high level 6G architecture consists of four levels starting namely central server, Base station with edge server, URLLC and the end devise getting access from the URLLC radio access network (Carroll et al., 2016). The central server is the core network which is responsible for performing network operation like routing, handover and network slicing. The core network is connected to the base stations which offers high data rate and low latency to the end devises which uses the state of art technology such as statics analysis, autonomous vehicle, robotics, remote communication, health care and virtual reality through the URLL radio access network (Nayak et al., 2021).

Figure 4. High level 6G architecture

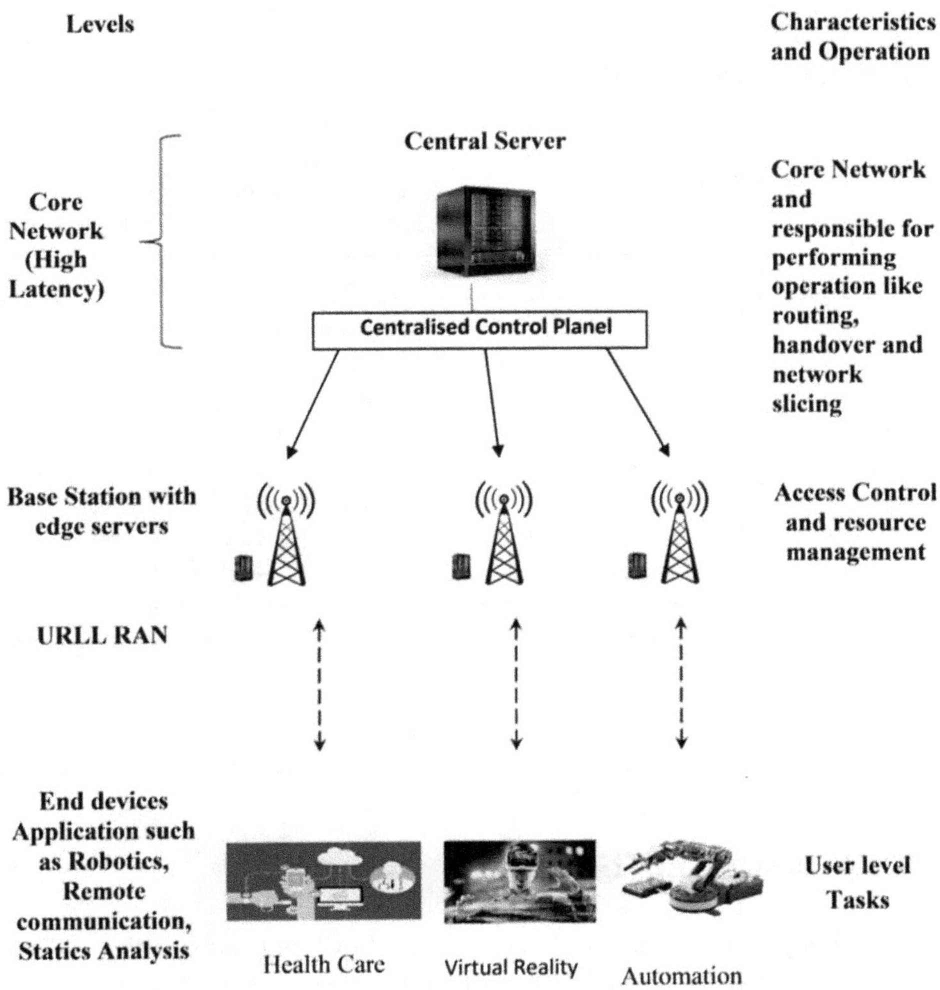

The 6G engineering incorporates building blocks across key compositional areas of a correspondence organization, beginning from the actual layer as far as possible up to the assistance layer in combination with a protected and computerized coordination design (Janjua et al., 2020). 6G engineering disintegration into building blocks, as made by Nokia Bell Labs, comprises of four significant interworking parts and the first part deal with open service. The second part does the specialised functions such as ensuring the flexible data flow. It ensures smooth flow of the data from the physical layer to the service layer. The third part deal with the mesh connectivity and RAN core convergence. The final part of the model deals with type of the platform deployed related to hardware acceleration, Het-cloud and data flow centric. The name Het-cloud is heterogeneous cloud and this name is given as the various classification are cloud such as public cloud, and private cloud are integrated. The advantage of the het-cloud is that this can be easily scaled and any new services can be easily incorporated. The model is represented in the figure 5 below.

Figure 5. Nokia Bell architecture model

Orchestration	Open Service Domain resource monetization Cognitive closed loop and optimization
Specialized	Flexible offload Extreme Slicing Sub-Networks
Functional	Cell free and Mesh Convergent Ran-Core Information Architecture and AI
Platform	Het-Cloud Data flow centric Hardware acceleration

6G FUNCTIONAL ARCHITECTURE

The key objective of any architecture model has to ensure flexibility and simplicity. The architecture model can be made simple by understanding the overall functions and there by elimination of any duplication function in the model.

RAN-Core Convergence

Key arrangement measures for 6G limits will consolidate objections of unraveling, versatility, time-to-exhibit and guarantee less likelihood of blunders. Isolating control plane (CP) and customer plane (UP) will continue to be a key arrangement rule and CP will presumably be organized as a chain of organizations. The Lower Layer capacities (LLF) will guarantee that there is least idleness in the air interface. It might be normal that vRAN and edge cloud will meet i.e., will share a comparable run-time environment for viability reasons. Since we expect various het-cloud circumstances of the 6G time frame to be server less, novel systems for instance, sandboxing to smooth out fire up delays for work execution and resource impression will be required. Meeting support will empower diligent capacities for constant information handling while various levelled stockpiling and informing layers give quick, versatile area mindful capacity access and work connections. The figure 6 shows the RAN Core Convergence Architecture.

Figure 6. RAN-core convergence

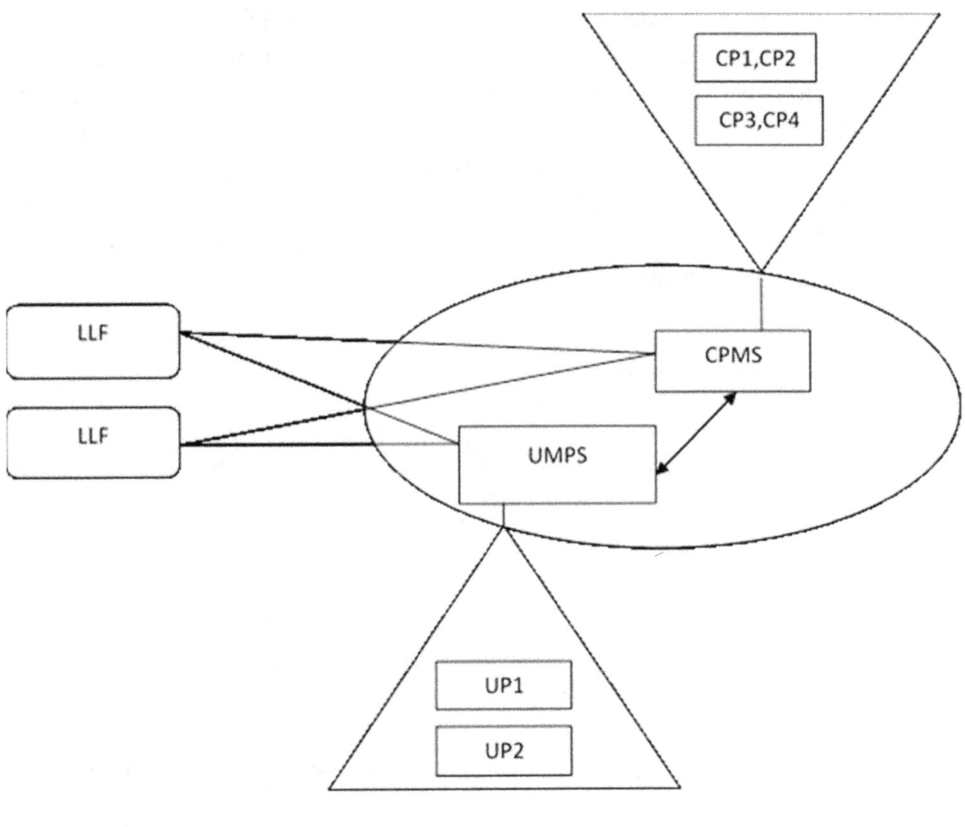

In 2G and 3G, the General packet radio Service (GPRS) network included four various hubs in the client plane namely Gated-GPRS Support node, Service-GPRS support node, Radio resource Management (RNC) and hub B. This was decreased to PGW, SGW, eNodeB in the client plane in 4G EPC and essentially to the UPF and gNB in the 5G Technology. Adaptability is accomplished by permitting free scaling, furthermore arrangement of various capacities and guaranteeing the capacity to rapidly make new administrations. A few standards can be utilized to advance practical arrangements like dormancy, security, strength and energy proficiency. The significant patterns that have worked with the adaptability have been the detachment of client plane and control plane. With the advancement of 5G RAN to vRAN, there is a near change in the RAN with the unit of the base station CU control and customer plane limits, cloud neighbourhood execution, and concentrated course of action. Furthermore, the assist-based designing with drawing closer of the 5G Centre will contact the RAN later on. As the Core customer plane limits move closer to the edge considering extending traffic volume and lower latency essential, there is an opportunity to fit the RAN and Core abilities to make a less intricate association.

The association will essentially have a Lower Layer Function (LLF) substance that joins all the inactivity fundamental air interface related RAN limits that are avoided from the radio unit, and User Plane Micro Services (UPMS) and Control Plane Micro Service (CPMS) helpful substances that fuse all the higher layer RAN and Core limits as smaller than normal organizations. The CPMS joins both RAN and Core organizations, for instance, radio resource control, radio insightful control (RIC), adaptability the board, approval, radio resource the leaders, etc. The UPMS consolidates higher layer RAN customer plane similarly to Core customer plane organizations like header pressure, encryption, QoS technique approval, and significant pack examination. The UPMS and CPMS will be established on a design that revealed APIs for new small-scale organizations to be added to the middle course of action of organizations that describe the two utilitarian substances. The small organizations that set up the UPMS and CPMS can be set in the het-cloud in a disaggregated plan. There may be an area and central event of the UPMS serving different plans of usage cases. RAN-CORE blending and functional improvement will enable astoundingly specific RAN and cut unequivocal RAN. The effortlessness of show of new organizations and new kinds of devices with different radio capacities and dedicated stacks will further develop time to showcase and orchestrate outright cost of ownership.

Figure 7. Cell free and mesh connectivity

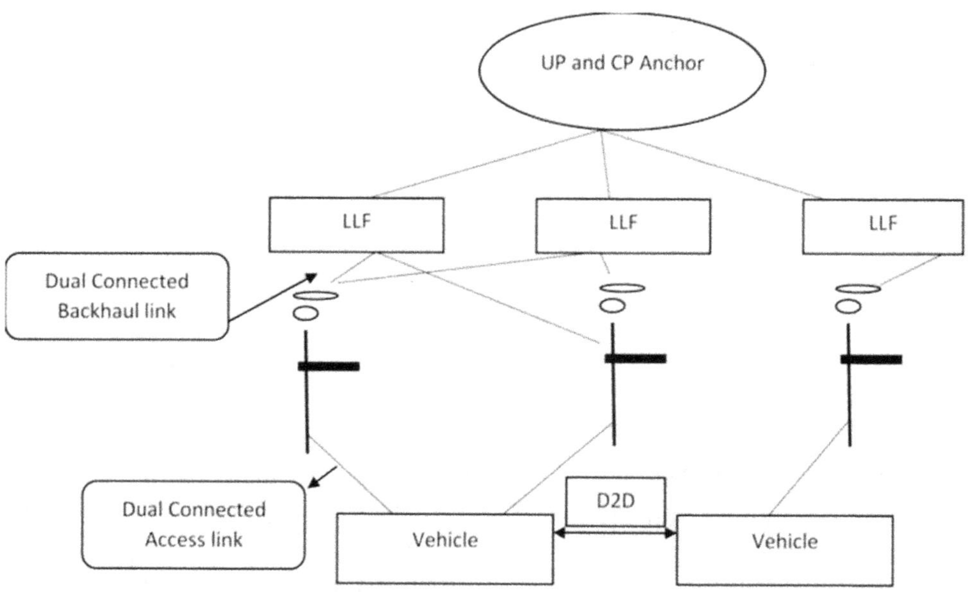

Cell Free and Mesh Connectivity

The game plan of non-autonomous 5G designing has decidedly set up the usage of a twofold organization in access where every contraption is related with LTE and NR cells. The unwavering quality of the organization can be improved by guaranteeing double accessibility of NR cells. In this sort of organization, a contraption isn't connected with only a lone cell but to an expert cell and a slave cell. All the while, fused Integrated Access and backhaul (IAB) has been standardized to extend the extent of the distant relationship in high gatherings. IAB center points are Layer 2 center points that essentially store and forward information from the sponsor center point and do whatever it takes not to stay aware of any UE control plane or higher layer customer plane state themselves. Engaging twofold organizations for IAB centers and end contraptions will achieve clear cross-section accessibility, where a device can communicate with the association through various courses. Furthermore, this will bring about the lattice network as displayed in the above figure 7. Cross area relationship of extensively higher thickness can be refined by loosening up twofold to multi-network at each ricochet. Also, with cloud execution of the 6G CP and UP limits tantamount to CU-CP and CU-UP and higher layer DU limits, devices can become cell-less with state stayed aware of simply the CP and UP got in the edge cloud. Bundles can be coordinated through the lower layer limits in the cross-section organization.

The data generated from various gadgets are sent at the same time from numerous radio units and thus creating MIMO and this result in achieving a high spectral efficiency. 6G gadgets may moreover be privately associated with a few proximal companion gadgets to frame 6G sub-organizations while keeping up with their wide region network. The sub-network availability will be very much useful in proving the necessary communication to Industry 5.0.

DEPLOYMENT DIFFICULTIES

6G will utilize TeraHertz (THz) recurrence band for transmission. The following are the challenges in deploying the 6G technology

- **Generation and Complexity of Hardware Design:** 6G transmission requires ceaseless THz signal (Rappaport et al., 2019). THz signal is troublesome on the grounds that it has more severe necessities in regard to the complexity in designing the hardware and it is exorbitant to produce of THz signal (Zhang et al., 2019). Transmission part of 6G must be low in cost to help the minimal expense advantage asserted by 6G.
- **Spectral Loss:** The electromagnetic signal always has spectral loss with the increase in the frequency and for the frequency of this range the spectral loss is quite significant. The experimental result of free space path loss for a reference distance of 10 m is found to be 100dB and it quite obvious as the loss is proportional to further increase in the communication distance.
- **Atmospheric Loss:** The presence of the polar gas molecule in the atmosphere causes the absorption loss in the signal transmitted and thus the signal is attenuated because of both spectral and atmospheric loss. The energy loss of signal happens due to atomic retention and spreading misfortune. The atomic ingestion misfortune happens because of transformation from THz signal energy to the inward active energy of the atoms present noticeable all around.

- **Reflection Loss:** As the wavelength is in sub millimetre range the possibility of scattering is more and which further cause the decrease in the magnitude of the signal.

APPLICATION

The future technology will be at its full swing around the year 2030 and the communication support to those technologies will be provided by 6G technology as it has the advantage of very high speed and low latency (Strinati et al., 2019). Thus the following are the application of the 6G technology which is certainly the state of art of technology of the future as represented in the figure 8.

- Autonomous vehicle
- Cloud computing
- Artificial Intelligence/Machine Learning/ Data Science
- 'e' Health Care
- Home/office Automation
- Robotics
- Under water Communication etc.

Figure 8. Application of 6G technology

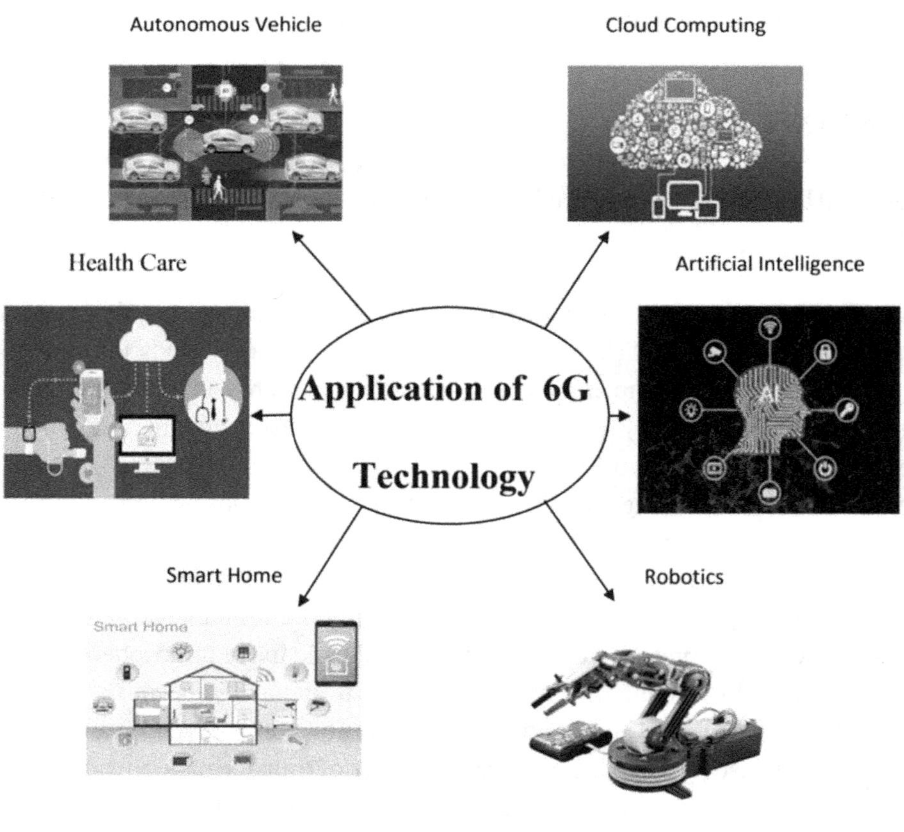

ROLE ON ML IN 6G

6th Generation (**6G**) is another remote innovation that numerous scholastics and analysts are leaving on (Rommel et al., 2018). The essential assurances of 6G are to grow Artificial Intelligence (AI) and Machine Learning (ML) benefits in distant associations and to the customers. 6G also will give impels in particular estimations like high throughput, supporting new prominence applications, further created usage of radio repeat gatherings and a great deal genuinely using AI and ML strategies. One of the huge ML advancements envisioned as a vital development for 6G will be Deep Learning(DL) considering its strong applications in achieving acquiring from circumstances that are more close to humans. Figure 9 beneath addresses the advancement of 6G and the possible job of Machine learning strategy which is bound to be prevailing around 2025 to 2030.

Figure 9. Evolution of 6G and potential role of ML techniques

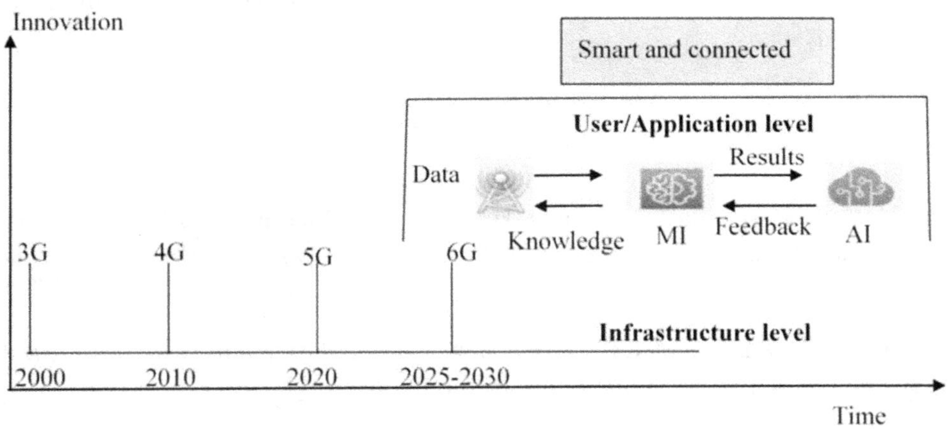

At various organization levels, ML calculations are to be prepared for certain given informational collections and after the data analysis the necessary control action are taken in the real time for the device at the remote location through the support of the Base station and network help (e.g., gadget programmability and setup). These new strategies drive the prerequisite for a data-driven, and ML-neighbourhood association as different association limits in the organization region satisfy the necessity. In continuation of the above motivation, it has, in addition, focused on describing 6G improvement towards ML and AI. Many existing researchers have zeroed in on explaining the need of organizing ML with AI and how this compromise has been applied in numerous splendid locales for achieving improvement in people's lives, including the clinical benefits region, business, clinical systems, and wise request motors. Simulated intelligence models are computational systems that are used to get comfortable with the discriminative components of a structure that can't be tended to by a mathematical model. ML is spine of 6G distant associations as it displays the structures that can't be tended to mathematically. Moreover, certain ML techniques are at this point being used to substitute savage power or heuristic estimations to find ideal courses of action for network issues. As ML makes its progression in 6G associations, it will be achievable for persistent noticing and robotized zero contact action and control. Additionally, ML figures can be done even in the advanced cell which has the necessary application and essential activity or proposal should be possible through the end gadgets as everything relating is associated with the cloud. One of the imperative necessities of any ML model is getting ready data that is used to set up the model to get the needed yield from the model.

CONCLUSION

With the advancement of remote innovation, 5G would not have the option to completely fulfil the developing need for remote interchanges in 2030. Hence, 6G should be rolled out. The use of 6G innovation in industry would help all the future generation state of art technology to come into reality and will certainly enhance the user experience. Also, 6G with Computational Intelligence could assist us with handling a lot of information gathered from the IoT devices connected to the cloud and can predict the future by analysing the data. This article broke down 6G specialized benefits, portrayed 6G empowering the modern applications, presented the premise of AI key innovations and summed up important use of AI in various situations.

REFERENCES

Akhtar, M. W., Hassan, S. A., Ghaffar, R., Jung, H., Garg, S., & Shamim Hossain, M. (2020). The shift to 6G communications: Vision and requirements. *Human-centric Computing and Information Sciences*, *10*(1), 1–27.

Arias, Salado, Medina, & Zambrano. (2021). *5G technology deployment in Latin America: An analysis of public policy and regulation environment.* Academic Press.

Babukarthik, R. G. (2021). 15 Lethal Vulnerability of Robotics in Industrial Sectors. In *Big Data Management in Sensing: Applications in AI and IoT* (pp. 227–238). River Publishers.

Carroll, N. (2016). Key success factors for smart and connected health software solutions. *Computer*, *49*(11), 22–28.

David, K., & Berndt, H. (2018). 6G vision and requirements: Is there any need for beyond 5G? *IEEE Vehicular Technology Magazine*, *13*(3), 72–80.

Elsayed, M., & Erol-Kantarci, M. (2019). AI-enabled future wireless networks: Challenges, opportunities, and open issues. *IEEE Vehicular Technology Magazine*, *14*(3), 70–77.

Janjua, M. B., Duranay, A. E., & Arslan, H. (2020). Role of Wireless Communication in Healthcare System to Cater Disaster Situations under 6G Vision. *Frontiers in Communications and Networks*, *1*, 6.

Nayak, S., & Patgiri, R. (2021). 6G communication technology: A vision on intelligent healthcare. In *Health Informatics: A Computational Perspective in Healthcare* (pp. 1–18). Springer.

Nayak, S., & Patgiri, R. (2020). *6G communications: A vision on the potential applications.* arXiv preprint arXiv:2005.07531.

Rappaport, T. S., Xing, Y., Kanhere, O., Ju, S., Madanayake, A., Mandal, S., Alkhateeb, A., & Trichopoulos, G. C. (2019). Wireless communications and applications above 100 GHz: Opportunities and challenges for 6G and beyond. *IEEE Access: Practical Innovations, Open Solutions*, *7*, 78729–78757.

Rommel, Raddo, & Monroy. (2018). Data center connectivity by 6G wireless systems. In 2018 Photonics in switching and computing (PSC) (pp. 1–3). IEEE.

Sivaramakrishnan, S. (2021). Terrance Frederick Fernandez; R. G. Babukarthik; S. Premalatha, "4 Forecasting Time Series Data Using ARIMA and Facebook Prophet Models. In *Big Data Management in Sensing: Applications in AI and IoT* (pp. 47–60). River Publishers.

Sivaramakrishnan, S., & Kesavamurthy, T. (2017). Emseep—Effective Modelling of Scalable Energy Efficient Protocol In Wireless Sensor Network. *Advances in Natural and Applied Sciences*, *11*(12), 18–26.

Sivaramakrishnan, S., & Kesavamurthy, T. (2017). Identifying Cluster Head and Data Transmission Through Them for Efficient Communication in Wireless Sensor Network. *Journal of Computational and Theoretical Nanoscience*, *14*(8), 4014–4020.

Sivaramakrishnan, S., & Kesavamurthy, T. (2017). Optimum Routing of Data through Cluster Head in Wireless Sensor Network. *International Journals of Control Theory and Applications*, *10*(31), 27–35.

Strinati, E. C., Barbarossa, S., Gonzalez-Jimenez, J. L., Ktenas, D., Cassiau, N., Maret, L., & Dehos, C. (2019). 6G: The next frontier: From holographic messaging to artificial intelligence using subterahertz and visible light communication. *IEEE Vehicular Technology Magazine*, *14*(3), 42–50.

Vanitha, U., Sivaramakrishnan, S., Karthika, S., & Babukarthik, R. G. (2020). Efficient Communication Routing Through WiMax Network During Disaster. In *IOP Conference Series: Materials Science and Engineering* (*vol. 994*, no. 1, p. 012035). IOP Publishing.

Yang, P., Xiao, Y., Xiao, M., & Li, S. (2019). 6G wireless communications: Vision and potential techniques. *IEEE Network*, *33*(4), 70–75.

Yastrebova, A., Kirichek, R., Koucheryavy, Y., Borodin, A., & Koucheryavy, A. (2018). Future networks 2030: Architecture & requirements. In *2018 10th International Congress on Ultra Modern Telecommunications and Control Systems and Workshops (ICUMT)* (pp. 1-8). IEEE.

Zhang, J., Kang, K., Huang, Y., Shafi, M., & Molisch, A. F. (2019). Millimeter and THz wave for 5G and beyond. *China Communications*, *16*(2), iii–vi.

Ziegler, V., Viswanathan, H., Flinck, H., Hoffmann, M., Räisänen, V., & Hätönen, K. (2020). 6G architecture to connect the worlds. *IEEE Access: Practical Innovations, Open Solutions*, *8*, 173508–173520.

Chapter 8
Role of Blockchain in Security of 6G Networks

Shikha Garg
Punjabi University, Patiala, India

Sonia Goyal
Punjabi University, Patiala, India

Abhinav Bhandari
Punjabi University, Patiala, India

ABSTRACT

Over the last decade, the demand for faster data rates has fueled the rapid growth of wireless communication systems that introduce the concept of new wireless technologies such as 6G or beyond 6G. The world is going through an ultimate transformation with the arrival of the intelligent information era. It integrates 6G, internet of things (IoT), blockchain, and other techniques to address the issues and challenges faced by 6G networks. Although an IoT-enabled application is crucial in this case, it faces security, privacy, latency, and reliability issues. The 6G networks can accommodate various heterogeneous devices and infrastructure to provide ultra-high data rates, high reliability, low latency, and secure communication processes. Motivated from these facts, this chapter explores the role of blockchain in addressing security and privacy issues in 6G networks and future application prospects and research objectives.

INTRODUCTION

6G wireless communication technology is currently one of the most in-demand research areas. Personal life, business, society, and communication systems will all be revolutionized by 6G. It's critical to consider the possible applications, use cases, methodologies, issues, and challenges of 6g networks. The 5G technology has yet to be fully deployed globally. Furthermore, there are various security issues in 5G technology to overcome. Thus, it inspired us to explore the security issues and challenges of 6G

DOI: 10.4018/978-1-6684-3804-6.ch008

technology. Noteworthy, 6G technology will create many research possibilities and facilitate several new technologies.

Furthermore, the 6G will be demonstrated to be a paradigm shifter in a variety of industries. As a result, creating a vision for 6G technology is required to alter the present world. Nevertheless, 5G and Beyond 5G (B5G) will not meet all of the Internet of Everything's requirements (IoE). As a result, there is a tremendous demand for 6G. Researchers have already addressed the issues of B5G mobile communication. It expects that 6G will be completely operational by 2030. For the next 10 to 25 years, 6G will be the most critical research topic. Many nations have already begun their research projects on 6G networks. The researchers have started solving the challenges of 6G mobile communication such as THz frequency, minimum end-to-end delay, reliability, capacity, global coverage, security, and many more. The "6genesis flagship project" is the 6G research project that started in Finland in 2018. 6G project was launched by China, the USA, and South Korea in 2019 (Dang, 2020). In 2020, Japan also introduced the 6G project. Also, Japanese telecommunications company Nippon Telegraph and Telephone Public Corporation (NTT) had presented a white paper in January 2020.

Several countries have begun the 6G project as we see that in every decade, a new mobile generation emerges. So, it expects that in 2030, 6G technology will be deployed as 6G networks have no standard functionalities or specifications, just many possibilities which empower various new technologies. The maximum data rate from 1G to 6G is increasing exponentially. It is expected that the 6G networks will provides a 1tbps data rate with a maximum wavelength of 300 micrometers to achieve the THz frequency band. Also expects that the prices will be 1000*cheaper in the 6G era. 6G will provide not only ground coverage but also underwater and space coverage. 6G networks can accommodate various heterogeneous devices. 5G Network softwarization technologies, like Software Defined Network (SDN), Network Slicing, Network Functions Virtualization (NFV), and Multi-access Edge Computing (MEC) are still relevant in 6G, and their security challenges will persist. Attacks on SDN controllers, southbound and northbound interfaces, and platforms for installing SDN controllers/applications with inherent vulnerabilities are all security issues related to SDN. Denial of Service (DoS) attacks and stealing data by compromising slices are two possible attacks for network slicing. Security vulnerabilities of NFV are attacks on Virtual Machines, Virtual Network Functions (VNF), hypervisor, VNF manager, and NFV orchestrator. MEC in 6G is vulnerable to physical security issues, such as man-in-the-middle attacks and Distributed Denial of Service (DDoS), due to the enormously distributed structure of 6G networks. The 6G network will not achieve the guaranteed dynamicity and full automation due to attacks on network softwarization technology. Softwarization, cloudification, and virtualization, of upcoming wireless generations, undoubtedly provide significant benefits such as micro-based enterprise models, flexible and effective management and network orchestration (MANO), different services using network slicing. Nonetheless, they try to intensify the issues such as data privacy, security vulnerability, immutability and network reliability (Nguyen & Pathirana, 2020), authentic Virtual Network Functions (VNFs), multiple access control, soft spectrum sharing, and authentic use of a resource. Virtual networks offer various services with varying levels of security. In general, Blockchain technology (BCT) and distributed ledger technology (DLT) have gained traction and have been accepted by industry and research organizations worldwide. Although the blockchain and the distributed ledger technology are very safe, and have numerous potential applications in 6G wireless network like network decentralization, ensuring integrating resource sharing promises to provide minimum processing delay and improve the security of spectrum sharing. The blockchain used in network decentralization and distributed ledger technologies has the great potential to improve network performance and authentication security, respectively. Spectrum sharing based on blockchain technology could solve the problem of spectrum monopoly and low spectrum utilization. It is expected that the 6G technology would also utilize Artificial Intelligence (AI). Many experts believe that the 6G network should have "AI-enabled" capabilities. The 6G network should have a high integration of newly developed AI technologies and networking capabilities. Furthermore, as network security and privacy have grown increasingly vital in the latest days, risk reduction should be an essential part of the design. Therefore, this chapter discusses the prospective enhancements to the 5G network that may be combined to form a 6G network, the security and privacy issues that each important technology raises, and future applications for 6G networks. Hence, AI capabilities on the 6G network should also be significantly improved. Figure 1 depicts the topics mentioned in this study and how they connect.

The four fundamental components of a 6G network are represented by the red and blue circles in Figure 1, which comprise 3D intercoms, intelligent real-time edge, intelligent radio, distributed AI. These four categories are the focus of this chapter since they contain the essential features of 6G research. They are also the ones who are most concerned about security and privacy. AI-based software, molecular communications (MC), quantum communications (QC), blockchain technology (BCT), TeraHertz (THz) technology, and Visible Light Communication (VLC) technology are some of the technologies used in this study. These technologies have a lot of potential for usage in 6G network applications like blockchain and distributed ledger technologies, wireless brain-computer interaction (BCI), connected robotics and automated systems, and multi-sensory X Reality (XR), as demonstrated in Figure 1 by the yellow color circles. Figure 1 is an example of 6G security and privacy issues in this chapter.

Figure 1. 6G security and privacy issues

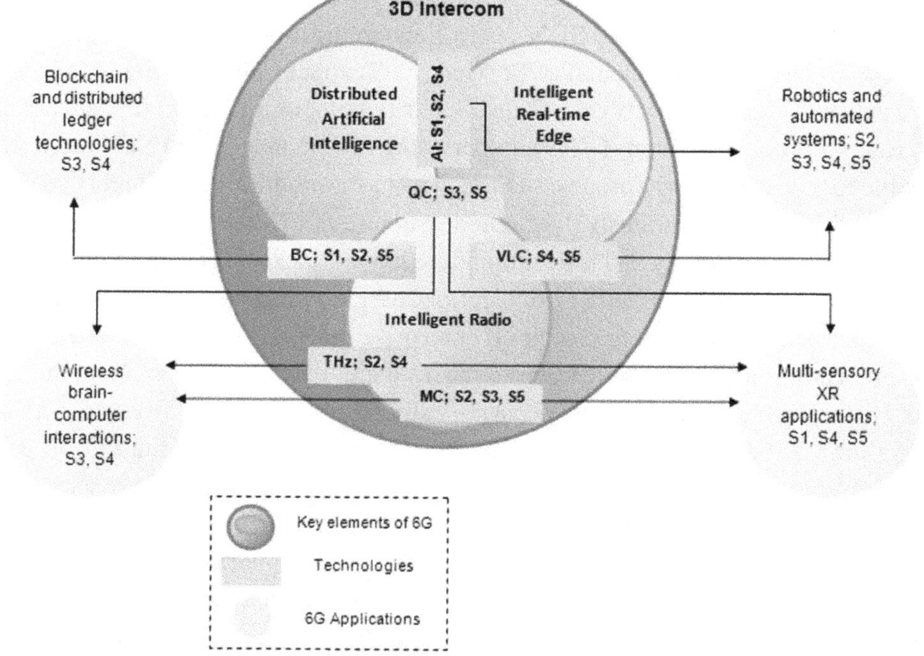

S1: Access Control, S2: Authentication, S3: Encryption, S4: Malicious behaviors, S5: Communication

The three sectors, Distributed AI, intelligent radio, and the intelligent real-time edge sector containing intersect technologies. Furthermore, AI resides at the intersection of these major sectors because 6G networks are expected to be AI-enabled.

The five most important security and privacy issues are outlined just below the diagram. In the diagram, authentication, access control, and malicious behavior are all more vulnerable. Certain technologies, on the other hand, are susceptible to specific challenges. For example, the VLC, in combination with both intelligent radio and intelligent real-time edge, is especially vulnerable to malicious behavior and data transfer processes. The THz and molecular communication technology supports intelligent radio. The MC technology has security and privacy concerns related to encryption, authentication, and communication, whereas the THz technology is particularly vulnerable to security and privacy concerns. BC and QC technology overlapped with intelligent radio and distributed AI. Access control, authentication, encryption, and data transmission are the primary security and privacy considerations here. Additionally, 6G applications have their own set of issues. The robotics and automated systems often depend on VLC and AI technology, where encryption, malicious behavior, and data transfer might be challenging. The multi-sensory X Reality application uses MC, QC, and THz technology, making it vulnerable to data transfer leakage, malicious behaviors, and access control threats. Wireless BCI employs similar concepts as the multisensory X Reality applications; however, they are concerned about their safety and privacy. The most significant flaws are encryption and malicious behaviors. As the 6G networks last application, the BC and DLT are generally very safe; however, they remain vulnerable to malicious attacks. There are five basic categories of security and privacy challenges in these new areas: Access control, authentication, access control, encryption, malicious behavior, and data transmission, which will be referred to as S1, S2, S3, S4, and S5. This chapter aims to provide more information on how these challenges impact the various parts of the 6G network and provide strategies to reduce the risks.

BACKGROUND

1G and 2G

Both the 1G and 2G systems are focused mainly on voice calls, with the 2G system providing more excellent speech quality and spectrum management than the 1G system. The data rates for 2G systems range from 9.6 to 14.4 kbps. China Mobile's annual report reported a ten times reduction in price from 0.1 to 0.01 US dollars/minute. More than half of the global population has embraced these services in only the past two decades (Gong, 2017).

3G and 4G

Third and fourth-generation cell phones, sometimes known as "3G or 4G Internet" phones, are a collection of wireless mobile communication system standards that offer high-quality voice transmission and multimedia services like video calling, online gaming, television services, etc. Data transmission is a primary function of mobile 3G and 4G networks. Users using 4G can expect speeds of up to 100 Mbps, whereas 3G offered only a peak rate of 14 Mbps. From 3G to 4G, technical advancements include orthogonal frequency division multiplexing (OFDM) and Multiple-input multiple-output (MIMO) to reduce the price a thousand times than the previous generations.

5G and 6G

The exponential expansion of 5G and beyond5G will be offered human-to-machine and machine-to-machine (M2M) communications. However, the current 5G still relies on enhanced mobile broadband (eMBB) and uses a pricing structure similar to 4G networks. Therefore, charging based on connectivity rather than data flow will be more acceptable. According to FTTH systems, China charges between $100 and $200 per terminal. To revolutionize 6G, 100 trillion sensors are planned to be built and connected to the internet by the end of 2030. As a result, thousand times price decrease will be required to create a long-term smart society. Therefore, it is a challenge for 6G to achieve the desired criteria listed in Table 1.

Table 1. Comparison between 5G and 6G

S.No.	Issues between 5G and 6G		
	Issues	5G	6G
1	AI	Half Supported	Fully supported
2	Satellite Integration	Partial	Fully
3	Autonomous Vehicle	Partially	Fully
4	Terahertz Communication	Very Limited	Widely
5	Operating Frequency	3-300 MHz	1THz
6	Data Rate	1 Gbps	1Tbps
7	Wavelength	3mm	300μm
8	Core Network	IoT	IoE
9	Architecture	Massive MIMO	Intelligent Surface
10	End to End Latency	10ms	1ms
11	Reliability	Good	Extreme
12	VLC	No	Yes
13	Centre	User	Service
14	Dimension	1D / 2D	3D
15	Privacy	Normal	Extreme
16	Real Time	No	Yes

Overview of Blockchain:

Blockchain is a kind of distributed ledger in which data is stored in a chain of data blocks. A new transaction on the blockchain is recorded in the logs of all users, regardless of whether or not a block is created. DLT is a decentralized database that a group of individuals manages. Blockchain is a distributed ledger technology (DLT) in which transactions are recorded using a hash, an immutable digital signature. This implies that if a single block in a chain is altered, it will be evident that the entire chain has indeed been tampered with. If hackers wanted to take down a blockchain system, they'd have to modify each block in the chain across all distributed copies.

Types of Blockchain:

There are two primary forms of blockchain: public (permissionless) and private (permissioned). The term "public blockchain" refers to a blockchain that is accessible to the whole people. In these blockchains, the roles of participant and validator are unlimited. This type of blockchain has the advantage of being uncontrollable, which implies that no single party will still have complete control over the network. This public blockchain will be truly dispersed because all nodes connecting to it will have equal authority. Bitcoin, Ethereum, and Litecoin are examples of public blockchain in practice. Private blockchains, on the other hand, are centralized, which means a single company controls them. All transactions are visible only to those who are members of the blockchain ecosystem. These blockchains are far more tightly regulated than public blockchains. The Hybrid or Consortium blockchain, which incorporates the benefits of both public and private blockchains. Some entities in this blockchain are public and can participate in transactions, while others are private and are controlled by the consensus mechanism. As a result, it's known as the hybrid blockchain, and it offers high-level security that may be used to protect the network. To fully utilize blockchain's potential in a 6G network, it is needed to grasp the fundamental concept, critical aspects of blockchain, and how blockchain can assist 6G applications. Table 2 shows the main features of blockchain with their applications in 6G networks.

Table 2. Characteristics of blockchain and their applications to 6G

Features of blockchain	Explanation	6G network and service applications
Decentralization	Transactions do not require the involvement of a central authority or a responsible third party. users have complete control over their personal information.	There is no requirement for trusted external authority in 6G networks, such as band managers, spectrum licensing, and database managers; mobile computing and D2D network central cloud/edge service manager; 6G UAV control center; and complicated cryptographic primitives in 6G IoT systems. Decentralized 6G networks might avoid single-point failure, guarantee data availability, and enhance customer service efficiency.
Immutability	The data recorded in the blockchain is difficult to modify.	Allow for great immutability in 6G services. Information sharing, virtualized network resource provisioning, spectrum sharing, and resource trade can all be immutably recorded in the only-appended blockchain. Furthermore, without modification, it can establish ubiquitous IoT networking, D2D communications, and large-scale human-centric interconnectivity via peer-to-peer networks of ubiquitous blockchain nodes. The great immutability is also beneficial for 6G networks performing accounting functions, such as logging session statistics and using data for billing, resource utilization, and trend monitoring.
Transparency	Every network member can see all transaction details on the blockchain (public ledgers).	Provide improved-localized visibility into 6G service utilization. For public verification, a single copy of blockchain records is distributed throughout a vast network. Users and service providers have equal access, validation, and tracking privileges over the network. Additionally, blockchains may provide open 6G infrastructures with transparent ledger solutions (i.e., distributed edge computing, distributed network virtualization, dispersed IoT networks). Blockchain ledgers can also support fair service trading applications (e.g., resource trading, payment) under the control of all network elements.
Security and privacy	Asymmetric cryptography is used in blockchain for security with strong authentication, integrity, and non - repudiation. Blockchain smart contracts can help with data auditability, data provenance for privacy and access control.	Provide high security for 6G networks using decentralized ledgers. Blockchain lets 6G networks safeguard themselves and their data with distributed trust models and strong access authentication. It is significantly more challenging for hackers to compromise data when stored over a network of computers (IoT metadata). Apart from that, smart contracts may enable 6G services, including user verification, data authentication, and resource preservation.

MAIN FOCUS OF THE CHAPTER

Basic Key Elements of 6G Networks:

There is a lack of 5G network compatibility for AI-based technology because of privacy, security, lightweight authentication, latency, etc. 6G networks are used to handle these issues. Hence, four major areas are advanced in 6G networks:

a. Intelligent real-time edge: 5G network failed to control the vehicle and unmanned aerial vehicle as it requires low network latency and real-time intelligence. However, it provides the possibility of autonomous driving. Hence, a 6G network enables such AI-powered services, allowing autonomous driving and responses to an unfamiliar environment. Summarization of key areas are presented in Table 3.

Table 3. Summarization of key areas

Key areas	Summary	Characteristics and their relationship to 6G
Intelligent real time edge	It facilitates autonomous driving response in real-time to the new environment	Real time response, capable to control 6G
Intelligent radio	It configures and updates itself dynamically based on hardware data	Self-adaptation, communication responsibility of 6G network
Distributed artificial intelligence	It is a decentralized system that can make intelligent decisions at several levels	Makes intelligent decisions for 6G network
3D intercoms	It offers services when and where they are required	Entire 3D coverage, responsible for coverage in 6G

b. Intelligent radio: Devices and transceiver algorithms were created simultaneously in prior network generations. As a result, the hardware capabilities of the networks, such as decoder processing, antenna count, and so on, have persisted in a quasi-static state (Letaief, 2019). Nevertheless, current developments in circuitry and antennas have facilitated decoupling the transceiver algorithms from the hardware, allowing intelligent radio to function as a single system. Based on the hardware knowledge, the transceiver mechanisms may dynamically configure and upgrade themselves under this paradigm.
c. Distributed artificial intelligence: Although 5G networks belong to the internet of things, 6G networks belong to everything. It provides a decentralized system that can make intelligent decisions at several stages (Huang, 2019) to satisfy multiple requirements. Every edge device becomes capable of controlling and accessing any part of the data, computing and storing data independently, and enhancing the security and privacy of the 6G network.
d. 3D intercoms: 6G networks have been used to analyze, plan, optimize capabilities that have gone from two dimensions to three dimensions throughout previous network generations. These networks have also supported satellite, UAV, underwater communications in 3D space. In addition to this, the 6G network also achieves routing, mobility, and resource management in the 3D intercom.

Motives for Integrating Blockchain with 6G Technology

(Kim & Huang, 2019) Presents a distributed machine learning (DML) paradigm relying on privacy protection. In several private settings, the suggested error-based aggregation rules outperform other anti-attack abilities and computational costs. It is a solution designed to address network performance, security, and privacy concerns. (Gai, 2019)Presented a privacy protection scheme for IoT edge computing based on blockchain technology that employs differential privacy technology in the blockchain system to achieve privacy protection, task allocation, and anti-tampering function, potentially preventing illegal threats. Furthermore, blockchain can be utilized to strengthen the safety of cloud medical data in a large context of medical data exchange. In addition to security, it is possible to build a direct link between network elements, decreasing management expenses and increasing spectrum efficiency. As a result, blockchain facilitates interoperability between network operators and service providers as well. This section discusses the motivation for the integration, which originates from the security issues faced by 6G networks and the promising opportunities presented by combining these two technology families.

6G and Blockchain Integration

To emphasis the motivation, this chapter will highlight essential features of the two technologies in terms of integration. Blockchain can store and manage 6G data with its secure distributed ledger. More crucially, blockchain can offer many security properties, including decentralization, transparency, privacy, and immutability, all of which promise to address existing 6G network security challenges effectively. Thus, the main advantages of blockchain for 6G networks and applications are its security and network management skills. 6G, on the other hand, refers to the most recent generation of wireless networks, which are designed to provide more capacity, lower latency, higher data rates, better end-user quality-of-experience (QoE), massive device connection, reducing operating expenses, and reliable service provisioning. Thus, the benefits of 6G in offering fast and high-quality services and the need for security and networking improvements are the most critical aspects of the technology.

According to the latest research, the rationale for combining blockchain and 6G originates from the potential benefits of blockchain in overcoming 6G network challenges related to networking, service management, security, and privacy. With the aid of blockchain technology, 6G is projected to overcome existing challenges and provide new potential for blockchain-based 6G services and applications. The following discusses the integration's rationale, current 6G challenges, and potential benefits of blockchain-6G integrations.

General and Security Challenges in 6G Networks

(Biral & Centenaro, 2015) Discusses some of the perceptible issues in 6G. Furthermore, this paper address issues related to M2M communications. General challenges faced by 6G networks are discussed below:

i. Scalability: Industrial IoT aficionados expects that billions of gadgets will be connected and controlled in the development of industrial environments as a result of the introduction of concepts like massive Machine Type Communications (mMTC). As a result, customizing the design of 6G networks to meet such an enormous traffic demand will be tough.

ii. Real-time communication with minimal latency: Real-time communication will be crucial in future computing ecosystems. For precision functioning, device-to-device (D2D) and machine-to-machine (M2M) communication need a high degree of precision and near-zero latency. AR-assisted healthcare and autonomous driving, for example, would necessitate relevant and valuable content.
iii. Higher throughput: 6G is expected to raise the flow of data as well as efficiency. Mission-critical systems which rely on 6G and even beyond communication networks require the continuous connection of billions of devices. The massive volume of transactions should be managed in real-time by base stations and other network equipment.
iv. Synchronization: Synchronization is a critical need in most industrial applications. For practical uses like transportation networks and electricity distribution, synchronization is essential for precise operation.

Requirements for Security in Future Computing Systems

i. Confidentiality: Future computer infrastructure, such as IoT, reveals large attacks surfaces due to wireless communication. Encryption solutions, such as symmetric essential encryption methods, must be lighter in weight for low-power IoT devices. However, because of computational constraints, the data may be vulnerable to privacy concerns. Because of their high computational requirements, light encryption techniques may expose data to privacy issues.
ii. Integrity: As future systems will generate a significant volume of data, authorized users will need to access and edit information while it is in transit. In transit, eavesdropping and data tampering will lead the system to react unexpectedly.
iii. Availability: Service availability is a critical requirement in upcoming networks. DDoS attacks are more likely in 5G ecosystems with many connected devices due to their complexity. The specializations of present network security technologies can't be applied instantly to 5G and beyond networks to identify attacks and intrusion attempts.
iv. Authentication and access control: Data in transit and storage must be safeguarded using access control mechanisms to prevent unwanted manipulations. Formal centralized identification and authentication solutions would be restricted in scalability in the face of the enormous futuristic demands predicted in 6G. Access control requirements that are resource-intensive to meet the variety of potential tenants in the 6G network will cause bottlenecks in the associated services.
v. Audit: An audit is needed to define whether the tenants are following the norms of the network ecosystem. To achieve the higher security criteria, deep packet-level monitoring may be required to detect and flag the activity of sure tenants. Auditing a large number of tenants will be problematic from the aspect of enforcing security.

6G Technologies: Security and Privacy Issues

Security issues have been a major consideration in the design and implementation process for 6G and beyond networks. The present 5G technology infrastructure faces networking, security, and computational performance degradation issues because of its centralized architecture. Single-point failures in this architecture threaten on-demand user access to edge/cloud services. Centralized systems fail when several users request data simultaneously or when servers are disturbed by cyberattacks or software flaws. Furthermore, NFV and service function chaining (SFC) bring additional security issues in 5G networks.

Data communication may be hacked by curious cloud entities causing data spillage problems if NFVs use in multi-cloud providers. Moreover, the cloud infrastructure is shared by all tenants in a virtualized environment. As a result, service providers' accountability and transparency may be damaged due to an increase in the likelihood of cloud-based attacks. A virtualization server in NFVs can provide unique features for multiple operating systems, like resource allocation or VM migration via orchestration protocols. Securing communication between the virtual machine manager and the orchestrator is a significant issue. Mobile data traffic is rapidly increasing, and rising user expectations on 6G infrastructure pose provide additional security and performance issues that must be addressed. For example, rising demand for bandwidth-hungry 6G applications like video streaming and big data processing necessitated an effective 6G spectrum resource management plan to minimize resource shortages and ensure service continuity. Thus, spectrum sharing is required between mobile subscribers and mobile network operators (MNO). However, spectrum sharing raises security concerns in such cases, and malicious users can utilize it as a central point of attack.

In the context of 6G IoT like smart cities, smart healthcare, large-scale data storage, heterogeneous networks, and cloud computing, security and privacy challenges become considerably more complicated to tackle. IoT devices will continuously generate IoT data in a large amount. It is extremely difficult to quickly identify the relevant information or detect suspicious activity in large-scale data exchange. Centralized control may not be practical because of long latency, network congestion, and third-party privacy risks. So, how might mobile services be made more efficient? (such as sharing and processing data and managing users) while maintaining high levels of security is a major challenge for the industry. Because of this, innovative solutions are needed to overcome the network performance and security limitations of future 6G networks. Some key technologies for the 6G network have already been proven effective, as outlined above. Low latency, high reliability, and secure transmission services are provided to 6G networks. However, as stated previously, most of these technologies come with additional security and privacy risks. This is the subject of this section. Table 4 provides the summary of the security and privacy concerns raised up by 6G technology.

i. Artificial Intelligence

Among all the technologies predicted to be employed in 6G networks, AI is widely recognized as a key component. It has created a lot of interest in the networking sector. Many new security and privacy issues have arisen due to increasing attention towards AI (Zhu & Philip, 2019). Artificial Intelligence (AI) is being used in the 5G network. However, it is supposedly being run in isolated places with enormous volumes of training set, powerful but private processing centers, whereas the 6G network's AI will become more integrated (Tariq, 2020). In terms of architecture, AI technologies can be separated into two categories: physical (Data lines and network architecture) and computing (such as SDN, NFV, and edge/cloud/fog computing).

ii. Molecular Communication

The natural phenomena of molecular communication have been discovered among biological beings with nanoscale structures (Akan & Ramezani, 2016). Nanoscale and micro scale technologies are becoming a reality due to advances in bioengineering, nanotechnology, and synthetic biology over the last decade (Farsad, 2016). Furthermore, the energy required for the formation and molecular signal's propagation is negligible. However, in biology, this phenomena has been extensively investigated over a long period of time and recently it has become a topic of interest in communication study. Molecular communication holds great promise for 6G communications. However, it is a relatively new multidisciplinary technique. The core concept of MC is the transmission of information via biochemical signals. (Nakano, 2019) Described a method for mobile MC in which the transmitter, receiver, and connected nodes can interact while moving. Though, many security and privacy vulnerabilities relating to the authentication, encryption, communication processes had been discovered. (Farsad, 2016) Noted that an adversary may disrupt this sort of communication and that only a few research have ever investigated the security of MC lines. (Lu & Higgins, 2015) Proposed a coding system that could increase data transmission security. MC could benefit from developing novel authentication techniques to protect data confidentiality and privacy, as suggested by (Loscri, 2014). Flooding attacks, jamming, and de synchronization are among the attack tactics discussed by these authors at various levels of molecular communication. Though MC mechanisms for the 6G networks need to be developed for practical use, they are projected to accomplish what conventional communication channels are incapable.

iii. Quantum Communication

QC is another promising technique for 6G networks. A real benefit is that it can significantly increase data transfer security and reliability. Quantum communication is vulnerable to eavesdropping, measurement, and replication by an adversary. As a result, the recipient cannot be oblivious to the interfering signal (Zhang & Ding, 2017). Encryption is a crucial feature of quantum communication, and it could be ideal for long-distance communication with the right breakthroughs. It provides a slew of new features and raises conversations to a level that standard communication systems can't match (Gyongyosi, 2018). In contrast, quantum communication does not yet address all security and privacy concerns. Though tremendous work has been done in establishing quantum cryptography for QC, long-distance quantum communication remains a substantial barrier due to fiber attenuation and operation mistakes. According to (Hu & Yu, 2016), Quantum secret sharing, distribution, teleportation, direct transmission, and quantum dense coding may be necessary to enable entirely secure quantum communication. (Zhang & Ding, 2017) Provide additional information on the protection of quantum secure direct contact that could send secret information directly across a quantum medium without the use of an encryption key. (Nawaz, 2019) Also examined several quantum procedures that protect fundamental security by using quantum key distribution models.

iv. Terahertz Technology (THz)

The high transmission rates needed for 6G make mm-wave bands ineffective, despite their widespread use in 5G networks. However, the Radio Frequency (RF) band is unavailable for the following technologies. These reasons have accelerated the advancement of THz technology. THz communication makes use of the 0.1–10 THz spectrum band, which offers enormous spectrum resources than the millimeter-wave band.). The THz band has various advantages, according to (Huang, 2019). THz technology might be capable of supporting data speed of up to 100 Gbps. Second, eavesdropping would be more difficult because of THz communication's shorter pulse duration and narrower beam, which would make it more secure. Third, THz waves have a very low attenuation through specific materials; thus, they can be used for many different applications. Furthermore, THz communication transmissions can be highly directional, reducing intercell influence significantly (Zhang & Ding, 2017). Perhaps most importantly, IEEE 802.15.3d has already defined two forms of THz physical layers. (Strinati, 2019) Observe that THz communication consumes a significant amount of energy, which is an evident challenge. 6G cells must be shrunk from "small" to "tiny," necessitating the development of more complicated hardware and designs (Saad & Bennis, 2019). THz, like all other technology, has its own set of security and privacy concerns. The majority of these are concerned with authentication and malicious conduct. Although eavesdropping is thought to be difficult with THz transmission, (Ma, & Shrestha, 2018) claim that even if a signal is being transmitted using narrow beams, an eavesdropper can still intercept it. They do, however, discuss a technique of preventing such eavesdropping.

v. Visible light communication (VLC)

VLC technology may be employed to address the growing demand for wireless connectivity (Islim, 2017). VLC has already been explored and applied in several sectors, including the Vehicular Ad Hoc Network (VANET) and the indoor positioning systems. For example, numerous papers on VLC-based positioning have been published by researchers such as (Luo, 2017). VLC is superior to RF in terms of bandwidth and latency since it can withstand electromagnetic interferences (Khan, & Balan, 2019). The advent of solid-state lighting has also aided the advancement of VLC technology. (Chen & Li, 2019) Created LiFi, a VLC system that allows for multiple access and can offer high-speed services to many connected mobile users. However, there are significant flaws in VLC technology that are preventing it from progressing. Because intense natural light will impact transmissions, the primary application cases for VLC should be indoors. Malicious behaviors and communication mechanisms are among the security and privacy concerns raised by VLC. (Pathak, 2015) Pointed out that there must be a direct line of sight between an attacker and a victim to initiate an attack on an active VLC operation. This would help attackers to be detected. A precoding strategy for visible light communication links (Mostafa, 2014) had been offered to enhance the physical layer's security.

Similarly, Sec VLC protocol had been developed by (Ucar, 2016) to safeguard the transmission data's security. The authors did it by using the protocol in a vehicle network and (Cho & Chen, 2019) Investigated that eavesdropper's cooperation can threaten VLC systems' security.

vi. Blockchain

Blockchain technology has numerous applications in a 6G network. Distributed ledger systems, network decentralization, and spectrum sharing are a few examples of blockchain technology. According to (Dang, 2020), the blockchain used in network decentralization and distributed ledger technologies has the great potential to improve network performance and the security of authentication. Spectrum sharing based on blockchain technology could solve the problem of spectrum monopoly and low spectrum utilization. Blockchain technology has the potential to be one of the disruptive IoE technologies. Authentication, access control, and communication processes are all related to blockchain security and privacy issues. A blockchain-based radio access network are explored that would securely access and authentication across trustless network elements, whereas using blockchain technology, (Kiyomoto, 2017) propose a novel mobile service authorization architecture. To acquire access to licensed spectrums, (Kotobi, 2018) presented a technique of enhancing the security of cognitive radio and media access protocols by using the blockchain.

Furthermore, although the 6G network's decentralized architecture means that a hacker could only update records if the hacker has control of more than 51% of the nodes. i.e., 6G networks do not have any reliable third party in charge of storing data and managing data in the event of security breaches (Puthal, 2018). (Ferraro, 2018) Also, point out that a blockchain-based system may compromise the hashing capabilities needed to process payments in security. Table 4 presents the various security and privacy issues occurring in 6G networks in different technologies.

Table 4. Security and privacy issues in 6G networks

Key technology	Research	Security and privacy issues	Key technology contribution
Artificial Intelligence	(Hong, 2019)	Communications	A machine-learning-based antenna design that could be used in the Physical layer to avoid data leakage.
	(Dang, 2020)	Malicious behavior	Detect network abnormalities
	(Loven, 2019)	Access control	Processes for fine-grained control
	(Sattiraju & Weinand, 2018)	Authentication	To improve the security of physical layers
Molecular communications	(Farsad, 2016)	Malicious behavior	An adversary interfering with MC
	(Lu, 2015)	Encryption	Coding scheme to improve the security of data
	(Loscri, 2014)	Authentication	To develop new authentication mechanism
Quantum Communication	(Hu, 2016)	Communication	Different modes of QC
	(Nawaz, 2019)	Encryption	To protect quantum encryption keys
Blockchain	(kiyomoto, 2017)	Authentication	New architecture for mobile service authorization
	(Kotobi, 2018)	Access Control	Technique to improve access protocols
	(Ferraro, 2018)	Communication	Validation of transactions with the use of hashing power
THz	(Akyildiz, 2014)	Authentication	Authentication using electromagnetic signatures
	(Ma & Shrestha, 2018)	Malicious behavior	An eavesdropper can detect a signal even if it is being transmitted across a narrow beam.
VLC	(Ucar, 2016)	Communication	A secure protocol for communication
	(Cho & Chen, 2019)	Malicious behavior	Security can be harmed by cooperating eavesdroppers.

SOLUTIONS AND RECOMMENDATIONS

Blockchain-Enabled 6G Networks and Services

Blockchain may deliver a new set of creative solutions for 6G networks and services that are more secure, decentralized, and revolutionize network management architecture for enhanced quality of service (QoS). As a result, 6G should use blockchain's advantages to provide greater security and flexibility in mobile network coverage and services. In a nutshell, we discuss the tremendous benefits that blockchain may provide to 6G networks, emphasizing three primary areas: security enhancement, improvement in system efficiency, and simplification of the network.

Security Enhancement

6G wireless networks provide a wide range of possible uses for blockchain technology, such as enhancing security and privacy, decentralization, immutability, transparency, and traceability. The block chain's decentralized network architecture eliminates the need for third-party authority, therefore eliminating the concept of centralized network management. For instance, cloud computing based on blockchain facilitates decentralization of edge/cloud 6G networks, eliminating centralized core network control and single point failure bottlenecks and significantly increasing system performance. 6G services (such as the exchange of data and the allocation of resources and sharing spectrum) are robust to data changes. One of the benefits of adopting blockchain is that it improves spectrum management in two ways: first, by allowing for more accurate spectrum access verification and, second, by making it more accessible. Furthermore, the usage of blockchain allows peer-to-peer ledge networks to enable scaled spectrum sharing without the requirement for spectrum license holders or band administrators. The immutability of blockchain ledger services ensures a high level of security and more excellent system protection from DoS attacks and threats. Blockchain could potentially bring new authentication methods for 6G cellular networks, thanks to smart contracts, allowing highly efficient and flexible user access control mechanisms using intelligent coding logic and access rules. Rather than depending on external public critical infrastructure, contracts can automatically authenticate user access, discard malicious access, and detect threats from networks without revealing personal data. Furthermore, blockchain solutions guarantee excellent data protection by publishing user data to ledgers signed using hash functions and added immutably to blocks. When sharing personal information across an untrusted network, Blockchain technology can provide complete data control, which is unlike any other traditional technique that prevents users from tracking their data.

Improvements in System Efficiency

Blockchain technology can boost the performance of 6G networks. Decentralized blockchain nodes can verify resource requests (such as data access) using intelligent smart contracts, which eliminates the need for centralized authority and, as a result, reduces the latency of the network. Furthermore, by removing decentralization, blockchain can create direct contacts between 6G service providers and phone users, which will lower management costs dramatically. This would allow 6G ecosystems to have significantly more efficient and flexible data delivery models while meeting strict security standards. For instance, by utilizing the computational capacity of all network participants, blockchain can allow users to create secure peer-to-peer communication (i.e., D2D communication) rather than relying on a third-party intermediary. This could improve overall system performance by reducing latency, transaction cost and providing worldwide access to all users. Although malicious attacks and threats might compromise the entire network, distributed ledger consensus ensures that no single point of failure vulnerabilities exist, resulting in increased security for the whole network.

Network Simplification:

Blockchain's decentralized architectures are supposed to simplify 6G network deployments. With blockchain, cellphone operators no longer need to worry about centralized control systems. The blockchain network is capable of delivering 6G services. Service responses, user access, and service trade (e.g., resource trading and payment) among network members, including phone users and network operators, can all be implemented on decentralized ledgers without additional management infrastructure requirements (Christidis, 2016). Hence, adopting blockchain technology minimizes network complexity, and thus it reduces operational expenses. Besides, the blockchain network controls the 6G transactions (data sharing, spectrum sharing) itself, where all entities have the same management and maintenance privileges over the network.

ROLE OF BLOCKCHAIN IN 6G APPLICATIONS:

Although several applications from existing network generations can still be used in 6G networks, there are still some limitations. As a result, the 6G technology future applications are summarized as follows in Table 5.

a. Multisensory XR: Multiple sensors are used to collect sensory data for which 6G networks in the XR combines classic ultra-reliable low-latency communication with improved mobile broadband. The security and privacy issues with ultra-reliable low latency communication in multi-sensory XR applications include malicious behavior.
b. Robotics and Automated systems: This is the 6G network's second application in which an automated system is required to drive alone efficiently. This system includes both intelligence and logic across the network, which enables autonomous control of all internal parts.
c. Wireless brain-computer interaction: Wbci is used to connect the brain to a computer or other device. With the 6G network wireless BCI is about to find more applications in which few are XR.
d. Blockchain: As the blockchain technology exchanges information with all participants, it is probable to be used for data and spectrum sharing, boosting the security of wireless 6G mobile networks. Still, several issues remain particularly malicious. There are three forms of malicious behavior attacks: most double-spending attacks, security flaws, and a breach of transaction privacy. They also offer solutions based on blockchain technology to address these problems in 6G mobile networks, including cryptographic algorithms, incentive strategies, etc. Blockchain technology also can prevent Distributed Denial of Service (DDoS) attacks, traffic diversion threats, and access control and data protection. There are a variety of methods by which DDoS attacks can occur. e.g., 51% of attacks may result in DDoS. It does not require making a botnet network and simply uses the internet of things network to attack. The combination of both 6G and IoT can result in new security challenges. It can also be said that traditional techniques can be easily broken, so more security is needed by replacing the previous methods with new techniques such as blockchain, China Block. On the other hand, Machine learning (ML), Artificial Intelligence (AI) technologies, and Blockchain can detect and prevent the network from various attacks. The chapter discusses the different types of blockchain of which the role of Consortium blockchain provides high-level security, which can be used to secure the network. Table 5 discusses a summary of the security and privacy issues in 6G applications.

Table 5. A summary of the security and privacy issues in 6G applications

Applications	Research	Issues of Security and privacy	Contribution
Multisensory XR applications	(Chen & Li, 2019)	Communications	Improves the security of the network by working on its dynamics
	(Hamamreh, 2017)	Malicious behavior	A novel anti-eavesdropping system has been developed.
	(Eryani, 2019)	Access control	Handling access with the DOMA multiple-access approach
Robotics and Automatic system	(Hooper, 2016)	Malicious behavior	Protect Wi-Fi attacks, explores eavesdropping attacks, hijacking attacks, etc.
	(Sanjab, 2017)	Encryption	Innovative mathematical framework to increase the security of autonomous drones
	(Sun, 2019)	Communication	The only one communication mechanism which could protect against eavesdropping.
	(Ni & Lin, 2019)	Authentication	Two-factor verification could be supported by a self-driving vehicle protocols.
Wireless brain computer interaction	(Ramadan, 2017)	Malicious behavior	Presents hacking application
	(Svogor, 2012)	Encryption	Preventing reply attacks with a password method
Blockchain technology and DLT	(Nguyen, 2020)	Malicious behavior	Three forms of malicious behavior in the 6G applications are discussed.
	(Nguyen, 2020)	Encryption	Discusses various cryptographic algorithms to improve security of blockchain

BENEFITS OF BC IN 6G:

Blockchain technology has shown its benefits in various fields such as infrastructure and resource management, network slicing and resource brokerage, authentication, international roaming, local area wireless networks, and cloud computing.

a. Infrastructure and resource management: The resource management functions, including decentralization, orchestration, spectrum sharing, caching, computation and communication, are challenging tasks in the vast connectivity requirements of next-generation telecommunications infrastructure. In various ways, blockchain technology addresses the issue of infrastructure sharing among several mobile network operators (MNOs). Initially, it observed each mobile network operator's infrastructure utilization and managed the related payment using cryptocurrencies, which can then be pre-booked using smart contracts. On the other hand, Telecom regulatory organizations cannot track real-time spectrum utilization and restrict unfair activity with current technologies. Furthermore, combining various spectrum bands (e.g., licensed, unlicensed) from multiple network operators and wireless standards adds complication.

Moreover, the complexity of interference handling and real-time agreements raises the stakes for dynamic spectrum sharing incentives. Combining an automated spectrum trading system based on blockchain technology with current mobile network infrastructure addresses the issue mentioned above of MNOs agreeing on smart contracts. Figure 2. Shows a detailed depiction of a resource management technique using blockchain, including computing architectures, communication, and energy infrastructure in 6G communication.

Figure 2. Framework of resource management using blockchain in 6G communication

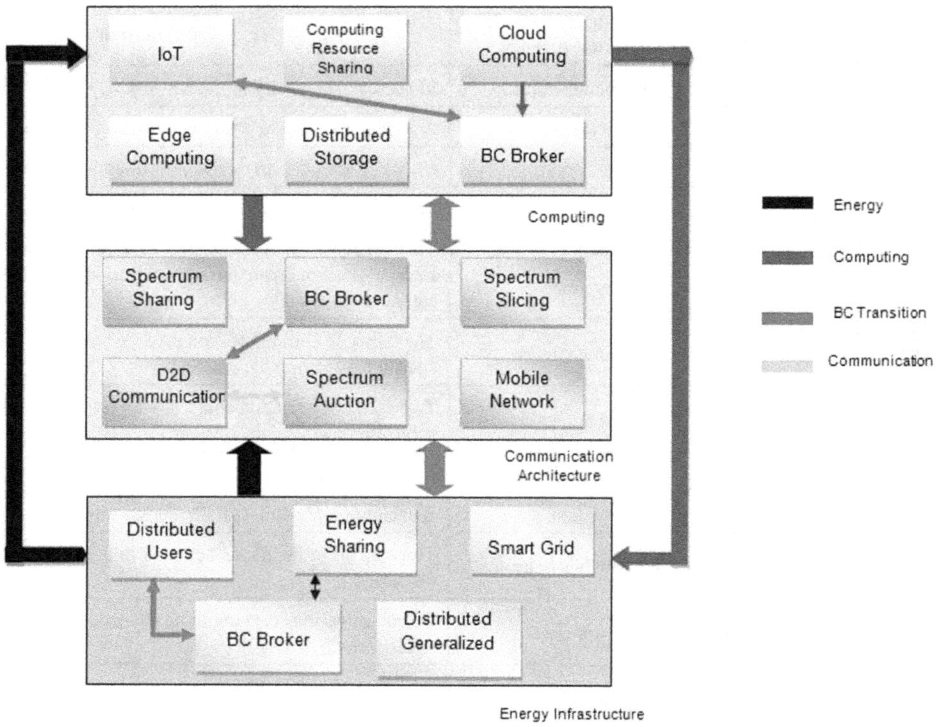

b. Network slicing and resource brokerage: Based on the service level agreement (SLA), the smart contract orders the network slice orchestration from the broker under the blockchain-based network slice trade. Blockchain technology is also used to automate the process of billing, leasing, payment, etc. In addition to this, it also provides some benefits such as reducing the operational cost, enhanced operational productivity for every network slice, agreement, uplifting the enforcement of straightforward agreements related to brokering operations. Moreover, it also supports cross-carrier payments and money transfers across countries and controls license spectrum access, accessible providing of roaming terms and agreement among MNOs, and network element authorization.

c. Authentication and Access control: Generally, a subscriber identity module (SIM) is used by the user for existing mobile networks. Hence, whenever a user has to change the mobile network operator, each end-user is given a unique wallet id, and the sim card is changed either manually or electronically. The blockchain technology was used for content-centric 6G wireless networks that stay transparent and immutable for each event that can be used during audits (Nguyen & Pathirana, 2020).

d. International roaming: Blockchain is primarily dependent on the availability of internet connections and does not depend on the administrative borders of countries. It removes the international roaming restrictions because the data transmission between domestic and international MNOs is simple. Furthermore, blockchain enables customizable service level agreements and smart contracts among end customers and MNOs in quasi-real-time by solving various concerns such as signal coverage package plan, service quality, etc. (Hewa, 2020).

e. Local area wireless networks: Blockchain smart contracts address the real-time local spectrum allocation problem. A small MNO acquires a new frequency band from significant operators across the desired coverage region and period. Furthermore, blockchain provides the central regulatory body with higher-level policies for controlling spectrum utilization by MNOs.
f. Cloud computing: Cloud computing allows sophisticated computations to be offloaded to remote servers for the resource-constrained user equipment. As far as its safety is concerned, security and trustworthiness are the primary considerations to protect sensitive information. Hence, Blockchain can provide the required level of trust between end nodes and smart devices, ensuring the integrity of remote servers and offloaded operations (Porambage & Liyanage, 2019).

FUTURE RESEARCH DIRECTIONS

With various combinations of telecommunication and computer science research routes, the study scope of 6G is vast. This section discusses the most significant 6G research potential using blockchain technology.

IOE

The Internet of Everything (IoE) is a broader concept than IoT, to intelligently connect people, processes, data, and objects. The role of IoE is defined. Business models and business processes are likely to be re-invented as a result of the IoE. First, digital technology allows for process optimization and automation. Second, new business models in several industries are becoming possible due to the use of digital technology. It will be interesting to study the business implications of the different options available when introducing IoE. There will be a strong need to compete with unprecedented corporate speed and agility. In addition, more research is needed to determine the effect of using blockchain-based technology for interoperability among various organizations, such as billing.

Data Storage and Analytics

Millions of devices and objects will continuously produce real-time streams of new information if the IoE is implemented. As a result, appropriate and effective centralized and decentralized data storage systems are necessary in the first place. Blockchain-based technology can play a significant role in this regard. It is still unclear how these technologies will be distributed and combined across diverse areas (fog, cloud, and edge).

Second, data analytics research will be critical for effective and precise decision-making to assess and extract the vital elements from this massive amount of data. Methods are classified into four types based on their application: diagnostic, descriptive, prescriptive, and predictive analytics. It will be fascinating to look at the possibility of combining these data analytics methodologies with a decentralized blockchain-based data storage system, where smart contracts can be used to automate procedures.

AI

AI was not used in 4G, but it is now being used in 5G to a limited extent. The ultimate goal of the 6G network is to make our country smarter, more efficient, and greener by integrating AI at every level of the network. First, AI and machine learning techniques have been proven to improve physical layer security (Weinand & Karrenbauer, 2017), channel coding, and range and obstacle detection (Sattiraju & Weinand, 2018). Each of these fields is still in the early stages of research and requires more exploration. The present 5G technologies, such as Network slicing, NFV, and SDN, will need to be improved further at the network layer to achieve a more flexible and self-learning adaptive architecture capable of supporting more complex and heterogeneous networks that are frequently also dynamically changing. The blockchain's purpose in this field will primarily be to make the machine learning decision-making process more intelligible and logical by allowing all of the underlying materials on which the choices are based to be tracked back.

Vehicle to Vehicle Communication

Intelligent Transportation System (ITS) will undoubtedly be one of the most critical applications to emerge in the coming decade, necessitating the technical abilities provided by a 6G network. In (Khan & Balan, 2019), a blockchain-based solution to define vehicle trust management was developed and validated through simulation. The critical flaw in their technique was that it was limited to ad hoc networks; therefore, more research is needed to ensure that it can be deployed autonomously in complex mobility situations like a multi-junction road network

CONCLUSION

With the research phase of 5G networks concluding, 6G networks have risen to the top of many academics' minds. Likewise, Blockchain has gotten a lot of attention and is being hailed as one of the primary enablers for the 6G network. The role design of such a wireless network, driven by the massive and heterogeneous demands of the hyper-connected existence of everything, would undoubtedly open up new commercial opportunities. Many security solutions are being proposed about the security and privacy of 6G networks using state-of-the-art techniques that include statistical techniques, machine learning, deep learning techniques, theory-based information approaches, etc. Among all of the Blockchain technology is going to be proved to be the most effective. As a result, this chapter emphasizes the significance of 6G, its security and privacy concerns, a comparison with 5G, the integration of Blockchain with 6G, and the characteristics, applications, and benefits of Blockchain in 6G.

REFERENCES

Aazhang, B., Ahokangas, P., Alves, H., Alouini, M. S., Beek, J., Benn, H., . . . Zhu, P. (2019). *Key drivers and research challenges for 6G ubiquitous wireless intelligence (white paper).* Oulu, Finland: 6G Flagship, University of Oulu.

Akan, O. B., Ramezani, H., Khan, T., Abbasi, N. A., & Kuscu, M. (2016). Fundamentals of molecular information and communication science. *Proceedings of the IEEE, 105*(2), 306–318. doi:10.1109/JPROC.2016.2537306

Akyildiz, I. F., Jornet, J. M., & Han, C. (2014). Terahertz band: Next frontier for wireless communications. *Physical Communication, 12*, 16–32. doi:10.1016/j.phycom.2014.01.006

Al-Eryani, Y., & Hossain, E. (2019). The D-OMA method for massive multiple access in 6G: Performance, security, and challenges. *IEEE Vehicular Technology Magazine, 14*(3), 92–99. doi:10.1109/MVT.2019.2919279

Biral, A., Centenaro, M., Zanella, A., Vangelista, L., & Zorzi, M. (2015). The challenges of M2M massive access in wireless cellular networks. *Digital Communications and Networks, 1*(1), 1–19. doi:10.1016/j.dcan.2015.02.001

Chen, C., Bian, R., & Haas, H. (2018, May). Omnidirectional transmitter and receiver design for wireless infrared uplink transmission in LiFi. In *2018 IEEE International Conference on Communications Workshops (ICC Workshops)* (pp. 1-6). IEEE. 10.1109/ICCW.2018.8403727

Chen, R., Li, C., Yan, S., Malaney, R., & Yuan, J. (2019). Physical layer security for ultra-reliable and low-latency communications. *IEEE Wireless Communications, 26*(5), 6–11. doi:10.1109/MWC.001.1900051

Cho, S., Chen, G., & Coon, J. P. (2019). Enhancement of physical layer security with simultaneous beamforming and jamming for visible light communication systems. *IEEE Transactions on Information Forensics and Security, 14*(10), 2633–2648. doi:10.1109/TIFS.2019.2904440

Christidis, K., & Devetsikiotis, M. (2016). Blockchains and smart contracts for the internet of things. *IEEE Access: Practical Innovations, Open Solutions, 4*, 2292–2303. doi:10.1109/ACCESS.2016.2566339

Dang, S., Amin, O., Shihada, B., & Alouini, M. S. (2020). What should 6G be? *Nature Electronics, 3*(1), 20–29. doi:10.103841928-019-0355-6

Farsad, N., Yilmaz, H. B., Eckford, A., Chae, C. B., & Guo, W. (2016). A comprehensive survey of recent advancements in molecular communication. *IEEE Communications Surveys and Tutorials, 18*(3), 1887–1919. doi:10.1109/COMST.2016.2527741

Ferraro, P., King, C., & Shorten, R. (2018). Distributed ledger technology for smart cities, the sharing economy, and social compliance. *IEEE Access: Practical Innovations, Open Solutions, 6*, 62728–62746. doi:10.1109/ACCESS.2018.2876766

Gai, K., Wu, Y., Zhu, L., Zhang, Z., & Qiu, M. (2019). Differential privacy-based blockchain for industrial internet-of-things. *IEEE Transactions on Industrial Informatics, 16*(6), 4156–4165. doi:10.1109/TII.2019.2948094

Gong, X., & Cortese, C. (2017, September). A socialist market economy with Chinese characteristics: The accounting annual report of China Mobile. *Accounting Forum, 41*(3), 206–220. doi:10.1016/j.accfor.2017.04.002

Gyongyosi, L., Imre, S., & Nguyen, H. V. (2018). A survey on quantum channel capacities. *IEEE Communications Surveys and Tutorials, 20*(2), 1149–1205. doi:10.1109/COMST.2017.2786748

Hamamreh, J. M., Basar, E., & Arslan, H. (2017). OFDM-subcarrier index selection for enhancing security and reliability of 5G URLLC services. *IEEE Access: Practical Innovations, Open Solutions, 5*, 25863–25875. doi:10.1109/ACCESS.2017.2768558

Hewa, T., Gür, G., Kalla, A., Ylianttila, M., Bracken, A., & Liyanage, M. (2020, March). *The role of blockchain in 6G: Challenges, opportunities and research directions. In 2020 2nd 6G Wireless Summit (6G SUMMIT)*. IEEE.

Hong, T., Liu, C., & Kadoch, M. (2019). Machine learning based antenna design for physical layer security in ambient backscatter communications. *Wireless Communications and Mobile Computing, 2019*, 2019. doi:10.1155/2019/4870656

Hooper, M., Tian, Y., Zhou, R., Cao, B., Lauf, A. P., Watkins, L., & Alexis, W. (2016, November). Securing commercial wifi-based uavs from common security attacks. In *MILCOM 2016-2016 IEEE Military Communications Conference* (pp. 1213–1218). IEEE. doi:10.1109/MILCOM.2016.7795496

Hu, J. Y., Yu, B., Jing, M. Y., Xiao, L. T., Jia, S. T., Qin, G. Q., & Long, G. L. (2016). Experimental quantum secure direct communication with single photons. *Light, Science & Applications*, *5*(9), e16144–e16144. doi:10.1038/lsa.2016.144 PMID:30167186

Huang, T., Yang, W., Wu, J., Ma, J., Zhang, X., & Zhang, D. (2019). A survey on green 6G network: Architecture and technologies. *IEEE Access: Practical Innovations, Open Solutions*, *7*, 175758–175768. doi:10.1109/ACCESS.2019.2957648

Islim, M. S., Ferreira, R. X., He, X., Xie, E., Videv, S., Viola, S., Watson, S., Bamiedakis, N., Penty, R. V., White, I. H., Kelly, A. E., Gu, E., Haas, H., & Dawson, M. D. (2017). Towards 10 Gb/s orthogonal frequency division multiplexing-based visible light communication using a GaN violet micro-LED. *Photonics Research*, *5*(2), A35–A43. doi:10.1364/PRJ.5.000A35

Khan, A. S., Balan, K., Javed, Y., Tarmizi, S., & Abdullah, J. (2019). Secure trust-based blockchain architecture to prevent attacks in VANET. *Sensors (Basel)*, *19*(22), 4954. doi:10.339019224954 PMID:31739437

Khan, L. U. (2017). Visible light communication: Applications, architecture, standardization and research challenges. *Digital Communications and Networks*, *3*(2), 78–88. doi:10.1016/j.dcan.2016.07.004

Kim, H., Kim, S. H., Hwang, J. Y., & Seo, C. (2019). Efficient privacy-preserving machine learning for blockchain network. *IEEE Access: Practical Innovations, Open Solutions*, *7*, 136481–136495. doi:10.1109/ACCESS.2019.2940052

Kiyomoto, S., Basu, A., Rahman, M. S., & Ruj, S. (2017, December). On blockchain-based authorization architecture for beyond-5G mobile services. In *2017 12th International Conference for Internet Technology and Secured Transactions (ICITST)* (pp. 136-141). IEEE.

Kotobi, K., & Bilen, S. G. (2018). Secure blockchains for dynamic spectrum access: A decentralized database in moving cognitive radio networks enhances security and user access. *IEEE Vehicular Technology Magazine, 13*(1), 32-39.

Letaief, K. B., Chen, W., Shi, Y., Zhang, J., & Zhang, Y. J. A. (2019). The roadmap to 6G: AI empowered wireless networks. *IEEE Communications Magazine*, *57*(8), 84–90. doi:10.1109/MCOM.2019.1900271

Loscri, V., Marchal, C., Mitton, N., Fortino, G., & Vasilakos, A. V. (2014). Security and privacy in molecular communication and networking: Opportunities and challenges. *IEEE Transactions on Nanobioscience*, *13*(3), 198–207. doi:10.1109/TNB.2014.2349111 PMID:25148668

Lovén, L., Leppänen, T., Peltonen, E., Partala, J., Harjula, E., Porambage, P., ... Riekki, J. (2019). EdgeAI: A vision for distributed, edge-native artificial intelligence in future 6G networks. *The 1st 6G Wireless Summit*, 1-2.

Lu, Y., Higgins, M. D., & Leeson, M. S. (2015). Comparison of channel coding schemes for molecular communications systems. *IEEE Transactions on Communications*, *63*(11), 3991–4001. doi:10.1109/TCOMM.2015.2480752

Luo, J., Fan, L., & Li, H. (2017). Indoor positioning systems based on visible light communication: State of the art. *IEEE Communications Surveys and Tutorials*, *19*(4), 2871–2893. doi:10.1109/COMST.2017.2743228

Ma, J., Shrestha, R., Adelberg, J., Yeh, C. Y., Hossain, Z., Knightly, E., Jornet, J. M., & Mittleman, D. M. (2018). Security and eavesdropping in terahertz wireless links. *Nature*, *563*(7729), 89–93. doi:10.103841586-018-0609-x PMID:30323288

Miraz, M. H., Ali, M., Excell, P. S., & Picking, R. (2015, September). A review on Internet of Things (IoT), Internet of everything (IoE) and Internet of nano things (IoNT). In 2015 Internet Technologies and Applications (ITA) (pp. 219-224). IEEE.

Mostafa, A., & Lampe, L. (2014, June). Physical-layer security for indoor visible light communications. In *2014 IEEE International Conference on Communications (ICC)* (pp. 3342-3347). IEEE. 10.1109/ICC.2014.6883837

Nakano, T., Okaie, Y., Kobayashi, S., Hara, T., Hiraoka, Y., & Haraguchi, T. (2019). Methods and applications of mobile molecular communication. *Proceedings of the IEEE*, *107*(7), 1442–1456. doi:10.1109/JPROC.2019.2917625

Nawaz, S. J., Sharma, S. K., Wyne, S., Patwary, M. N., & Asaduzzaman, M. (2019). Quantum machine learning for 6G communication networks: State-of-the-art and vision for the future. *IEEE Access: Practical Innovations, Open Solutions*, *7*, 46317–46350. doi:10.1109/ACCESS.2019.2909490

Nguyen, D. C., Pathirana, P. N., Ding, M., & Seneviratne, A. (2020). Blockchain for 5G and beyond networks: A state of the art survey. *Journal of Network and Computer Applications*, *166*, 102693. doi:10.1016/j.jnca.2020.102693

Ni, J., Lin, X., & Shen, X. (2019). Toward privacy-preserving valet parking in autonomous driving era. *IEEE Transactions on Vehicular Technology*, *68*(3), 2893–2905. doi:10.1109/TVT.2019.2894720

Pathak, P. H., Feng, X., Hu, P., & Mohapatra, P. (2015). Visible light communication, networking, and sensing: A survey, potential and challenges. *IEEE Communications Surveys and Tutorials*, *17*(4), 2047–2077. doi:10.1109/COMST.2015.2476474

Porambage, P., Kumar, T., Liyanage, M., Partala, J., Lovén, L., Ylianttila, M., & Seppänen, T. (2019). Sec-EdgeAI: AI for edge security Vs security for edge AI. The 1st 6G Wireless Summit.

Puthal, D., Malik, N., Mohanty, S. P., Kougianos, E., & Yang, C. (2018). The blockchain as a decentralized security framework. *IEEE Consumer Electronics Magazine*, *7*(2), 18–21. doi:10.1109/MCE.2017.2776459

Ramadan, R. A., & Vasilakos, A. V. (2017). Brain computer interface: Control signals review. *Neurocomputing*, *223*, 26–44. doi:10.1016/j.neucom.2016.10.024

Refaey, A., Hammad, K., Magierowski, S., & Hossain, E. (2019). A blockchain policy and charging control framework for roaming in cellular networks. *IEEE Network*, *34*(3), 170–177. doi:10.1109/MNET.001.1900336

Saad, W., Bennis, M., & Chen, M. (2019). A vision of 6G wireless systems: Applications, trends, technologies, and open research problems. *IEEE Network*, *34*(3), 134–142. doi:10.1109/MNET.001.1900287

Sanjab, A., Saad, W., & Başar, T. (2017, May). Prospect theory for enhanced cyber-physical security of drone delivery systems: A network interdiction game. In 2017 IEEE international conference on communications (ICC) (pp. 1-6). IEEE. doi:10.1109/ICC.2017.7996862

Sattiraju, R., Kochems, J., & Schotten, H. D. (2017, May). Machine learning based obstacle detection for Automatic Train Pairing. In *2017 IEEE 13th International Workshop on Factory Communication Systems (WFCS)* (pp. 1-4). IEEE. 10.1109/WFCS.2017.7991962

Sattiraju, R., Weinand, A., & Schotten, H. D. (2018, December). Performance analysis of deep learning based on recurrent neural networks for channel coding. In *2018 IEEE International Conference on Advanced Networks and Telecommunications Systems (ANTS)* (pp. 1-6). IEEE. 10.1109/ANTS.2018.8710159

Sattiraju, R., Weinand, A., & Schotten, H. D. (2019). *AI-assisted PHY technologies for 6G and beyond wireless networks.* arXiv preprint arXiv:1908.09523.

Strinati, E. C., Barbarossa, S., Gonzalez-Jimenez, J. L., Ktenas, D., Cassiau, N., Maret, L., & Dehos, C. (2019). 6G: The next frontier: From holographic messaging to artificial intelligence using subterahertz and visible light communication. *IEEE Vehicular Technology Magazine*, *14*(3), 42–50. doi:10.1109/MVT.2019.2921162

Sun, X., Yang, W., Cai, Y., Ma, R., & Tao, L. (2019). Physical layer security in millimeter wave SWIPT UAV-based relay networks. *IEEE Access: Practical Innovations, Open Solutions*, *7*, 35851–35862. doi:10.1109/ACCESS.2019.2904856

Švogor, I., & Kišasondi, T. (2012, June). Two factor authentication using EEG augmented passwords. In *Proceedings of the ITI 2012 34th International Conference on Information Technology Interfaces* (pp. 373-378). IEEE.

Tariq, F., Khandaker, M. R., Wong, K. K., Imran, M. A., Bennis, M., & Debbah, M. (2020). A speculative study on 6G. *IEEE Wireless Communications*, *27*(4), 118–125. doi:10.1109/MWC.001.1900488

Ucar, S., Coleri Ergen, S., Ozkasap, O., Tsonev, D., & Burchardt, H. (2016, November). Secvlc: Secure visible light communication for military vehicular networks. In *Proceedings of the 14th ACM International Symposium on Mobility Management and Wireless Access* (pp. 123-129). 10.1145/2989250.2989259

Weinand, A., Karrenbauer, M., Lianghai, J., & Schotten, H. D. (2017, June). Physical layer authentication for mission critical machine type communication using Gaussian mixture model based clustering. In *2017 IEEE 85th Vehicular Technology Conference (VTC Spring)* (pp. 1-5). IEEE. 10.1109/VTC-Spring.2017.8108527

Zhang, W., Ding, D. S., Sheng, Y. B., Zhou, L., Shi, B. S., & Guo, G. C. (2017). Quantum secure direct communication with quantum memory. *Physical Review Letters, 118*(22), 220501. doi:10.1103/PhysRevLett.118.220501 PMID:28621985

Zhu, T., & Philip, S. Y. (2019, July). Applying differential privacy mechanism in artificial intelligence. In *2019 IEEE 39th International Conference on Distributed Computing Systems (ICDCS)* (pp. 1601-1609). IEEE. 10.1109/ICDCS.2019.00159

Chapter 9
Role of Machine Learning in 6G Technologies:
Healthcare and Education Sectors

Dhaya R.
https://orcid.org/0000-0002-3599-7272
King Khalid University, Saudi Arabia

Kanthavel R.
King Khalid University, Saudi Arabia

ABSTRACT

Artificial intelligence is enhancing the standards and quality of healthcare and educational sectors aggressively. Wearable healthcare innovations additionally utilize 6G-AI to serve patients. 6G-AI machine learning in edification is a type of customized knowledge to be used to give every understudy a personalized educational encounter. The next-generation wireless network, known as 6G, will require a new paradigm in connectivity increasing efficiency, involving the use of efficient and effective resource organizational structures. Understudies are likewise getting to gadgets at an expanding rate. Obviously, it would be difficult to examine innovation and education without referencing 6G-AI. Computer-based intelligence can reshape education with its capability to affect educational approaches on nearby, public, and worldwide scales. Man-made intelligence makes it conceivable to mine that information and figure it out. Hence, this chapter, in its first part, presents the essential 6G-enabled machine learning applications in the education field in an elaborated manner.

INTRODUCTION

In healthcare applications, machine learning plays a significant role in serving patients and physicians commencing a variety of health related viewpoint. The nearly all renowned use cases have been robotizing experimental incrimination of clinical consideration systems. Human prepared br6G-AIns have nonstop submissions in healthcare portfolios regardless of its utilization to hit upon joining among transmissible

DOI: 10.4018/978-1-6684-3804-6.ch009

set of laws in order to have power over watchful robots by applying 6G-AI. It can affect medical clinics and health frameworks in improving productivity, while lessening the expense of care." The healthcare area has for some time been a promoter of and profited extraordinarily from innovative proceeds (Y.LeCun et al,2015) Nowadays, 6G enabled ML assumes a critical job in numerous health- connected dom6G-AIns, including the improvement of new operations, the treatment of patient information and reports and the therapy of persistent sicknesses. Figure 1 declares the required items to make high-quality ML schemes.

As machines prepared human being muscle multiple times more grounded, machines will create the human being mind multiple epochs all the extra remarkable. In spite of alerts from cert6G-AIn specialists that effects are poignant too quick, the pace of development continues to increment. Furthermore, for some, that is as it ought to be. A definitive objective is improved consideration at a lower cost (Sathyaraj et al, 2022). KenSci utilizes ML to anticipate and curing to assist doctors and payers mediate prior, foresee populace health hazard by recognizing examples and surfacing high danger markers and model sickness movement. Ciox Health utilizes ML to upgrade "health data the executives and trade of health data," with the objective of new work processes, encouraging admittance to medical information and civilizing the precision and stream of healthiness data (J. Qiu et al,2016)

Figure 1. Required items to create good for machine learning systems

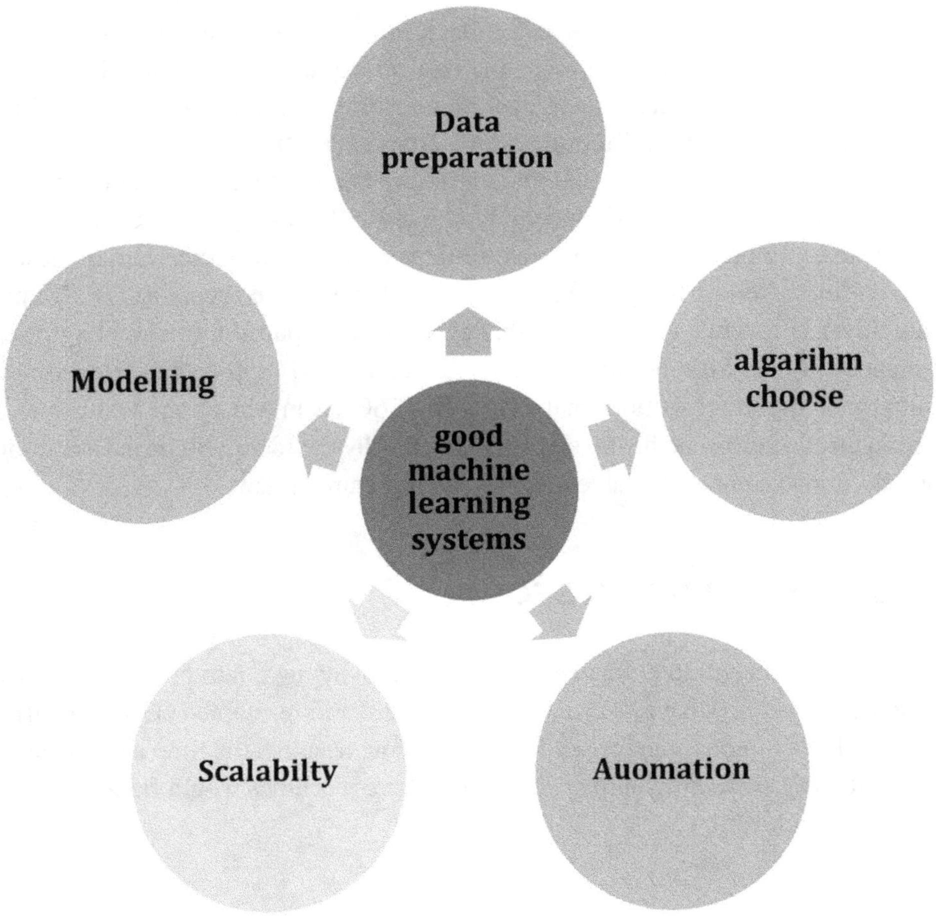

6G-AI's innovation utilizes ML to assist pathologists construct speedier and extra precise findings just as recognize patients that may profit by new kinds of medicines or treatments. Quantitative imminent need to get better the velocity and precision of bosom malignancy determination with its PC helped bosom MRI workstation Quantx. The objective is to improve outcomes for patients by means of improved conclusions by radiologists. Microsoft's Project Inner Eye utilizes ML to separate among cancers and fit life systems utilizing 3D radiological pictures that help clinical specialists in radiotherapy and careful arranging, in addition to other things (N. Wang et al, 2017). With the assistance of IBM's Watson 6G-AI innovation, Pfizer utilizes ML for immuno-oncology investigates concerning how the human body resistant framework can battle malignant growth (M. I. Jordan and T. M. Mitchell, 2015). Machine learning and information science joined with cutting edge research center innovation are serving late establish insitro create medicines with the objective of all the more rapidly relieving patients at a minor price. Through its ML stage Augusta, Biosymetrics, "enables customers to perform motorized ML and data prehandling," which progresses precision and executes a dull endeavor that is regularly finished by individuals in different zones of the clinical space, tallying drugs, exactness prescription, development, crisis centers and wellbeing systems (S.B. Kotsiantis et al,2007).

Concerto Health 6G-AI utilizes the machine learning to dissect oncology information, giving bits of knowledge that permit oncologists, drug organizations, payers and suppliers to rehearse accuracy medication and health (R Dhaya et al,2019). Beta Bionics is building up a wearable "bionic" pancreas it names the iLet, which oversees glucose point nonstop in persons have Type 1 diabetes." For machine learning designers, the utilization of expectation devices in genuine clinical settings can be a far off objective. As of late distributed rules for revealing clinical exploration that includes machine learning will help associate clinical and software engineering networks, and understand the maximum capacity of the machine learning apparatuses (Heuser and M. Zohner, 2012).

In the previous decade, numerous great outcomes have been accounted for machine learning (ML) instruments that are created to 6G-AId clinical dynamic. Be that as it may, deciphering these to great extent hypothetical accomplishments to sensible settings rem6G-AIns a considerable test and requires the dynamic coordinated effort between ML analysts and healthcare specialists (P. Domingos, 2012) . A significant advance towards encouraging this cycle is the expansion of revealing rules in clinical and health sciences to fuse ML and artificial intelligence (6G-AI) approaches. Hence, this chapter is segmented into three important sections, namely, need of 6G-AI in education, health first, followed by Machine learning in education and health sectors. Then thirdly the focus will be in case studies in 6G-AI of health and education sectors. The table 1 shows the literature review.

NEED OF 6G- AI IN EDUCATION SECTOR

6G-AI has recently been useful to guidance basically in convincing contraptions that assist make aptitudes and difficult structures (Dhaya, R et al, 2020). The networking enabled 6G-AI can drive adequacy, personalization and smooth out manager endeavors to allow teachers the time and chance to give cognizance and adaptability, especially human capacities where machines would fight. The figure 2 shows the Need of 6G-AI in Education Sector.

Figure 2. Need of 6G-AI in education sector

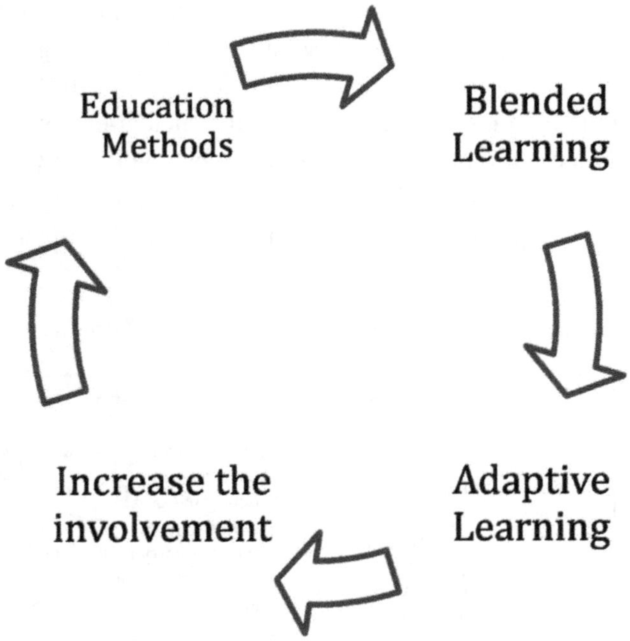

As 6G-AI takes on to a greater extent a showing job by furnishing understudies with essential data, it will change the job of instructors in homeroom. Educators will move into the part of study hall facilitator or learning inspiration. 6G-AI will likewise give customized mentoring to understudies outside of the homeroom (C. Zhang and Y. Ma,2012).

NEED OF 6G-AI IN HEALTH SECTOR

It places purchasers in charge of health and prosperity. Additionally, 6G-AI assembles the limit with respect to medical services specialists to all the almost cert6G-AIn appreciate the ordinary models and necessities of the people they care for, and with that understanding they can give better analysis, bearing and sponsorship for rem6G-AIning solid. In the exceptionally mind boggling universe of healthcare, 6G-AI gadgets can m6G-AInt6G-AIn human providers to offer snappier help, dissect issues and explore data to recognize designs or genetic information that would slant someone to a particular affliction. Table 1 shows the Pros, Cons, Challenges and Applications of 6G-AI in Health care.

6G-AI VS. MACHINE LEARNING

Artificial Intelligence is the further widespread design of machinery cont6G-AIning the choice to complete undertakings such that we would believe "smart". ML is a current usage of 6G-AI based around the likelihood that we ought to just have the choice to give machines admittance to data and let them find out on their own. DL is a division of ML, and ML is a division of 6G-AI, which is an umbrella phrase for any PC list that achieves something sharp. As such, all ML is 6G-AI, however not all 6G-AI is ML, thus forth.6G-AI alludes to a machine with cleverness. This does not mean the machine is mindful or like human intelligence; it just implies that the machine is fit for tackling a particular issue. Machine learning alludes to a specific kind of 6G-AI that learns by itself (V. Devedži, 2004). 6G-AI based learning, another sort of machine learning, is the group of ML calculations, which have 6G-AI utilizes in example discovery and distinct displaying. These calculations don't have yield classes or marks on the information (M. Chassignol et al, 2018) . 6G-AI is the discipline and designing of causing PCs to carry on in manners that, as of not long ago, we thought required human intelligence.

Table 1. Need of 6G-AI in health sector-pros, cons, challenges and applications

PROS	CONS	CHALLENGES	APPLICATIONS
• Better decisions • Fast disease diagnosis • Uninterrupted treatment • Information integration • Reduce hospital visits • Time saving process	• Privacy • Security • Lack of healthcare information • More Capital investment • Lack of interoperability • Increased unemployment • Adaptability of the staff	• Capital Money deployment for infrastructure • Difficulty in 6G-AI Deployment • State and country regulations • Difficult to handle medical software	• Robot Assisted Surgery • Virtual Nursing assistant • Preliminary diagnosis • Automated image diagnosis • Checking health through wearable devices • Drug development

MACHINE LEARNING VS. DEEP LEARNING

ML utilizes computations to parse information, 6G-AI from that information, and build instructed choices subject to what it has figured it out. Profound knowledge configurations estimations in layers to make an "fake neural association" that can find out and build insightful choices isolated (14. Kanthavel, R., and Dhaya, R.,2013) As a rule, the learning cycle of these calculations can either be directed or solo, contingent upon the information being utilized to take care of the algorithms. The field has been getting lots of thought of late and considering current conditions: Recent improvements have prompted results that were not idea to be conceivable previously (R. C. Sharma et al, 2019) Deep learning portrays calculations that break down information with a rationale structure like how a human would make determinations. Note that this can happen both through regulated and 6G-AI based learning. To accomplish this, deep learning applications utilize a covered configuration of calculations describes an Artificial Neural Organization (ANN). The plan of such an ANN is propelled by the natural neural organization of the human 6G-AI, prompting a cycle of learning that is definitely more proficient than that of standard machine learning models. Table 2 illustrates the dissimilarity between ML and DL.

The major differences among machine learning and deep learning are defined. First and principal, while the conventional machine learning calculations have a somewhat basic structure, for example, 6G-AI relapse or a choice tree, deep learning depends on an artificial neural organization. This multi-layered ANN is, similar to a human-brain 6G-AI , complex and entwined. Also, deep learning calculations require significantly less human intercession. Recall the Tesla model? In the event that the STOP sign picture acknowledgment was a more customary machine learning calculation, a programmer would physically pick highlights and a classifier to sort pictures, check whether the yield is as required, and change the calculation if this isn't the situation (Dhaya, R. and Kanthavel, R, 2020). When a deep learning figuring, in any case, the features are isolated subsequently, and the computation g6G-AIns from its own slip-ups .

Table 2. Difference between ML & DL

Machine Learning	Deep Learning
ML is a super ordinate of DL	DL is a division of ML
The information spoke to in ML is very unique when contrasted with Deep Learning as it utilizes organized information	The information portrayal is utilized in DL is very unique as it utilizes neural networks (ANN).
ML is a development of AI	DL is a development to ML. Essentially it is the way profound is the ML.
Input+ Decision Tree® Output	Input+ Future extraction + Classification® Output
Machine learning comprises of thousands of information focuses.	Large Data: Millions of information focuses.
Yields: Numerical Value, similar to characterization of score	Anything from mathematical qualities to freestyle components, for example, free content and sound.
Utilizations different kinds of mechanized calculations that go to display works and anticipate future activity from information.	Utilizations neural organization that goes information through preparing layers to the decipher information highlights and relations.
Figuring's are distinguished by data agents to investigate unequivocal variables in enlightening assortments.	Calculations are generally self-portrayed on information investigation whenever they're placed into creation.
ML is exceptionally used to stay in the opposition and study novel thing.	DL addresses multifaceted ML problems

MACHINE LEARNING IN HEALTH SECTORS

ML is helping with smoothing out administrative cycles in centers, guide and treat powerful diseases and customize clinical medicines."The worldwide healthcare industry is blasting. There is expanded mindfulness about living organ gift. Robots are being used for gallbladder removal, hip replacements, and kidney moves. An early investigation of skin tumors with least human slip-up is a reality. Bosom reconstructive medical procedures have empowered bosom cancer survivors to participate in modifying their organs. Figure 3 shows the applications of Machine learning in Health Sectors (Kun-Hsing Yu et al, 2018) . This mechanical headway has begun to establish the framework for closer coordinated effort among industry partners, affordable and less intrusive medical procedure alternatives, all encompassing treatments, and new consideration conveyance models. Here are five instances of current and arising ML developments:

Drug Discovery: From the fundamental showing of prescription merges to calculating the accomplishment speeds of a specific recommendation subject to anatomical segments of the long suffering humans – the Knight Cancer Institute in Oregon and Microsoft's Project Hanover are correct presently pert6G-AIning this improvement to redo medicine mixes to fix blood disease. Machine learning has likewise brought forth new approaches, for example, accuracy medication and the new advancement that can guarantee a drug has the correct impact on the patients. For instance, today, clinical experts can create calculations to comprehend illness measures and imaginative plan treatments like Type 2 diabetes (Nariman Noorbakhsh-Sabet et al, 2019) .

Optimized Patient Recruitment for Clinical Trials: Joining volunteers for clinical trials isn't simple. Numerous channels must be applied to see who is good for the investigation. With machine learning, gathering patient information, for example, past clinical records and mental conduct. The innovation is additionally used to screen natural measurements of the volunteers and the conceivable mischief of the clinical trials over the long haul. With such convincing information close by, clinical experts can diminish the time for testing, subsequently lessening generally costs and expanding the test viability [14].

Figure 3. Machine learning in health sectors

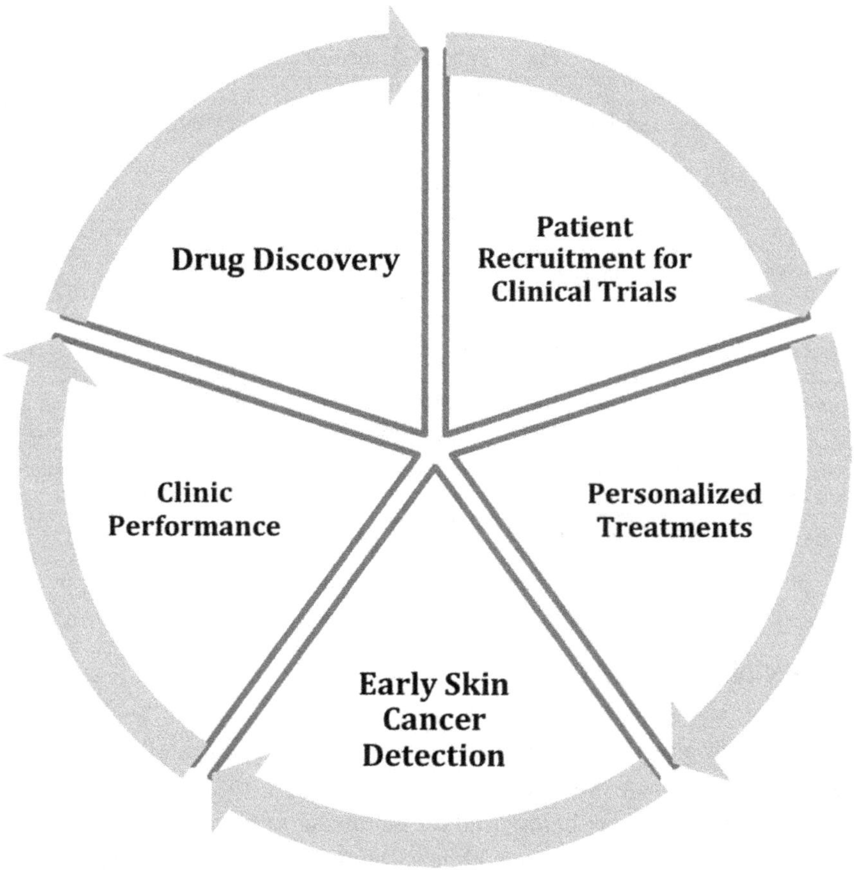

Customized handling of Healthcare actions: Every person body functions in a startling manner. 6G-AI helps clinical specialists with choosing the risk for each patient, dependent upon their signs, past clinical records, and family lineage using smaller than expected bio sensors. These second contraptions screen tolerant wellbeing and flag abnormalities without tendency, in this way engaging more mind boggling limits of assessing wellbeing.

Early Skin Cancer Detection: Machine learning is m6G-AInly used to mine and investigate patient information to discover examples and do the determination of such countless 6G-AIlments, one of them being skin cancer.

Clinic Performance: Medical care providers need to ideally submit reports to the public authority with central patient records that are treated at their facilities. Consistence plans are endlessly creating, which is the explanation it is fundamentally more essential to ensure crisis facility objections to mind the remote possibility that they are being reliable and working inside as far as possible. With the machine learning, it is definitely not hard to assemble data from different sources, using different techniques and designing them precisely (D Peteiro-Barral, B Guijarro-Berdiñas, 2012).

MACHINE LEARNING IN EDUCATIONAL PART

ML in guidance is a type of customized education that could be utilized to give each understudy an individualized educational encounter. Here, the understudies are directed for their individual education, can track the speed they need and settle on their own choices concerning what to realize (M. Stamp,2017).

Versatile Education

Versatile education is a novelty supported or online informative structure to dissect a performance continuously and changes encouraging methods reliant on that sequence. Imagine 6G-AI convenes dedicated numerical coach assembles personalized dedication. Versatile education is the conveyance of the personalized learning encounters that address the one of a kind necessities of people through modified substance, continuous criticism, and assets (instead of giving a standard learning experience to all).It tries to give a special yet personalized insight for every client. Versatile learning is one strategy for giving personalized learning, which 6G-AIms to give proficient, viable, and tweaked learning ways to connect every understudy. Versatile learning frameworks utilize information driven way to deal with change the way and speed of learning, empowering the conveyance of personalized learning at scale (Dhaya R et al,2021).

How to Actualize Versatile Learning?

1. Implement dynamic learning in class with the AI of advancing basic reasoning and critical thinking.
2. Help 90% of understudies improve to ensure they have authority over a point.
3. Reduce understudy dropouts in courses underneath 5% to expand understudy AI intendance (Dhaya, R. and Kanthavel, R, 2022).

Escalating Efficiency

Machine learning as artificial intelligence can probably build teachers more efficient by concluding assignments, for instance, study hall the slat, booking, and so forth Thusly, teachers are permissible to zero in on assignments that can't be cultivated by 6G-AI, and that require a human touch. ML has the capacity of improved substance and instructive arrangement affiliation and the board. It assists by bifurcating the function as requirements be and grasp the ability of everyone. This assists with investigating what work is most appropriate for the instructor and what works for the understudy. It makes crafted by educators and understudies simpler and that makes them glad and comfortable with schooling. This likewise expands the contribution and their affection towards cooperation and learning (I Arel et al, 2010) . It additionally can possibly make teachers more productive by concluding undertakings, for instance, study hall the panel, booking, and so forth, in this manner, teachers are permissible to zero in on assignments that can't be cultivated by 6G-AI, and that require human contact (Kanthavel R et al,2022).

Knowledge Analytics

Ordinarily it happens that the instructor additionally stalls out while educating. Along these lines, the bits of knowledge and essence aren't appropriately perceived by the understudies. With learning investigation, the educator can 6G-AI knowledge into information and can perform deep jumps into information. She/he can move through an enormous number of pieces of substance, decipher it and afterward can make relations and ends (D. Vernon et al, 2014) . This can forcefully affect the teaching and knowledge measure. Aside from this, the learning investigation recommends ways the understudy should take. Understudies can 6G-AI benefits by accepting recommendations concerning materials and other learning approaches from this product (X. Du, Y, 2017).

Prescient Investigation

Prescient investigation in 6G-AI is tied in with knowing the outlook and requires of the understudies. It assists with making conclusion regarding the effects that might happen afterward on. With class tests and half-yearly outcomes, it very well may be perceived which understudies will perform well in the test and which understudies will make some intense memories (K. B. Letaief et al,2019) . Through this, an understudy can benefit from outside assistance in a superior manner and can deal with his feeble subjects.

Modified Learning

This is the best usage that machine learning gives. It is adaptable and singular supplies are being managed with through this. Throughout this instructive reproduction, the understudies can manage their personal learning. They can have their individual tempo and can settle on choices about what to realize and how to learn (Dhaya R et al, 2022) . They can pick the subject matter they are keen on, the instructor they need to 6G-AI as of and what educational plan, norms and example they need to follow.

Estimating Assessments

Innovation during my educational time was limited to OMR answer sheets. It was made by utilizing the OMR Sheet Design Software. It had a dark shading record point which is at the side of the sheet. Additionally, machine learning as artificial intelligence is used to audit understudy errands and tests more exactly than a person be able to yet, the best outcomes will have higher legitimacy and dependability when a machine accomplishes the work as there are higher unwavering quality and low odds of mistake (A.S. Wale and S. S. Sonawani, 2018) . Table 3 shows the profit and confines using ML in Education part.

Table 3. Benefits and limitation of using ML in the Education sector

Benefits	Limitations
• An Automatic grading system gives the chance for totally unbiased grading • saves instructors time and gives a more practical review of a youngster's accomplishments • Help educators to seem headed for the future. • Specify if a student is at hazard of falling out or getting bigger punitive action. • Predicting Career Paths	• Enchanting personal communication missing from the students. • Innate cost. • It very well may be very expensive to obt6G-AIn the entire new technology and projects to build customized education a viable answer for instructors and learners. • Data Collection and Transparency

CASE STUDIES OF 6G -AI IN HEALTH CARE APPLICATIONS

Coming up next are the utilization instances of artificial intelligence in healthcare. Physical layer encryption, which is planned to be supplied in 6G for keeping the data transmission safe, is another advanced tool, especially for e-health application. However, new technology uptake, particularly among the elderly, is a significant impediment to attaining this goal. Furthermore, digital modern medical dose modification makes life easier for people who require regular care, such as diabetics (Sangeetha, S. K. B., & Dhaya, R., 2016). To protect the patient's privacy and safety, these technologies require a secure and trustworthy connection. As a result, the 6G aim is to transform the globe into a safe and fully connected digital society by combining communication technologies with intelligence. The development models for 6G vision are divided into four categories: I large spectrum frequency band (VLC and THz); (ii) integrated terrestrial-aerial systems (UAVs, HAPs, and satellites); (iii) energy efficiency and levels of engagement intellectual ability (ambient backscatter communication (AmBC), multifunctional thoughtful substrate (RIS); and (iv) holis (BC).

Virtual Help with Healthcare

There is an enormous measure of health applications av6G-AIlable on the App Store and Play Store. As per Statista, before the finish of 2019, the absolute number of cell phone clients will arrive at more than $2.7 billion all around the world. What's more, every one of them will have both of the voice collaborators referenced previously. A mix of these voice 6G-AI coddes and healthcare applications can assist clients with the conveying medicine cautions, instructive material, and the sky is the limit from there (K ShG-AIlaja et al,2018) . This degree of individual help can assist individuals with carrying on with a healthier existence without a human guardian.

Exploration and Progressed Investigation

A larger part of today, healthcare has gotten reliant on machines. Regardless of whether it is for a normal exam or critical tasks, machines have become a best approach for specialists to help their patients carry on with healthier lives. For a larger part of the time, machines and hardware turn out great, can assist specialists with distinguishing the flaw beforehand so that machines don't at the hour of crises. Aside from this, AI can likewise help specialists to perform a medical procedure by giving them information about the patient's body part, which requires a medical procedure(Dhaya, R. and Kanthavel, R, 2022) . It likewise causes specialists to make profoundly personalized treatments for their patients. For instance, IBM Watson's 6G-AI has the capacity to handle unstructured just as organized patient information and present proof based therapy options for cancer patients.

Life Teach for Individual Health

Subsequent meet-ups are a fundamental piece of healthcare, particularly if a patient is experiencing a persistent infection. In a lion's share of clinics and clinics, specialists offer standard development or life instructing as a piece of their treatment. Nonetheless, such administrations can be very exorbitant, and only one out of every odd patient can afford it. Be that as it may, novel wearable innovation and 6G-AI-empowered versatile application can help take care of the issue (N Marline et al, 2018). It will catch the information coming from the wearable gadget and recommend the necessary drug, activities, exercises, and even propensities, which will help them carry on with a healthier life.

Healthcare Bots

It is obvious that very soon chat bots will supplant call-based client support out and out. Chat bots in healthcare can assist patients with settling their interests quicker than conventional advising meetings with health care specialist co-ops. Bots can assist a patient with planning a development momentarily. Other than this, bots can likewise assist patients with covering their clinic tabs in a couple of snaps (P. S. Mung and S. Phyu, 2020) . To close, bots can help healthcare specialist co-ops to improve day in and day out client care and help patients by handling each assistance demand quicker, beginning from booking an arrangement to taking care of tabs.

Diagnostics Assist and Medical Imaging

Clinical imaging has gotten progressive changes giving healthcare benefits by giving a superior image of different oddities. However, at times, an abnormality is so little in size that a natural eye gets incapable to recognize them. Specialists not having the option to recognize such anomaly can even cost a patient's life. 6G-AI helps in identifying such irregularities that the natural eye can't distinguish (S. S. Nikam, 2015).

Communication Help

We have seen health care experts perusing different archives constantly with the goal that they don't miss any 6G-AI while diagnosing patients. Missing a solitary AI can cost life to a specialist. NLP can assist specialists with sifting through significant information from long reports. NLP not just permits specialists to sift through significant information yet additionally encourages them by portraying that information to them so it gets simpler for specialists to purchaser information (R. Kanthavel et al,2022) .

Medicine Creation

The drug business is a billion-dollar business today. Expensive meds are frequently an aftereffect of thorough innovative work. 6G-AI can help decrease the expense of medication by diminishing the expense of R&D. In 2014, Atomise dispatched 6G-AI-fueled program, which was utilized to discover elective prescriptions to the ebola infection. This program found two drugs that could lessen the impacts of Ebola inside a day. It diminished a long time of R&D work, which saved clinical organizations time, cash, and efforts to give medications to the individuals out of luck (Devi, M et al,2019).

Network Protection

Clinics and clinics hold a great deal of classified information. This information needs assurance from digital assaults. Indeed, even the littlest information break can make huge damage the two patients just as healthcare focuses/experts (Kanthavel, R., & Dhaya, R.,2022). It probably won't be feasible for the emergency clinics' network safety groups to sort out each expected danger to their frameworks. 6G-AI can help network safety groups to sort out each possible issue or danger. The calculation can likewise rank them dependent on their need and present it to the group (V. Efrati, C et al, 2014) .

Misrepresentation Discovery

6G-AI can help keep these false exercises by learning from past occasions. The machine learning calculation proactively looks for false 16G-AI and sends cautions to the concerned individual when they distinguish one (Kanthavel, R., & Dhaya, R.,2022).

Healthcare Structure Examination

Superfluous hospitalization has driven numerous individuals to run into obligation. Nonetheless, new 6G-AI-based frameworks can assist patients with recognizing work process shortcomings and errors in treatment.

Some of the issues in healthcare that have been raised as a result of the use of 6G-enabled AI include:

- Medicinal inaccuracy
- Agreeable Death mortality
- Deficient in intelligibility
- It's hard to find a suitable physician.

These 5 Steps Will Help to Improve Healthcare Security

- Endpoint Security to Protect Mobile Devices... Email Security Measures to Limit Unauthorized Access.
- Validating Data Access and Ownership with a Zero-Trust Framework...
- All Healthcare Employees Should Receive Cyber security Training.
- Disaster Recovery Solutions to Recover from the Unexpected

THE IMPACT OF 6G-AI IN EDUCATIONAL SECTORS

ML and 6G-AI have the ability to change the 6G-AI area for good by tending to probably the most basic difficulties it is right now confronting. Not exclusively will it sway the general turn of events and execution of understudies, yet additionally help diminish the pressure a lot on teachers (C. Romero and S. Ventura, 2010). Simulated intelligence arrangements deliver understudy execution investigation, customized learning, and understudy commitment devices to improve the customary study hall experience. Table 4 shows the advantages and disadvantages of utilizing 6G-AI in Education area.

Shortening Administrative Works

Simulated intelligence can be used to smooth out numerous regulatory errands and obligations that are presently taken care of by instructors, for example, getting ready plans, assessment reviewing, checking schoolwork, and so on Robotizing these cycles can facilitate the tension on the educators, furnish them with a feathered creatures eye see over every understudy's advancement, and let them center around improving their instructive strategies (Banumathi, J., et al,2022).

Table 4. Pros and cons of using 6G-AI in education part

PROS	CONS
• Game based learning • Digital textbooks • Predictive analysis • Enhanced course quality • Real time problem solving • Virtual learning and grading system • Personalized learning	• Cost • dependence • shortage of personal connection • unemployment • competent decision making • loss of data

Smart Data

Smart substance is an essential piece of the digitized educational plan. Despite the fact that even during the beginning period, numerous educational institutes are endeavoring to formulate the virtual educational program accessible all the way through various computerized stages to construct in an intelligent manner. In the midst of 6G-AI-based devices, educational resources, for example, electronic mode of manuscripts, sound, and record of moving pictures could be effectively ordered and gathered by the instructive plans and even understudy inclinations (Sangeetha SKB et al,2022) . This method the education learning cycle and the educational plan are completed extra understudy driven.

Custom-Made Learning

Not at all like the conventional tr6G-AIning framework that typically focuses on the center 79% of understudies that state that students of 10%, frameworks guarantee in the direction of serve up by 6G-AI, everything being equal, and find their maximum capacity (Pardo,2006). Schools and colleges are currently using 6G-AI in the homeroom for customized proposals on the meetings, their exhibition, everyday tasks, and even in assessments. Additionally, understudies currently approach 6G-AI-controlled smart mentoring frameworks anywhere they cont6G-AIn the advantage of customized study plans, open from any client gadget or area (Dhaya, R et al,2022).

Voice Associate

The intuitive voice associates can help with various learning material even without an educator, and have made learning conceivable anyplace and whenever. Aside from supporting learning, reconciliation of such voice collaborator innovation inside the school grounds framework can make following timetables a fun and drawing in experience for the students[27].

Global Learning

Innovation is crossing over and associating individuals of various topographical areas. This has ended up being a help for the understudies as they can now effectively convey and impart thoughts to their companions from around the globe. With the assistance of cutting edge 6G-AI arrangements, a few schools and colleges are making taking in stages for understudies from various pieces of the globe to contemplate together (Dhaya, R., & Kanthavel, R.,2022) .6G-AI-based apparatuses are assisting them with getting sorted out classes dependent on educational plan and redo the experience to coordinate the understudy's nationality, on condition that right to use to teaching resources inside their enlightening environment (Dhaya Ramakrishnan, Kanthavel Radhakrishnan,2022) . Verbal message conversion and computerized description construction allows instructors and scholars to exchange a few words virtually devoid of any restrictions. The expectations of worldwide education look brilliant with the proper utilization of 6G-AI tools.

Competent Educational Administration

Above and beyond set of courses and instruction processes, 6G-AI equipments are capable of optimizing and reformatting organization processes for learning organizations. For example, the setting up of modern interactive virtual as well as physical classes in terms of competing international standards, I tools can be used effectively (Y. Xiao,2020). The institutions also can have the capability to distribute the learning materials without any extra money and human effort by means of 6G-AI based applications at ease.

Following and Evaluating Student's Ability

Assembling enormous datasets on understudies' presentation for various courses, 6G-AI-based stages can investigate and give important knowledge to instructors and tutors (J. Du, et al,2020). Joining the pointers of time taken to finish a specific course and execution on individual tasks, educators can change exercise designs appropriately to coordinate the requirements of any understudy.

Generating a Gamified Practice

Appreciating extreme ideas in science, physical science, and other complex subjects will turn out to be simple with gamification methods controlled by 6G-AI and computer generated reality. By giving understudies direct insight of such marvels, instructors can help them comprehend the material as well as improve m6G-AIntenance. These devices can likewise be utilized to impart ethics and build up understudies' enthusiastic knowledge which could somehow be difficult to achieve with customary techniques (Sangeetha, S. K. B et al,2019).

Shared Learning

In a course group of 20-30 understudies, everyone has various attributes, characters, and learning abilities. With the assistance of 6G-AI-based instruments in the homeroom, educators can distinguish and connect the information holes of each understudy (H. Yang et al,2020) . Likewise, such data can help tweak courses, tasks, and so forth this likewise establishes an open climate for understudies to work together with one another on undertakings with no limits. Even though we are using 6G-AI in education section in a positive manner, some limitations also there. The biggest flaws are malicious conduct and encryption in 6G. The blockchain and distributed ledger technologies (mostly based on blockchain) are generally safe as the last applications of the 6G network, although they may still be the target of malicious action.

CONCLUSION

This chapter presented the role of machine learning technology in healthcare and education sectors in a det6G-AIled manner. In the first section, the difference between 6G-AI and ML have been discussed, followed by the vital role and the current scenarios of education and health sector where these edging technologies involvements in taking the decision accurately. The different case studies in various health care applications around have been elaborated with the assistance of 6G-AI and ML. Then, in the second part of this chapter analyzed the effective utilization of Machine learning in Educational Sectors through its processes from adaptive learning to evaluating assessments. From the inference, it is concluded that the healthcare industry is consistently transforming through inventive advancements like 6G-AI and ML. The last will before long get incorporated into tr6G-AIning as an indicative 6G-AId, especially in essential consideration. It assumes a significant part in molding a prescient, personalized, and preventive future, making treating individuals a breeze. This chapter is concluding with the analytical evidences of the effective utilization of 6G-AI as a credible player in the health and educational sectors in the current and the near feature too.

REFERENCES

Ailaja, Seetharamulu, & Jabbar. (2018). Machine learning in healthcare: A review. *2018 Second International Conference on Electronics Communication and Aerospace Technology (ICECA)*, 910-914.

Arel, I., Rose, D. C., & Karnowski, T. P. (2010). Deep machine learning-a new frontier in artificial intelligence research. *IEEE Computational Intelligence Magazine*, *5*(4), 13–18.

Banumathi, J., Sangeetha, S. K. B., & Dhaya, R. (2022). Robust Cooperative Spectrum Sensing Techniques for a Practical Framework Employing Cognitive Radios in 5G Networks. *Artificial Intelligent Techniques for Wireless Communication and Networking*, 121-138.

Chassignol, M., Khoroshavin, A., Klimova, A., & Bilyatdinova, A. (2018, January). Artificial intelligence trends in education: A narrative overview. *Procedia Computer Science*, *136*, 16–24.

Devedžic, V. (2004). Web intelligence and artificial intelligence in education. *Educ. Technol. Soc.*, *7*(4), 29–39.

Devi, M., Dhaya, R., Kanthavel, R., Algarni, F., & Dixikha, P. (2019, May). Data Science for Internet of Things (IoT). In *International Conference on Computer Networks and Inventive Communication Technologies* (pp. 60-70). Springer.

Dhaya, Kanthavel, Devi, & Algarni. (2019). Assessment on Recurrent Applications of Machine Learning and its Behaviors. *International Journal of Engineering and Advanced Technology, 8*(6S3).

Dhaya, Ujwal, Sharma, Singh, R., Selvan, & Krah. (2022). Energy-Efficient Resource Allocation and Migration in Private Cloud Data Centre. *Wireless Communications and Mobile Computing.* doi:10.1155/2022/3174716

Dhaya, R., & Kanthavel, R. (2020). Accomplishment Of Intellectual Supply Managing Algorithm For Ubiquitous Cloud Computing. *Gedrag & Organisatie Review*, *33*(3), 236–246.

Dhaya, R., & Kanthavel, R. (2022). Dynamic automated infrastructure for efficient cloud data centre. *Cmc-Computers Materials & Continua*, *71*(1), 1625–1639.

Dhaya, R., & Kanthavel, R. (2022). Energy Efficient Resource Allocation Algorithm for Agriculture IoT. *Wireless Personal Communication*. doi:10.1007/s11277-022-09607-z

Dhaya, R., & Kanthavel, R. (2022). IoE based private multi-data center cloud architecture framework. *Computers and Electrical Engineering, 100*. j. compeleceng.2022.107933 doi:10.1016/

Dhaya, R., Kanthavel, R., & Ahilan, A. (2021). Developing an energy-efficient ubiquitous agriculture mobile sensor network-based threshold built-in MAC routing protocol (TBMP). *Soft Computing*, *25*(18), 12333–12342.

Dhaya, R., Kanthavel, R., Algarni, F., Jayarajan, P., & Mahor, A. (2020). Reinforcement Learning Concepts Ministering Smart City Applications Using IoT. In *Internet of Things in Smart Technologies for Sustainable Urban Development* (pp. 19–41). Springer.

Dhaya, R., Kanthavel, R., & Mahalakshmi, M. (2021). Enriched recognition and monitoring algorithm for private cloud data centre. *Soft Computing*, 1–11.

Domingos, P. (2012). A few useful things to know about machine learning. *Communications of the ACM*, *55*(10), 78–87.

Du, Ai, Wang, & Zhang. (2017). Overview of deep learning. *Proceedings - 2016 31st Youth Academic Annual Conference of Chinese Association of Automation YAC 2016.*

Du, J., Jiang, C., Wang, J., Ren, Y., & Debbah, M. (2020, December). Machine Learning for 6G Wireless Networks: Carrying Forward Enhanced Bandwidth, Massive Access, and Ultrareliable/Low-Latency Service. *IEEE Vehicular Technology Magazine*, *15*(4), 122–134.

Efrati, V., Limongelli, C., & Sciarrone, F. (2014). *A data mining approach to the analysis of students' learning styles in an e-learning community: A case study. In Universal Access in Human-Computer Interaction. Universal Access to Information and Knowledge*. Springer International Publishing.

Heuser & Zohner. (2012). Intelligent machine homicide. *Proc. Int. Workshop Constructive Side Channel Anal. Secure Design*, 249-264.

Jordan, M. I., & Mitchell, T. M. (2015). Machine learning: Trends perspectives and prospects. *Science*, *349*(6245), 255–260.

Kanthavel, R., & Dhaya, R. (2013). Reliable Hierarchical Security Routing For Delay Tolerant Sensor Networks. International Information Institute (Tokyo).

Kanthavel, R., & Dhaya, R. (2022). Wireless Underground Sensor Networks Channel Using Energy Efficient Clustered Communication. *Intelligent Automation And Soft Computing*, *31*(1), 649–659.

Kanthavel, R., & Dhaya, R. (2022). Prediction Model Using Reinforcement Deep Learning Technique for Osteoarthritis Disease Diagnosis. *Computer Systems Science and Engineering*, *42*(1), 257–269.

Kanthavel, R., Dhaya, R., & Ahilan, A. (2022). AI-Based Efficient WUGS Network Channel Modeling and Clustered Cooperative Communication. ACM Transactions on Sensor Networks, 18(3), 1–14. doi:10.1145/3469034

Kanthavel, R., Indra Priyadharshini, S., Sudha, D., Sundara Velrani, K., & Dhaya, R. (2022). Multi-hoped cooperative communication-based wireless underground sensor network design. *International Journal of Communication Systems*. doi:10.1002/dac.5174

Kotsiantis, S. B., Zaharakis, I., & Pintelas, P. (2007, July). Supervised machine learning: A review of classification techniques. *Emerg. Artif. Intell. Appl. Comput. Eng.*, *160*, 3–24.

LeCun, Y., Bengio, Y., & Hinton, G. (2016). Deep learning. *Nature*, *521*(7553), 436.

Letaief, K. B., Chen, W., Shi, Y., Zhang, J., & Zhang, Y. A. (2019, August). The Roadmap to 6G: AI Empowered Wireless Networks. *IEEE Communications Magazine*, *57*(8), 84–90.

Marline, Kumari, & Krishna. (2018). Prognosis of diseases using machine learning algorithms: A survey. *2018 International Conference on Current Trends towards Converging Technologies (ICCTCT)*, 1-9.

Mung, P. S., & Phyu, S. (2020). Effective analytics on healthcare big data using ensemble learning. *2020 IEEE Conference on Computer Applications (ICCA)*, 1-4.

Nikam, S. S. (2015). A comparative study of classification techniques in data mining algorithms. *Oriental Journal of Computer Science & Technology*, *8*(1), 13–19.

Noorbakhsh-Sabet, N., Zand, R., Zhang, Y., & Abedi, V. (2019). Artificial intelligence transforms the future of healthcare. *The American Journal of Medicine*.

Pardo. (2006). Designing learning analytics experiences. In Learning Analytics. From Research to Practice (pp. 15–38). Springer.

Peteiro-Barral, D., & Guijarro-Berdiñas, B. (2012). A survey of methods for distributed machine learning. *Progress in Artificial Intelligence*, *2*(1), 1–11.

(2016). Qiu, Wu, Ding, Xu, & Feng. (2016). A survey of machine learning for big data processing. *EURASIP Journal on Advances in Signal Processing*, (1), 67.

Ramakrishnan, D., & Radhakrishnan, K. (2022, March). Applying Deep Convolutional Neural Network (DCNN) Algorithm in the Cloud Autonomous Vehicles Traffic Model. *The International Arab Journal of Information Technology*, *19*(2), 186–194.

Romero, C., & Ventura, S. (2010). Educational data mining: A review of the state of the art. *IEEE Transactions on Systems, Man and Cybernetics. Part C, Applications and Reviews*, *40*(6), 601–618.

Sangeetha, S. K. B., & Dhaya, R. (2016). Fuzzy integrated gaming approach for relay selection in co-operative communication. *Journal of Applied Sciences Research, 12*(3), 45–49.

Sangeetha, S. K. B., & Dhaya, R. (2022). Deep Learning Era for Future 6G Wireless Communications—Theory, Applications, and Challenges. *Artificial Intelligent Techniques for Wireless Communication and Networking,* 105-119.

Sangeetha, S. K. B., Dhaya, R., & Kanthavel, R. (2019). Improving performance of cooperative communication in heterogeneous manet environment. *Cluster Computing, 22*(5), 12389–12395.

Sathyaraj, R., Kanthavel, R., Leonardo Cavaliere, L. P., Vyas, S., Maheswari, S., Gupta, R. K., Ramkumar Raja, M., Dhaya, R., Gupta, M. K., & Sengan, S. (2022). Analysis on Prediction of Covid-19 with Machine Learning Algorithms. *International Journal of Uncertainty, Fuzziness and Knowledge-based Systems*. Advance online publication. doi:10.1142/S0218488522400049

Sharma, R. C., Kawachi, P., & Bozkurt, A. (2019). The landscape of artificial intelligence in open, online and distance education: Promises and concerns. *Asian J. Distance Educ., 14*(2), 1–2.

Stamp, M. (2017). *Introduction to Machine Learning with Applications in Information Security*. Chapman and Hall/CRC.

Vernon, D., Metta, G., & Sandini, G. (2014, April). A survey of artificial cognitive systems: Implications for the autonomous development of mental capabilities in computational agents. *IEEE Transactions on Evolutionary, A247*, 529–551.

Wale & Sonawani. (2018). A Survey on Deep Learning Approaches in Healthcare. *International Journal of Innovative Research in Computer and Communication Engineering, 6*(5).

Wang, N., Jiang, T., Lv, S., & Xiao, L. (2017, July). Physical-layer authentication based on extreme learning machine. *IEEE Communications Letters, 21*(7), 1557–1560.

Xiao, Y., Shi, G., Li, Y., Saad, W., & Poor, H. V. (2020, December). Toward Self-Learning Edge Intelligence in 6G. *IEEE Communications Magazine, 58*(12), 34–40.

Yang, H., Alphones, A., Xiong, Z., Niyato, D., Zhao, J., & Wu, K. (2020, November/December). Artificial-Intelligence-Enabled Intelligent 6G Networks. *IEEE Network, 34*(6), 272–280.

Yu, K.-H., Beam, A. L., & Kohane, I. S. (2018). Artificial intelligence in healthcare. *Nature Biomedical Engineering, 10*, 719–731.

Zhang, C., & Ma, Y. (2012). *Ensemble Machine Learning: Methods and Applications*. Springer.

Chapter 10
Security and Privacy of Unmanned Aerial Network Communication Systems in 6G Networks

Kayalvizhi M.
SRM Institute of Science and Technology, India

Ramamoorthy S.
SRM Institute of Science and Technology, India

ABSTRACT

The massive deployment of unmanned aerial vehicles (UAVs) is growing in various fields in wireless network of 5G and 6G. It supports huge deployment of UAVs that offer secure, reliable, and low-cost data communications. They are widely deployed in surveillance and military applications for collecting sensitive and critical data, transferring the collected data to the data center, procuring high-definition images of disaster areas, surveying natural disaster zones, and other critical missions. Although, the UAV applications are growing, the risk and security factors associated with it are increasing parallelly. So, a need of secure and reliable network communication system for UAV is mandatory. In this chapter, the performance of UAV network system in 6G environment and the technological challenges faced with respect to security and privacy are discussed. Then a possible solution using blockchain and security policies that can be provided for a secure communication in a 6G network is presented.

INTRODUCTION

While the UAVs communication network are growing it needs to operate effectively amidst the type of wireless network. While the most promising cellular network, supports UAV to ground communication effectively, but deploying UAVs for robust missions needs additional secure framework to be operated effectively. And the fifth generation 5G networks are still in study, the use of 6G network have taken

DOI: 10.4018/978-1-6684-3804-6.ch010

initiative. The wider range of UAVs applications in various fields will be next a major part when supported by 6G wireless network. But the existing challenges of 6G network needs to be considered when deploying UAV swarms. And the provision of data privacy and network security with a cost-effective approach needs to be addressed when considering 6G networks. And the support of 6G network in UAV swarm needs to assure that the trust, security and privacy of the participating UAVs are assured and not compromised. And the 6G network needs to support new security policies to take the UAV swarm to next level of operation. The application of Unmanned Arial Vehicles (UAVs) is increasing with irrespective of the fields. And with the reinforcement of 6G network, UAVs can be tweaked to wider and variety of new range of application fields such as automated logistics, effective stealthy military operations. And considering the mission and payload capacity of the UAVs, the security framework needs to be cost-effective and light weight that support higher success rate in UAV missions. But still many other factors need to be considered such as energy level of UAV, heterogenous network setup, operating in high level dynamic network and its continuous high mobility. Moreover, collaborating AI technologies with the UAV swarm supports real time decision making and path navigation, still the security related attacks need to be addressed when using AI technologies in UAV 6G network model. Since in the most of the UAV operations, the UAVs are unmanned and navigated from ground control station, they are highly exploitable from security attacks. Especially the CIA (confidentiality, Integrity, Availability) of the network needs to be secured. In UAV missions, the attack can be launched by a new UAV device are security attacks can be launched by compromising the UAV. So, the attacks can be launched by an UAV and to the UAV swarm. Mostly the UAV devices are targeted with hardware level security attacks where the attackers take advantage of the default configurations of the UAV devices. And nowadays many open-source tools are available to perform vulnerability scanning which can lead to zero-day attacks.

Protecting privacy is a basic need in wireless communication in the era of sixth generation 6G. The main challenge in 6G network will be the data communication which will be huge, that poses a threat to privacy of public. When the data are easily accumulated and transferred in the 6G network it leads to greater security risk on regulating the security of user's data. With the evolution of 6G network, the UAV's operation can be controlled autonomously which exposes it to security attacks. And the 6G network needs to protect the data that is transmitted in the network.

Although 6G network can support with many features like higher frequencies, frequency divisor multiple access (FDMA), overall throughput and low-level latency, still the security and privacy of the network needs to be addressed. And the urge to support machine to machine communication is another major research area that needs to be discussed. Since with the upcoming growth of technologies, the UAVs are deployed based on the need and requirement. To support cost-effective missions, the UAV's deployed can be of different ranges and it needs to be operated in a heterogeneous network. So, amidst of varying high performance computing features in 6G network, our main aim is to operate the UAV swarm in a secured environment. And there need to be a secure framework that will be reliable and feasible based on the requirements of the UAV operations. Since the cyber-attacks are increasing and the attacker's behavior cannot be predicted when it comes to zero-day attacks. So, the existing security frameworks and the encryption mechanism can't support the CIA of UAV swarm when operated in 6G network.

And by the time the pros and cons of 6G network is discussed, the picture of 7G network will come and take over the smart devices to next level. So, the technologies are constantly evolving and the security mechanism always needs to be ahead of the evolving technologies. So currently to support smart integration and automation of intelligence, 6G networks is evolving rapidly. The features of 6G network can support the automation of UAV networks which support to experience better quality and cost-effective operations. So, by the combination of blockchain in 6G network, can enhance the effective operation of UAV swarm.

The blockchain technology can support heterogenous UAVs that can be coordinated and operated as a single network in the ear of 6G. And the data transmission can be secured where the UAV-to-UAV communication and UAVs to base station communication can be secured. And the decentralized concept of blockchain supports the upload and download of data from the UAV payloads. And in the surveillance operation, the UAVs registered in the blockchain network can be authenticated whereas the malicious UAVs will be automatically rejected from the network.

In this chapter, the earlier evolution of technologies are discussed and an overview of Unmanned Aerial vehicles and its applications is elaborated. And, how 6G network can support the UAV swarm operations effectively and increase the applicability on various new dimensions are discussed. And amidst of the wider range of application, how the privacy and security of the UAVs are overcome by incorporating blockchain technology and blockchain assisted UAV swarm in 6G is presented. And an overview of how blockchain enabled UAV swarm network in 6G can mitigate security attacks and can operate effectively is discussed.

RELATED WORKS

In this section, works of various authors are discussed which focusses on the deployment of unmanned aerial vehicle in 6G supported network environment and the various existing network challenges and risks faced in providing a secure and reliable network functioning environment.

Rakesh Shrestha, Shiho kim (Shrestha, R., 2021) studied on the traffic management of Unmanned Aerial Vehicles in a 6G enabled environment and discussed on the future perspective of UAV traffic management for collision free operations of UAV. And a perspective of 6G parameters are discussed that support air transportation in 3D space. And the issues are discussed in UTM system deployed in 6G.

Cheng-Xiang Wang, Jie Huang, Haiming Wang, Xiqi Gao, Xiaohu You, and Yang Hao (Wang, C. X., 2020) presented the performance of 6G network and its various characteristics and discussed on the various frequencies of 6G and its technological benefits and its application in vehicular communications and Internet of Things. They are provided an overview of the risk and gaps involved in 6G networks.

Lv, Z., Qiao, L., & You, I., (Lv, Z., 2020) discussed on the existing models of 6G wireless network, its channel of communication and its models and summarized on the various methods existing in the research. And discussed on the space, time needed for vehicle communication. And also suggested and provided new solutions for the existing gaps in the 6G networks.

Yu Liu, Cheng-Xiang Wang, Hengtai Chang, Yubei He, Yubei He (Liu, Y., 2021) proposed a 3D model to achieve space-air-ground-sea communication for 6G channel where UAVs are deployed as mobile base stations. And the proposed model was analyzed on various statistical scenarios and its performance was verified.

Mustafa A. Kishk, Ahmed Bader, and Mohamed-Slim Alouini (Kishk, M., 2020), discussed on the coverage capacity of UAV and its enhancements and how the performance can be improved from tethering UAV (tUAV). And its advantages are improved by using tUAV. And the authors discussed on some use cases about how tUAV is mounted to a base station will be a benefit in terms of coverage and network strength. And also they discussed on the risks and challenges involved in the performance of the proposed model.

Run Liu, Anfeng Liu, Zhenzhe Qu, and Neal N. Xiong (Run Liu, 2021) proposed a UAV enabled offloading scheme that delegates the task with maintain the energy, where the nodes with large data capacity is selected to collect task from the neighbor nodes which saves the distance and the energy. And the managing of the nodes are splitted into high data and low data nodes and managed effectively. And the collection of the data is optimized by saving UAV energy where the simulation results show better performance of the proposed model when compared with the existing method.

Erik Matti, Oliver Johns, Suleman Khan, Andrei Gurtov, Billy Josefsson (Matti, E., 2020) presented a survey of 5G/6G in the aviation communication sector. The 5g/6G can provide better positioning and will be energy conservative with good internet connection. And the data exchanges can be done effectively with the user of 5G/6G. They provide reliable and secured data exchange and surveyed on several scenarios specific to aviation that also focused on the security parameter.

Rajesh Gupta, Aparna Kumari, Sudeep Tanwar, and Neeraj Kumar (Gupta, R., 2020) proposed a blockchain incorporated UAV swarm based our on 6G network to contain disease outbreaks like COVID-19. And also the proposed model focused on the data security while exchanging critical data. And the experimental results showed that the proposed software model is better in throughput when compared to the existing 4G/5G systems.

Mohd Zuhair, Pronaya Bhattacharya, Farnazbanu Patel, Deepti Saraswat, Dhruvin Navapara (Zuhair, M., 2021) authors proposed a method namely blockchain based in 6G network by incorporating unmanned aerial vehicles to surveillance group of peoples in crowded areas. The thermal sensors in UAV will monitor the body temperature and the collected data is processed to the ground station by using 6G network communication service. This helps to tract the information of the COVID patients and their locations. The proposed model uses smart contracts in blockchain to initiate trust. The experimental results showed that the 6G network achieved low latency and better than 5G.

Chengxiao Liu, Wei Feng, Yunfei Chen, Cheng-Xiang Wang, Ning Ge (Liu, C., 2020) investigated on resource allocation in a UAV swarm to attain maximum coverage area. The authors proposed a process-oriented framework to allocate channels, power and time that assist the whole UAV operation in a mission. And the authors presented an iterative algorithm to allocate resource that improved network efficiency. And the results shows that the coverage is better and promising in the 6g era.

Antonio Brito, Pedro Sebastiao, Nuno Souto (Brito, A., 2019) authors developed a radio jamming technique to mitigate unauthorized UAV operations. And the study of the jamming techniques helped them to understand the factors involved in deterring a communication link and they used BladeRF x40 board as hardware platform and the experimental setup showed that the jamming was successfully launched and it leads to an anti-UAV system.

Bomin Mao, Fengxiao Tang, Yuichi Kawamoto, and Nei Kato (Mao, B., 2021) authors provided a deep learning task offloading strategy to address dynamic energy harvest. And the authors studied on 6G Iot services and the limitations in remote areas. And the proposed AI based method perrformace results are improved in the factors of computation rate. And the authors also discussed on the future directions of 6G in IoT.

Mohammad Javad-Kalbasi, Shahrokh Valaee (Javad-Kalbasi, 2021) authors derived an algorithm for refreshing the route and resource allocation in UAV networks to address the limited link capacity. And also the proposed scheme focused on providing collision free communicating service. And the experimental results showed the performance of the resource utilization in a sample UAV network.

Abhishek Gupta, Dr. Xavier Fernando, Dr. Olivia Das (Gupta, A., 2021) authors investigated on the various 5G and 6G communication technologies and its various modelling techniques. They surveyed on various IoT network models and studies on the factors affecting reliability of IoT. And how offloading of the tasks helps in achieving better reliability and availability.

Helin Yang, Jun Zhao, Jiangtian Nie, Neeraj Kumar, Kwok-Yan Lam, and Zehui Xiong (Yang, H., 2021) authors proposed a resource allocation model to improved the capacity of the network and also by utilizing the wireless service to the maximum. This will allow the UAVs to track their real time locations and the allocation of the resource will be done efficiently. And the experimental results showed that the proposed method provides better learning rate and provides better network capacity.

Jinna Hu, Chen Chen, Lin Cai, Mohammad R. Khosravi, Qingqi Pei, and Shaohua Wan (Hu, J., 2021) authors presented on the challenges in the 6G internet of vehicles network and how it affects the optimization of the network. And they also studied on the solutions by using Artificial Intelligence and how UAV can be deployed to support 6G network. And also the authors discussed on the future directions of UAV in 6G networks.

Antonino Masaracchia, Yijiu Li, Khoi Khac Nguyen, Cheng Yin, Saeed R. Khosravirad, Daniel Benevides da Costa, Trung Q. Duong (Masaracchia, A., 2021) authors surveyed on the UAV networks and the challenges faced in a reliable and low latency communication. And they also discussed on the classifications of UAV networks and the challenges in the practical large-scale implementations with security factors.

EVOLUTION OF TECHNOLOGIES

From past few years, the evolution of wireless communication has shown steady growth from first generation technology to the fifth-generation technology and the changes in the mobile communication it has provided is rapid in short time. And the use of fifth generation 5G in mobile technology provided far more better benefits when compared to fourth generation 4G. The fifth generation 5G provides greater communication benefits in vehicular communications and Internet of Things (IoT). And also, the technological benefits it is expected to be provided are enormous and promising. Together the both 5G and sixth generation 6G will provide industrial level applications and also in ten years the communication standard will be improved in terms of reliability and security. And also, the mobile usage is increasing rapidly day by day. And in ten years, the count of mobile users will be more than 16 billion. So, the base technology should be strong and supportive to adopt the increasing count of the smart devices and its supported applications.

The evolution of the communication channel has played a greater role in supporting new technologies and enhancing its performance. The Figure 1. shows the timeline of communication channel that supports different technologies over years. In the earlier days, the first generation was used to transfer voice signals and it was promising and it supported 2.4 kb/s of data transfer. But however, it couldn't operate in wider range. Then 2G, second generation came in picture where it supported mobile communication system. And also, it supported security in the system of communication. 2G had a data transfer of 64 kb/s by using digital technology. But 2G didn't support cloud technologies and multimedia. It handled only short message service. And the evolution of third generation 3G allowed data transfer of up to 2 Mb/s that supported multimedia operations and browsing internet faster. But the 3G also faced issues like buffering, intermittent disconnection from network which pushed forward for the evolution of fourth generation, 4G which had a data transfer of 1000 Mb/s. Then in later 2020s, evolution of fifth generation supported data transfer of 10 Gb/s which will be more promising to support industrial level applications which also incorporated Artificial intelligence and Internet of Things. The use of 5G supported multimedia to next level and the quality in terms of data transfer and the data capacity increased. The main functionality in 5G is it provides low latency which makes it applicable to wide range of applications.

Figure 1. Technology evolution and its applications

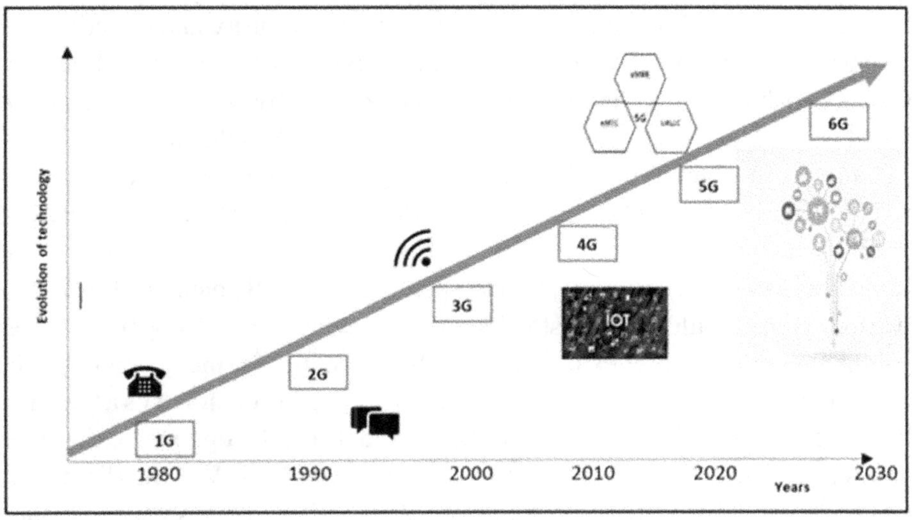

The 5G technology is still evolving and its behavior and characteristics are still in research but the era of 6G has started evolving. But nowadays everything is moving on a fast phase and all the systems focusses on data security and availability of data. And the people expect features like low latency, low consumption of power, high data capacity and speed which paves the way for the era of sixth generation, 6G. The growth of 6G in the next years will be promising that can mitigate the drawbacks in 5G.

When it comes to vehicular communications, the support of 4G, 5G is promising and still many research is ongoing in support of 5G in unmanned Aerial Communications (UAV). The UAVs are autonomous system that is widely deployed in military and surveillance operations. So, when a swarm of UAVs are deployed in a remote area, they need to perform effectively for successful mission. There are many factors that needs to be addressed for an effective coordination among UAVs and between the base station and the UAVs. So, in the next section, an introduction of UAVs is given and its challenges and gaps are discussed and how 6G can bridge the gap.

Unmanned Aerial Vehicles Applications

Unmanned aerial vehicles (UAV) is also known as drone where it can be operated either with human control or can be operated autonomously from base station. UAVs are widely being deployed in all fields based on the requirements. They are widely used for their cost effective, adaptability in any fields, availability based on the convenience and various UAV models available for various functionalities. They are being used widely in all fields like medical amenities transportation, surveillance, military operations, agriculture and many. Since the UAV's are available based on the requirements like size, payload capacity, power they are chosen widely. And recently they are widely deployed in communications but there are many external factors and challenges its facing like power capacity, security and the surveillance time.

And the recent growth in communication technologies have helped the UAV development in various aspects and have increased its range of applications. And if UAVs are properly managed and operated in an operation, it can provide cost effective reliable results in various real time cases. And also, UAVs can also be used as base stations which can effectively control the military and surveillance operation. So, since UAVs can be operated based on the requirements, the security and effective communication methods needs to be provided. UAVs when used as aerial equipment's, they can widely support broadband networks. They are capable of providing line-of-sight and on-the-fly communications. UAVs can also provide additional network capacity in existing cellular networks. And the major growing field of UAV is Internet of Things (IoT) where the devices with small payload and low power cannot fly for longer duration. The UAVs can server as relay that improves the network connectivity which will be of great support for surveillance operations.

Understanding the UAVs function is required to operate it to its full potential. The UAV fundamental structure and various types should be understood before deploying it into any operation. Many UAVs are manufactured light weight to support extra payload that is essential for the operation such as attaching high-definition camera, Global Positioning System, sensors, etc., Different UAVs are available cost wise based on their size, function and capacity. Many UAV models can be operated either manually or via remote or from ground control stations. The main component of the UAV is the flight controller which is also known as the heart of the Unmanned aerial vehicle. The UAV comes with an escape component which will be effective for supporting natural disaster operations.

For instance, when there is a natural disaster scenario like earthquake or forest fire, a speedy pre-assessment is needed so the deployment of UAVs comes in handy. And the UAV's can able to monitor the disaster area and send data to the base station. So, when deploying UAV, it can be either deployed as a single or a swarm of UAV. The swarm of UAV will be able to cover the disaster area with less time than a single UAV.

Flood is a natural disaster. The concerned officials need to inspect the area and need to take quick recovery actions to control the situation. And the deployment of UAVs will be of great benefit to cover the restricted area and the area where human intervention is not possible. And as a pre-disaster measurement, UAVs can monitor the rivers and flood and collect the status of the dam. And in case of unusual events, the authority will come to know the situation and take proper action. But the major problem is existing scenarios are the communication link. There needs to be a effective communication range to monitor the situation. Since the UAVs can be affected by weather and power conditions, need to make use of the flying time to the maximum. So, the support of 6G in these kinds of scenarios can support better communication link.

For forest fire monitoring which is unpredictable, UAVs can be deployed to collect the data and the exact locations of the fire outbreak to the rescue team. By pre-evaluating the area, many human and animal lives can be saved. And it also helps the officials to limit the damages caused from fire. But the challenges it faces is, due to smoke and other weather conditions, UAVs needs high-definition sensors and cameras to navigate itself through the disaster zone and collect the data. And in order to send the data real-time to the base station, the communication link needs to be up all the time. And the UAVs need to communicate among themselves to balance the load. So, for inter-communication the communication channel needs to be supportive where the 6G should provide high speed and capacity to communicate.

In many areas of medical sciences, UAVs are deployed as first aid to dispense medical services (Nguyen, D. C., 2021) to remote locations to reduce the time of service. This will assist in tracking the civilians that needs high priority medical support before the medical team arrives. These kinds of UAVs are designed light weight to carry the medical supplies. Since these medical kits are critical to be carried, extra support is needed to aid the UAV operation. In the era of 6G, these medical kits delivery can be enhances in terms of delivery speed and tracking.

The UAVs are commonly deployed in the areas of military and surveillance, normal usage by peoples, for research study, agriculture and medical. Each of the application area is discussed as follows:

UAV in Military Operations

UAVs deployed in Military operations (Porambage, P., 2021) is growing nowadays due to its cost-effective. There are many variants of UAV models that are engaged for different military operations. The main aim is to avoid human loss and to collect accurate data from the area. Military operations can be categorized based on the requirements which rely on the weight, payload and power of the UAV model. UAVs are deployed on the target location to collect sensitive data which can be used in their missions. UAVs are also involved in collecting data for defusing bomb and in air strike. And the UAV swarms are deployed to cover larger area and collect data in short time.

UAV in Monitoring Natural Disasters

The main idea of autonomous flying of drones is to monitor the disaster zones where the human intervention is difficult. So, UAVs play an important role in monitoring these disaster zones and provide detailed information of the current situation and stream real-time data. And UAVs are deployed (Luo, C., 2019) to provide food to the needy and also deliver medicines. So their main goal is to provide the sufficient things to the people who are in need during the time of any natural disaster.

UAV in Agriculture

When it comes to Agriculture which is a major growing field, the farming can be improved by deploying swarm of UAVs (Maddikunta, P. K. R., 2021). The UAVs can help to monitor the crops health status, monitor the fertility level of the soil, monitoring the irrigation systems and spraying of fertilizers to the fields at low cost. In the analysis of field soil, the data can be used to monitor the level of nitrogen and measurements can be taken to initiate proper fertilizer dosage. So, the soil fertility level can be improved and proper steps can be initiated to maintain it. And to monitor the crop health, it can help to spot the fungal and bacterial infections in the crops. It will help in providing immediate assessment as the common existing methods by monitoring using infrared light individually. And the device used to perform scanning of the land that supports spraying of fertilizers comes handy in the UAV devices that can crop spray effectively in the correct altitude and distance. And the crops sprays done by UAVs are more effective than the manual method. For monitoring crop health, UAVs are engaged to take high-definition quality images which will support to provide proper care beforehand. Assessment by UAV swarms is a cost-efficient way to monitor the life of crops.

Table 1. Summary of unmanned aerial vehicle applications

S.No	Application Area	UAVs Role
1	Natural Disaster	Pre-assessment of disaster area
		Delivery of medical goods
		Real time monitoring
		Collecting geographical locations
		For Rescue operations
2	Military Surveillance	Air Strike operations
		Collecting sensitive data from remote areas
		Bomb diffusing
		Area surveying
		Stealth operations
3	Agriculture	Monitoring Crop health
		Monitoring field soil
		Crop fertilizer spray
		Monitoring irrigation system
		Collecting high-definition images of crops
4	Commercial usage	Goods delivery
		Forecasting weather
		Monitoring human health
		Stock monitoring
		Photography

EVOLUTION OF SECURITY ATTACKS

Unmanned aerial vehicles (UAV) when operating as a swarm needs to function securely and the data upload and download needs to be done subjecting to security. Since the UAVs deployed are in military operations the data's that are circulating needs to be secured. The UAV's main aim is to collect data from the disaster zones and send the data securely back to the base station. The data's collected can be of images, geographical data or the status of the monitoring zones. So there needs to be a secure mechanism for data transfer that is suitable irrespective of data types. When it comes to security and privacy, the CIA (Confidentiality, Integrity, Availability) needs to be provided.

The 6G network is generally in a perception that it will be faster than the 5G network with improved features such as there shouldn't be any coverage issues. It needs to provide a full coverage and it shouldn't be constrained to only ground level communications. And the 6G should also possess capabilities of Artificial Intelligence (AI) which will be a more prominent featured of 6G. The AI should be deeply integrated into 6G as it should able to apply and provide all the functionality of AI tools and functions of networking. As currently the security and the privacy issues in the network communications are increasing and the attackers are always a way ahead in getting into the network system, there needs to be a standard and reliable communication setup to tackle and mitigate the security attacks.

Figure 2. Security attacks evolution

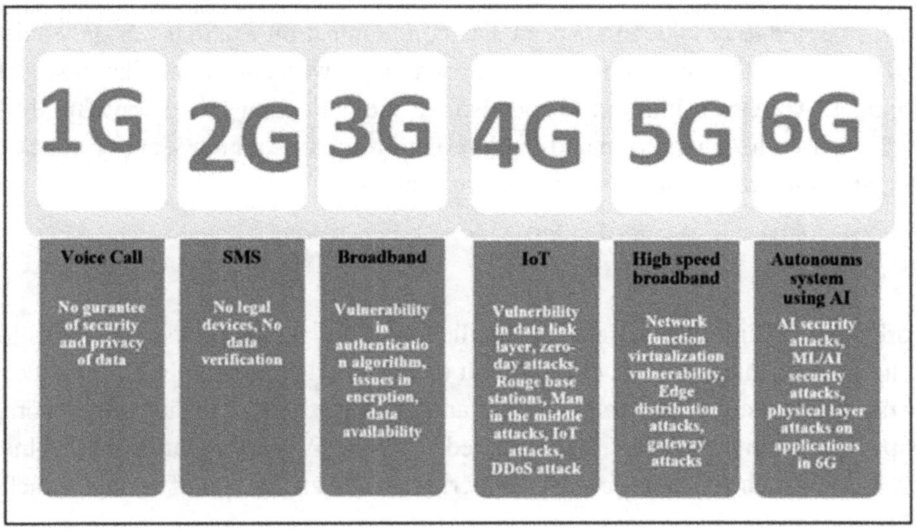

The evolution of security and privacy issues are depicted in the Figure.2 which is varying on the basis of communication network. The security attacks are based on both application and technology level. Most of them are vulnerable to authentication attacks, access control issues and abnormal behavior of network devices. And the applications involved in also possess security vulnerabilities.

1G

The evolution of 1G started in the year 1980 and it was dedicated for voice services. It relied only on analog signal. And it has many drawbacks and lacked security standards. There was no encryption of data where the signals are not secured and the phone conversation was not private and secured. As a result, the users face privacy gaps including eavesdropping and masquerading attacks.

2G

The 2G was mainly based on digital signals where the voice and SMS service was supported by Code Division Multiple Access and by Time Division Multiple Access. And the main standard used for communication was Global system for communication, GSM. It made the 2G network secured where the services involved are user identity, protection of user data, authentication of the source. But the 2G still had some security gaps where the authentication was one-way. There was only a way where the network will verify the user but the user cannot verify the network which led to masquerading attacks and information stealing.

3G

The 3G network emerged to provide high speed data communication where it was only able to meet the gaps of 1G and 2G. The GSM involved in 2G was used in 3G with additional features. Authentication and Key Agreement (AKA) was introduced and also the authentication is two-way. But the 3G network faced issues in keys used for encryption and the Internet Protocol. And the security attacks it faced are access data illegally, integrity issues, denial of service attack.

4G

The 4G network supports internet devices and applications. So, the 4G is more vulnerable to security attacks since the devices interact more closely with the terminal equipment's. So, the security vulnerabilities in the 4G network are more open to threats and risks. And due to the increased storage capacity, the malicious payloads can be injected and executed. And the Wi-Fi standards have vulnerabilities in data link layer, MAC which are more prone to integrity attacks and denial of service attacks.

5G

The evolution of 5G aim was to provide faster communication, more scalability and secure communicating architecture. The main feature of 5G is to support many devices and to provide more quality service irrespective of the increase in the number of devices. And the network architecture is a heterogeneous model. The use of Network Function Virtualization made the network dynamic and which also led to various security risks and challenges. And also, the data privacy issues in 5G network led to increase risk of data leaks.

6G

While the evolution of 5G is in research, the 6G network raised to support the gaps in all the earlier networks. The main key technologies that are supporting the 6G networks are Artificial Intelligence, Blockchain, Quantum communication and Molecular communication. Already 5G network have incorporated Artificial Intelligence technologies and in the various layers of network applications. But the deployment of AI technologies in the 5G network is not utilized since the 5G base is of traditional network setup and the features of the AI tools are not completed incorporated. And many vehicular communications network exist in the 5G network but the communication range is minimum, but more real-time use cases cannot be handled due to issues of latency.

The key areas of 6G network supports intelligent agents, Artificial Intelligence, 3D intercoms and intelligent radios. These technologies support real-time response, and make intelligent decisions, and are adaptive to the networks, and supports full 3D coverage. The 6G era will have more control to the communications, have a intelligent decision making capacity. But still the 6G network is under research and it's a growing field. It will take ten more years for the 6G network to completely evolve and to function.

And to support Unmanned aerial networks the 6G network has its own features but still there are many risks and gaps that are discussed in the next section.

6G ASSISTED UAV SWARM

When it comes to vehicular communications the 6G evolution has incorporated wide range of key areas where the support of Unmanned Aerial Vehicles is its key area. The Swarm of UAV operations faces many network challenges in the aspect of communication and coverage area. By the use of key technologies of 6G like Artificial Intelligence and Machine Learning, the process of UAV swarm can be boosted up. It can support positioning of the UAVs and planning of trajectory and power consumption in UAV communications. The 6G assisted Unmanned Aerial Vehicle (UAV) network will be used at its full potential and it provide 3D coverage of the network. And the optimization of the network such as radio signals, resource allocation and power consumption, energy level will be managed effectively. And the UAV operations involve to operate in a swarm and also it can act as a base station and as a server. And the 6G technologies such as AI can support efficient usage of network capacity and to helps to find the routing path and assist in optimizing other parameters. This will enable dynamic support of the network setup under real-time scenarios, and also the traversal of network device from one area to other will be smooth and doesn't need manual intervention. And the best base station support will be automatically selected from the available range of communications. And this will make the UAV operations to operate autonomously and will overcome the cell limitations.

The pioneer communication technologies supported 2D coverage of network and limited to ground equipment. The ear of 6G leads a path to heterogeneous network that can support 3D coverage that can cover air traffic such as unmanned aerial vehicles, satellites. And it can support continued and reliable communication of services which will also be cost-effective. The 6G network have a capability to bridge the ground and air level communications in remote areas.

6G network can support decentralized services that can provide efficient network coverage and can support Space-ground-sky communications parallelly. In the upcoming years, research of 6G can support global coverage and support continuous network coverage over mountain and ocean areas. So, the era of 6G have a great scope in multi-dimensional network coverage and can provide many benefits in 6G wireless network. And the scalability of the network can be improved by the use of 6G which has a greater network capacity and improves the traditional network setup.

6G SECURITY AND PRIVACY ISSUES

The 6G network applications require more secure and standard operations when compared to the 5G networks. The key technologies involved in 6G will support various range of use cases and the UAV operations will be functioned effectively in assistance of 6G network. The Figure 4. portraits the main key features involved in 6G network and its supported applications and technologies.

In this section, the security and privacy gaps involved in 6G architecture (Nayak, S., 2021) are discussed. Even though, the 6G comes in a package with technologies such as Artificial Intelligence, Machine learning and Quantum, its exposed to security and privacy attacks and more exposed to security attacks. Lets discuss on the security gaps and challenges involved in each technology.

Figure 3. 6G assisted UAV swarm network

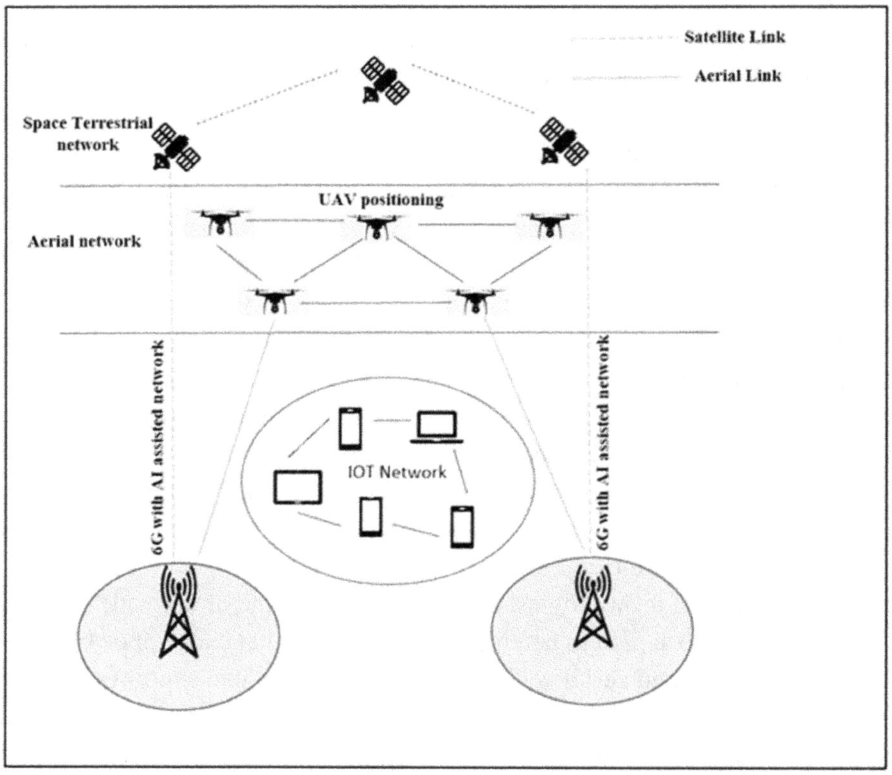

Figure 4. 6G architecture composition

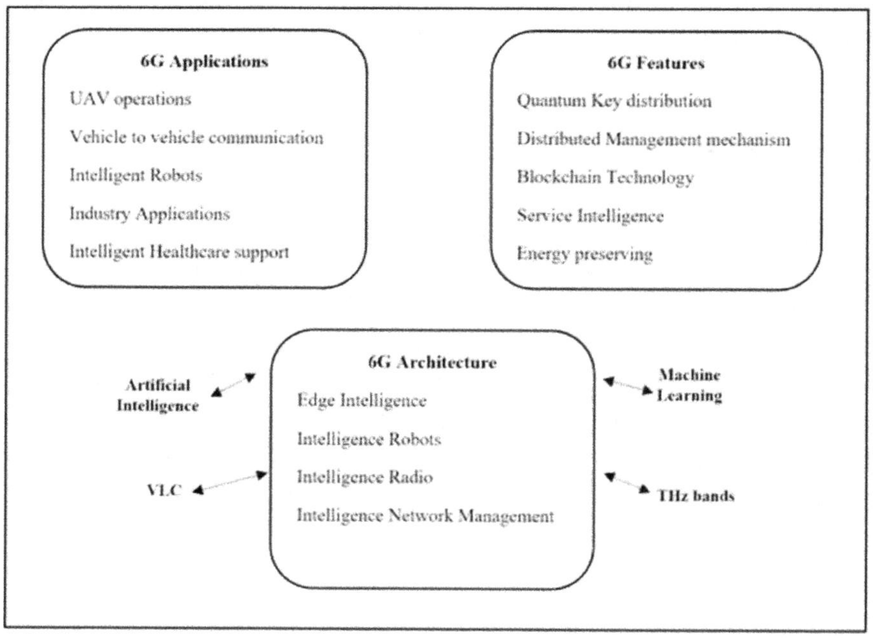

Artificial Intelligence

AI assisted 6G network will support resource management and allocation in a smart way, will ensure adjusting the network setup to the real time scenarios, and the devices are equipped with high intelligence which can support optimized path planning among the nodes. But Artificial Intelligence also faces with security risks that needs to be addressed for efficient communication. The more prominent security attacks are tampering data with malicious payloads which is also known as poisonous attacks. By tampering the network with false data can lead to poor outcomes and failure of the mission. And the communication tampering can leads to service interruption and compromise the whole network setup. And the Artificial Intelligence and Machine learning models and frameworks used in the network once compromised can lead to miscalculated decisions and results and can shutdown the whole network. So before using AI and Machine learning models, the open vulnerabilities in these models needs to be studied and a proper solution needs to be addressed.

Quantum Communications

The Quantum networks support transmission of data in quantum bits. It works in similar to the traditional networks, but it is able to solve a real-time scenario more efficiently than a traditional system. This quantum network is more exposed to security attacks such as cloning and collision attacks. The cloning attacks is nothing but cloning an entire block without altering any information. And the quantum collision attack provides same output with two different hash functions, when given as an input.

TeraHertz (THz)

The Thz technology is the more prominent key technology of 6G network. It is capable of offering stronger security in the physical layer which identifies the legitimate users in the network and align them in a straight path which is in line of sight. But attackers are capable of hacking the data which is done by scattering the signals with an obstacle, so a malicious node an easily intercept the signal and can gain advantage. And it can also steal user sensitive and critical data to gain access to system resources and alter the system configurations.

Visible Light Communication (VLC)

It is an optical technology that support wireless networks and offers additional features when compared to radio signals such as robustness, security, high data capacity. It can provide enhanced security than the radio frequencies system, since the light is not capable of penetrating through walls. But if a malicious node is present in the coverage area of transmitters, attackers can eavesdrop the signal and gain access to the network.

Molecular Communication

This technology is observed from the natural phenomenon of events where the nano and micro scale level improves the security and privacy of the network. By using distributes edge technology, the devices have no need to exchange data which can enhance the security. And the power consumption is low in molecular communication. The main aspect is to use biochemical signals to transmit the information in the network. And it also supports dynamic communication in real time scenarios that will support the communication of source, destination and all the participating nodes parallelly.

The above security and privacy issues in 6G are still in research and many standard security and protocols are designed to support the efficient communication of 6G network.

Table 2. Security and privacy issues in 6G

6G Technology	Security issues	Security issues outcome
Artificial Intelligence	Authentication	Learning model that can be trained to improve the authentication
	Encryption	Machine Learning models can be used to provide encryption
	Abnormal behavior	ML models can be trained to alert the system, is abnormal behavior observed
Quantum Communication	Communication network	Quantum communication modes can be engaged
	Encryption issues	Encryption keys needs to be protected
THz	Abnormal behavior	Signals can be interpreted by attacker when it hits an obstacle
	Authentication	Signatures can be used to ensure the source identity
VLC	Communication	Security protocols used for communication system
	Abnormal behavior	Eavesdropping can compromise the whole network
Molecular Communication	Encryption	Use of coding blocks can be used to train the model and provide encryption
	Authentication	New authentication mechanism models need to be trained
	Abnormal behavior	An attacker can interrupt the molecular communication

BLOCKCHAIN ASSISTED UAV SWARM IN 6G

When it comes to UAV swarm operations in 6G network, blockchain technology (Je, D., 2021) can be used to mitigate various security gaps in the existing 4G and 5G network communication setup. The 6G main key feature is DLT which can support decentralization of the node operations.

The existing challenges in UAV swarm in 4G/5G networks is that it operated in a centralized manner where the command and control rely on a single point that acts as a base station. And the coverage and the commands are issued from the base station that manages the whole UAV operation. So, when the UAV is out of coverage from the base station and when the UAVs are out of alignment, the communication can be interrupted. And the UAVs needs to be authenticated when participating in a network communication. The identity of each UAV should be authorized by the base station in the network. So, there are many gaps and challenges in UAV swarm. So, the 6G network with its technologies can be incorporated to mitigate these issues.

Blockchain Technology in 6G

The Unmanned Aerial Vehicles when assisted by blockchain network can overcome the identity issue and the authentication mechanisms since they are decentralized and it supports the network scalability.

Blockchain network is a distributed ledger where the records are recorded with the timestamp when it enters the network. So, this allows to authenticate the data and avoid tampering of the data. Here each block of data is provided with a hash value and the next block is interlinked with previous block hash. So, this allows to maintain the integrity of the data. Since, even if that unauthorized user tries to modify the data in the block, the hash value gets changed which will be automatically rejected by the network.

And there are consensus algorithms provided by blockchain which allows the peer nodes to authorize a new node when it tries to join the network. And whenever a new node tries to enter the network, the existing nodes in the network throws a puzzle to the new node and it needs to solve the puzzle to join the network. If the new node is malicious, the attacker needs massive computation power to solve the puzzle and to join the network.

Figure 5. Blockchain assisted UAV swarm

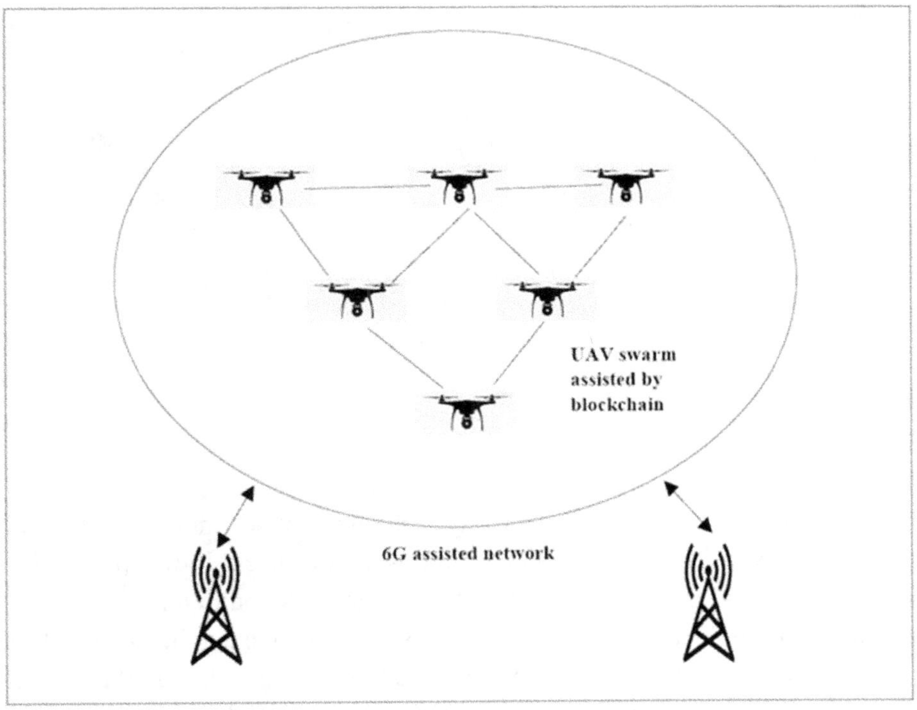

Unmanned aerial vehicles when operating as a swarm needs to communicate with each other and transfer the data among themselves. When the UAV swarm is deployed in 6G network, the DLT technology makes the network decentralized and each node is responsible to authenticate the other nodes, And the decentralization concepts support the UAV operations to be carried out effectively even in a real-time scenario. In a trusted UAV swarm network, each UAV is provided with a valid encryption key which allows it to authenticate the network and communicate securely. So, when a malicious UAV node try to enter the network, it cannot join the network since it doesn't have a valid key pair to communicate. So, it is automatically rejected by the network. The scenario is depicted in Figure.6. The availability of the network is ensured since the records are distributed in each UAV nodes and the availability of the network is enhanced by using blockchain technology. So, the 6G network aided blockchain UAV swarm can operate securely and can mitigate the security attacks. But some of the issues are still prominent is 6G network and its difficult to mitigate. Since the blockchain also suffers with few security attacks where if the attacker gain access to 50% of the network, there is a chance to compromise the network. And in blockchain there are various categories that supports security since it uses consortium blockchain.

Figure 6. Trusted UAV swarm

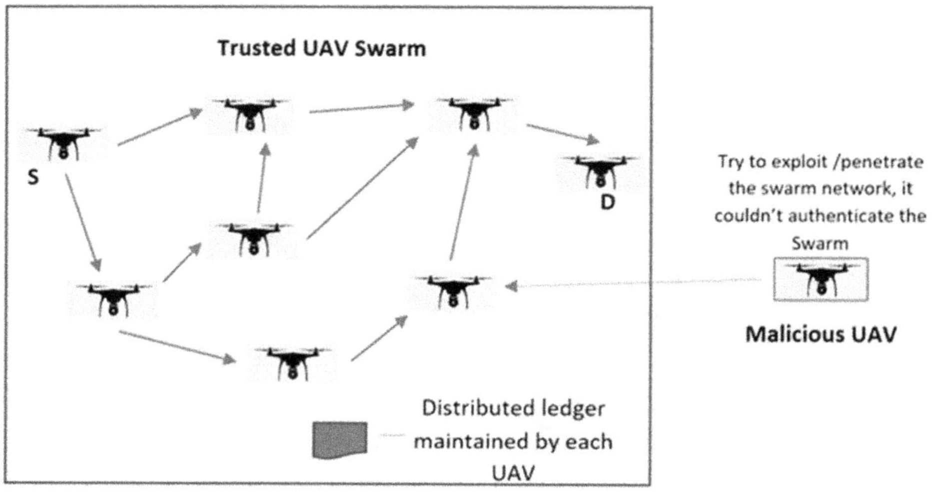

CONCLUSION

In the era of 6G network though it possesses many technological advantages when compared to the other communication network such as 4G/5G, still it has its drawback. In this paper, the evolution of communication networks and the security issues involved in each communication era was discussed. And the enhancement of UAV swarm in 6G network was elaborated in detail. And the discussion of 6G technologies and its behaviors are discussed. And the study shown how the limitations of UAV swarm can be fixed by using blockchain technology in 6G network. And since the 6G network is under research study, still there are many potential areas that needs a growth and in next ten years the full potential of 6G will be evolved and it will be capable of mitigating the gaps in existing network.

REFERENCES

Brito, A., Sebastião, P., & Souto, N. (2019, May). Jamming for Unauthorized UAV Operations-Communications Link. In *2019 International Young Engineers Forum (YEF-ECE)* (pp. 94-98). IEEE.

Gupta, A., Fernando, X., & Das, O. (2021, June). Reliability and availability modeling techniques in 6g iot networks: A taxonomy and survey. In 2021 International Wireless Communications and Mobile Computing (IWCMC) (pp. 586-591). IEEE.

Gupta, R., Kumari, A., Tanwar, S., & Kumar, N. (2020). Blockchain-envisioned softwarized multi-swarming uavs to tackle covid-i9 situations. *IEEE Network*, *35*(2), 160–167.

Hewa, T., Gür, G., Kalla, A., Ylianttila, M., Bracken, A., & Liyanage, M. (2020, March). *The role of blockchain in 6G: Challenges, opportunities and research directions. In 2020 2nd 6G Wireless Summit (6G SUMMIT)*. IEEE.

Hu, J., Chen, C., Cai, L., Khosravi, M. R., Pei, Q., & Wan, S. (2021). Uav-assisted vehicular edge computing for the 6g internet of vehicles: Architecture, intelligence, and challenges. *IEEE Communications Standards Magazine*, *5*(2), 12–18.

Javad-Kalbasi, M., & Valaee, S. (2021, June). Re-configuration of UAV Relays in 6G Networks. In *2021 IEEE International Conference on Communications Workshops (ICC Workshops)* (pp. 1-6). IEEE.

Je, D., Jung, J., & Choi, S. (2021). Toward 6G Security: Technology Trends, Threats, and Solutions. *IEEE Communications Standards Magazine*, *5*(3), 64–71.

Kishk, M., Bader, A., & Alouini, M. S. (2020). Aerial base station deployment in 6G cellular networks using tethered drones: The mobility and endurance tradeoff. *IEEE Vehicular Technology Magazine*, *15*(4), 103–111.

Liu, C., Feng, W., Chen, Y., Wang, C. X., & Ge, N. (2020). Cell-free satellite-UAV networks for 6G wide-area Internet of Things. *IEEE Journal on Selected Areas in Communications*, *39*(4), 1116–1131.

Liu, R., Liu, A., Qu, Z., & Xiong, N. N. (2021). An UAV-Enabled Intelligent Connected Transportation System With 6G Communications for Internet of Vehicles. *IEEE Transactions on Intelligent Transportation Systems*, 1558–0016.

Liu, Y., Wang, C. X., Chang, H., He, Y., & Bian, J. (2021). A Novel Non-Stationary 6G UAV Channel Model for Maritime Communications. *IEEE Journal on Selected Areas in Communications.*

Luo, C., Miao, W., Ullah, H., McClean, S., Parr, G., & Min, G. (2019). Unmanned aerial vehicles for disaster management. In *Geological disaster monitoring based on sensor networks* (pp. 83–107). Springer.

Lv, Z., Qiao, L., & You, I. (2020). 6G-enabled network in box for internet of connected vehicles. *IEEE Transactions on Intelligent Transportation Systems*.

Maddikunta, P. K. R., Hakak, S., Alazab, M., Bhattacharya, S., Gadekallu, T. R., Khan, W. Z., & Pham, Q. V. (2021). Unmanned aerial vehicles in smart agriculture: Applications, requirements, and challenges. *IEEE Sensors Journal*.

Mao, B., Tang, F., Kawamoto, Y., & Kato, N. (2021). Optimizing Computation Offloading in Satellite-UAV-Served 6G IoT: A Deep Learning Approach. *IEEE Network*, *35*(4), 102–108.

Masaracchia, A., Li, Y., Nguyen, K. K., Yin, C., Khosravirad, S. R., Da Costa, D. B., & Duong, T. Q. (2021). UAV-enabled Ultra-Reliable Low-Latency Communications for 6G: A Comprehensive Survey. *IEEE Access*.

Matti, E., Johns, O., Khan, S., Gurtov, A., & Josefsson, B. (2020, October). Aviation Scenarios for 5G and Beyond. In *2020 AIAA/IEEE 39th Digital Avionics Systems Conference (DASC)* (pp. 1-10). IEEE.

Nayak, S., & Patgiri, R. (2021). 6G communication technology: A vision on intelligent healthcare. In *Health Informatics: A Computational Perspective in Healthcare* (pp. 1–18). Springer.

Nguyen, D. C., Ding, M., Pathirana, P. N., Seneviratne, A., Li, J., Niyato, D., & Poor, H. V. (2021). *6G Internet of Things: A Comprehensive Survey. IEEE Internet of Things Journal.*

Porambage, P., Gür, G., Osorio, D. P. M., Liyanage, M., Gurtov, A., & Ylianttila, M. (2021). The roadmap to 6G security and privacy. *IEEE Open Journal of the Communications Society*, *2*, 1094–1122.

Shrestha, R., Bajracharya, R., & Kim, S. (2021). 6G Enabled Unmanned Aerial Vehicle Traffic Management: A Perspective. *IEEE Access: Practical Innovations, Open Solutions*, *9*, 91119–91136. doi:10.1109/ACCESS.2021.3092039

Wang, C. X., Huang, J., Wang, H., Gao, X., You, X., & Hao, Y. (2020). 6G wireless channel measurements and models: Trends and challenges. *IEEE Vehicular Technology Magazine*, *15*(4), 22–32. doi:10.1109/MVT.2020.3018436

Wang, H., Cheng, H., & Hao, H. (2020, October). The use of unmanned aerial vehicle in military operations. In *International Conference on Man-Machine-Environment System Engineering* (pp. 939-945). Springer.

Yang, H., Zhao, J., Nie, J., Kumar, N., Lam, K. Y., & Xiong, Z. (2021, May). UAV-Assisted 5G/6G Networks: Joint Scheduling and Resource Allocation Based on Asynchronous Reinforcement Learning. In *IEEE INFOCOM 2021-IEEE Conference on Computer Communications Workshops (INFOCOM WKSHPS)* (pp. 1-6). IEEE.

Zuhair, M., Patel, F., Navapara, D., Bhattacharya, P., & Saraswat, D. (2021, April). BloCoV6: A blockchain-based 6G-assisted UAV contact tracing scheme for COVID-19 pandemic. In *2021 2nd International Conference on Intelligent Engineering and Management (ICIEM)* (pp. 271-276). IEEE.

Chapter 11
Technological and Non-Technical Challenges Associated With 6G Networks

Anantha Sunil Maligi
BMS College of Engineering, India

Bhavana H. T.
BMS College of Engineering, India

Sanjana T.
BMS College of Engineering, India

Archana H. R.
BMS College of Engineering, India

ABSTRACT

Wireless cellular communication systems are witnessing tremendous improvement in QoS, new features, and new technologies from the past four decades. This leads to realization of advanced mobile phone systems (AMPS/1G), 2G, 3G, 4G, and 5G networks. Each generation, networks have faced potential challenges to cater large coverage, huge connectivity, high data rates, etc. Even though 5G has provided high standard infrastructures, the demand for a huge volume of data is exploding due to billions of IoT devices and multimedia services over mobile networks. Because of remarkable expansion in data traffic and improved digital makeover of industry and society, 6G is focused to provide ultra-broadband and ultra-low latency connectivity for wide range of digital devices and to address several 5G constraints. The implementation of 6G networks faces many technological and non-technical issues and challenges. Some anticipated challenges and potential solutions are discussed in this chapter.

DOI: 10.4018/978-1-6684-3804-6.ch011

INTRODUCTION

A new generation communication systems will be introduced in almost every ten years, these new systems emerge with high Quality of Service (QoS), increased bandwidth, less latency, provides new features and technologies to users. There is huge demand for global mobile Internet and telecommunication traffic due to increased high-definition multimedia applications, simultaneous communication between new featured devices connected over internet and immense demand for data users. Cisco has predicted that the demand for mobile traffic may reach 77 Exabyte per month in 2022 which is seven times greater than 2017 scenario (Tsiropoulos et al., 2017). The important performance requirements for massive machine-type communication (mMTC), enhanced mobile broadband (eMBB) and ultra-reliable and low latency communication (URLLC) are extensively researched in 5G networks (Zaidi et al., 2018). 5G succeeded in catering high standard infrastructure which enables various technologies such as AI, IoT, Mobile broadband communication, smart cities, self driving car etc. However 5G cannot accomplish requirement of increasing usage of smart devices and exponentially growing traffic every year shown in fig. 1 (Tariq et al., 2019). The demand for huge volume of data is exploding due to billions of IoT devices and multimedia services over mobile networks. 5G may not meet the future demands for following reasons

- Massive growth of IoT device deployment require improvement in connection density and 5G network coverage (10 million connections per km^2) (Jiang & Liu, 2017)
- The booming IoT services such as extended reality (XR), telemedicine systems, mind-machine interface (MMI) and flying cars require high transmission rates, reliability and low latency. These criteria cannot be fulfilled by 5G networks (Gupta et al., 2020).
- Architecture of current network systems (4G and 5G) cannot cope up ultra-large scale, high dynamic and complex systems (Zhang, Liang, & Niyato, 2019).
- 5G network cannot support intelligent and deep connectivity (Lu & Zheng, 2020).

Figure 1. Increase in Global Mobile data forecast by ITU

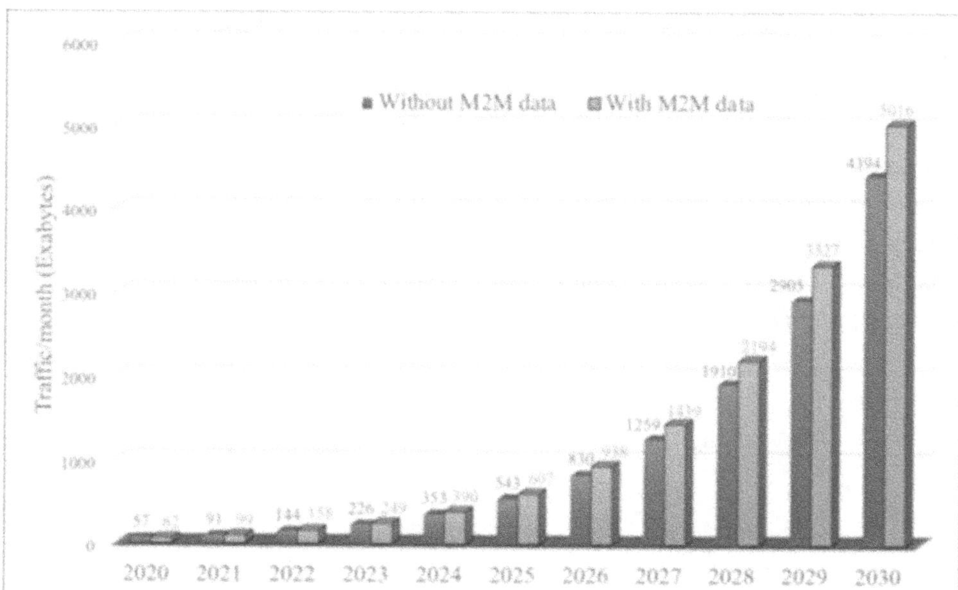

5G communication systems cannot cater to Internet of Everything (IoE). Because of remarkable expansion in data traffic and improved digital makeover of industry and society, 6G network systems are pondered that could provide ultra broadband and ultra low latency with connectivity for wide range of digital devices and to address several 5G constraints. Table 1 hypothesizes possible requirements of 6G networks against 5G networks.

Table 1. Key performance indicators of 6G against 5G (Tariq et al., 2019)

Characteristics	5G	6G
Individual data rate	1 Gbps	100 Gbps
DL data rate	20 Gbps	>1 Tbps
U-plane latency	0.5ms	<0.1 ms
C-plane latency	10 ms	< 1ms
Mobility	Up to 500 km/hr	Up to 1000 km/hr
DL spectral efficiency	30 bps/Hz	100 bps/Hz
Operating frequency	3-300 GHz	Up to 1 THz

6G systems implants novel protocols, architectures, solutions to support to provide intelligent, holographic, ubiquitous and deep connectivity and hold up internet of everything, nano-things and bodies(Marco, 2020; Nawaz et al., 2019). The implementation of 6G networks faces many technological and non technical issues and challenges, some of anticipated challenges and potential solutions are discussed in this article.

Technological Challenges

TeraHertz Band

6G network relies on THz frequency band which ranges from 0.1THZ to 10THz, where the wavelength is typically 300μm and supports Tbps for high speed communication. Even though THz frequency band is exceptional communication in transmission, it faces many challenges (Akyildiz et al., 2014; Elmeadawy & Shubair, 2020; Nagatsuma, 2019; Sengupta et al., 2018).

a) Generation: To generate THz frequency signals, strict and complex design specifications need to be followed. Generation of THz frequency is expensive.
b) Large scale fading: The THz frequency is very sensitive to shadow and can get attenuated easily. The effect of rainfall/humidity attenuation increases with increasing frequency, however after 100GHz attenuation is relatively constant. Frequency dispersion problem can be addressed by designing new multipath channel modules for broad bandwidth.
c) Coverage and directional communication: It need large number of antennas to provide directional communication and cater to larger coverage area. The relevant techniques and mechanisms need to be redesigned and optimized.

d) Fast channel fluctuation: The channel coherence time in the THz band is very small and rate of change of toppler spread is large than cellular networks. Fast adaptive mechanism need to be incorporated to overcome rapid change in service beam and cell association.
e) Energy loss: THz signals attenuates soon to zero after travelling a short distance. Molecular absorption (increases with moisture in the air) and spreading loss (due to extension of Electro Magnetic (EM) waves) are the main reasons for attenuation of THz waves. Amplifying the signal for a short distance is not economical.

Device Capabilities

Apart from the traditional mobile phones and laptops, 6G focuses new communication equipment such as wearable devices, integrated headsets and implantable sensors. The devices supporting 6G network must endure 1 Tbps data rate and high operating frequency. These devices must be incorporated with device to device communication, Artificial Intelligence, Extended Reality to inter operate with different devices (Chowdhury et al., 2019). Energy efficiency at the device level will be a major challenge as 6G user equipment will be power hungry than any other generation devices. Conventional transceiver components made up of Silicon (Si) and Gallium Arsenide (GaAs) are not suitable for 6G devices as they dissipate more heat and not energy efficient. CMOS transistor's design for frequency 300GHz and above operation is extremely difficult and may not support 6G communication features. New materials which can support the need of emerging systems should be exploited through research. Efficient utilization of energy consumption, and wireless/wired charging technologies, device weight, health and safety are the key concerns. These features of devices are challenging and costly (Pon et al., 2019).

Network Security and Privacy

Since 6G networks comprise of huge number of automated smart devices, data security, secrecy and privacy is vital requirement. Authentication, Access control, Malicious behaviour, Encryption and Data Transmission are the main security and privacy problems. The technologies supported by 6G networks such as Visible Light Communication (VLC), real time intelligent edge and radio are vulnerable against malicious behaviours and data transmission process. THz communication suffers from authentication and malicious behaviour issues. 6G applications running on robotics and autonomous systems rely on AI and VLC,where encryption and data transmission could create problems. The advanced technologies such as multi-sensory XR applications have their own security and privacy issues. Block-chain techniques can be implemented to realize complete anonymized, decentralized and untraceable network.Novel security systems with innovative cryptographic techniques need to be considered which also includes physical layer security with low cost, complexity and high security (Yang et al., 2019; Zhang et al., 2017).

Transceiver and Antenna Design

Each wireless communication requires specific transceiver and antenna design. Designing devices of millimetre component was challenging in 5G technology. Since 6G deals with THz high frequency signal, antenna size should in in range of nanometre to micrometre to support holographic beam formation (Al-Hajri et al., 2015; Goian et al., 2015). Meta-surface transceivers can be used which increases throughput and QoS, but integration of OFDM-MIMO with met-surface is more challenging (Tang et al., 2019).

Aesthetic signal processing algorithms should be designed to reduce huge propagation loss at THz frequency. Some of the transceiver parameters such as high power, sensitivity, noise level must be handled at high frequencies. Current transceiver architecture are not capable to deal with medium to high part of the THz signals (>300GHz). A novel transceiver architecture considering nonlinear amplifier, phase noise, and modulation index is essential and graphene nanomaterial used for new terminal architecture.

Artificial Intelligence (AI)

6G network comprises of intelligent devices which run on AI-based algorithms. Knowledge sharing will be carried by federated AI and performance of 6G will be improved by quantum machine learning algorithms (Alsharif et al., 2020). AI algorithms require high computation which requires longer execution times and power. High intelligence leads to utilization of complex devices which in turn increases the cost of network operator and device manufacturing. Since the communication systems are dynamic in nature, the prediction and inference about the future incoming data by AI algorithms become obsolete quickly. AI algorithms require frequent training which will be costly to bear by 6G networks. Implementation of 6G networks in physical layer is also very complex. User privacy, data security, wireless link latency, and capacity are the key concerns of AI based applications running on the edge devices. Decentralized ML algorithms can be used to provide security to user data (Albreem et al., 2020).

Distributed ML like Federated Learning (FL) are used to perform some special tasks such as Edge Intelligence(EI) in 6G networks. ML algorithms in centralized models can work as decision makers for establishing automated network infrastructure and maintenance (resource management) for 6G networks.

Energy Consumption

In 6G network devices at every node has to implement and utilize sophisticated signal processing mechanisms, AI-ML algorithms and have to process Big data which needs high energy. Also network devices will be dealing with huge transmission of data. 6G network has to solve the issues regarding efficient utilization of energy, charging, improvising battery life and energy harvesting by different means. Improvement of energy efficiency of low power devices further and wireless information and power transfer mechanisms (WIPT) are the potential solutions to tackle the energy related issues in 6G (Ng et al., 2019). In 6G communication low energy consumption and long battery life are the main challenges to make communication equipment free from daily charging constraints. Smart Base stations with reliable power supply can be used to in network which reduces the computational load on user gadgets. The power consumption can also be controlled by cooperative relay communications, network densification since these reduces the per-hop signal propagation distance. Energy harvesting of consumer nodes to utilize radio, vibratory and solar energy from surroundings would be applied in 6G. Long distance wireless power charging is also a promising solution to extend battery life of network nodes.

3D Coverage Challenges

The 6G network integrates terrestrial and non-terrestrial technologies, hence collaborative research on different aspects of 3D architectural network needs to be carried. 3D networks are considerably different from 2D network due to new degrees of freedom and dimension of altitude. The low-earth-orbit (LEO) is being used to provide global coverage in 6G communication. Issues related to LEO such as more path loss, longer transmission delay variation in Doppler need to be addressed. Since LEO travels much faster than rotation speed of earth, signal measurement, detection and synchronization creates problems which needs to be solved (Chen et al., 2020). 3D coverage aims to facilitate seamless communication from deep sea to high altitude. Providing quick, broader, reliable and continuous network coverage from telecommunication operator in rural areas and during natural disasters is most challenging job. Efficient Air-to-ground channel modelling, network topology planning and optimization methods for BSs or communicating nodes are to be designed. Energy management, mobility management, resource and communication track management in 3D network are crucial in 6G communication.

Tactile Internet Challenges

In shared real time devices of 6G network, control, touch, communication, communication and sensing capacities have to integrated together. This incorporation is an important issue since tactile internet is still in preliminary stage to address physical layer challenges. Some of the physical layer challenges include user plane separation, intelligent control, modulation shames, waveform selection have to be addressed. The signalling overhead and air interface latency issues can be addressed by finest waveform selection algorithms, intellectual control plane and strong modem techniques. Implementation of Adaptive network coding technologies and scalable routing algorithms can reduce end to end delay. Efficient security mechanisms to deal with malicious activities and adaptive network coding schemes should be designed (Dang et al., 2020; Gui et al., 2020).

Network Management

6G is an important platform for IoT devices, extensive IoT applications make 6G networks more challenging to deal with Management and Orchestration (NMO) and Network Traffic Monitoring and Analysis (NTMA) techniques (Shahraki et al., 2021). These techniques organizes network infrastructures such as fault management, network configuration management, performance and security management. The network QoS and QoE will be affected by these techniques. Major issues such as D2D facilitated 6G and considerable IoT supported 6G networks need to be addressed in future research works.

Peak Rate

Peak rate is an important measuring index for wireless communication systems, it is major technical indicator from first generation systems itself. 6G networks will further increase the peak rate and the Terabit (Tb/s) starts. Two applications require large peak rates to operate efficiently.

i) Smart applications need huge amount of data transmission requirements. Intelligent applications which run on Big data are key driving forces of next generation communication systems and need high speed in data transmission.
ii) AR/VR and holographic communication are the significant applications of 6G networks which require fast data transmission than any other wireless applications (Zhang et al., 2020; Zhang, Xiao, Ma et al, 2019).

Nontechnical Challenges

Industry Barriers

6G reaches to diverse aspects of social life and involves various Intelligent devices/gadgets with novel hardware and software technologies. 6G technology has to be integrated with many industries other than telecommunication based industry as it serves various phases of social production and life. Since traditional mobile communication based industries are restricted to its own field and do not open up for new industries, this may create industrial barriers (Lua & Zheng, 2020).

Policies and Spectrum Allocation

To 6G architecture needs different frequencies in unlicensed bands and utilize high frequency spectrum. It requires high speed spectrum in multiple bands such as combination of 1-3 GHz band, millimetre wave (30-300GHz) and TeraHertz Band (006-10THz). Unified spectrum allocation should be carried across different nations and areas in the world. The spectrum usage rules and coordination with customers of other phases of spectrum link weather radars has to be considered (Yang et al., 2019). The policies and guidelines with trajectory and spectrum resources of satellite communication should be sorted among the countries with proper consultation. Global roaming switching, coordination between satellite communication constructed and deployed by other countries is a challenging issue.

Green 6G

The primary goal of next generation wireless network is green computing and green communication. This is basically reduce overall power consumption, operating cost and to create positive effect on ecological and economical environment. Because of huge data rate, high power consumption, large number of operating devices and network covering equipments achieving green 6G network is a big challenge. Usage of TeraHertz waves may affect the human safety as these waves are non-ionizing due to insufficient photon energy. The waves may affect the highly sensitive organs of human body and must meet the International Commission on Non-Ionizing Radiation Protection Standards. Consideration of Electromagnetic radiations from 6G communication gadgets also need to be considered.

Geographical Constraints

Huge number of small/smart devices will be connected to internet in 6G networks which require service from multiple Access points and Base stations. Providing service to millions of devices within a small geographical region, overlapping coverage, multipoint transmission is challenging. In this scenario. This can be addressed by frequency allocation management, interference management and efficient handoff techniques.

Political and Social Factors

6G networks worldwide is creating geopolitical issues of raise in power over future recourse planning and utilization. Some of the questions related to new generation technology owning the accumulated data, deciding on technology, setting the rules and regulations need to be sorted. People of developing countries are not likely to be united because of lack of content relevance, language barriers and usage of modern technologies.

REFERENCES

Akyildiz, I. F., Jornet, J. M., & Han, C. (2014). Terahertz band: Next frontier for wireless communications. *Physical Communication*, *12*, 16–32. doi:10.1016/j.phycom.2014.01.006

Albreem, M., Alsharif, M. H., & Kim, S. A. (2020). Low Complexity Near-Optimal Iterative Linear Detector for Massive MIMO in Realistic Radio Channels of 5G Communication Systems. *Entropy (Basel, Switzerland)*, *22*(4), 388. doi:10.3390/e22040388 PMID:33286162

AlHajri, M., Goian, A., Darweesh, M., AlMemari, R., Shubair, R., Weruaga, L., & Kulaib, A. (2015). Hybrid rss-doa technique for enhanced wsn localization in a correlated environment. In *2015 International Conference on Information and Communication Technology Research (ICTRC)*. IEEE. 10.1109/ICTRC.2015.7156466

Alsharif, M. H., Kelechi, A. H., Yahya, K., & Chaudhry, S. A. (2020). Machine Learning Algorithms for Smart Data Analysis in Internet of Things Environment. *Taxonomies and Research Trends. Symmetry*, *12*, 88.

Chen, S., Liang, Y., Sun, S., Kang, S., Cheng, W., & Peng, M. (2020). Vision, requirements, and technology trend of 6g: How to tackle the challenges of system coverage, capacity, user data-rate and movement speed. *IEEE Wireless Communications*, *27*(2), 1–11. doi:10.1109/MWC.001.1900333

Chowdhury, M. Z., Shahjalal, M., Ahmed, S., & Jang, Y. M. (2019). *6g wireless communication systems: Applications, requirements, technologies, challenges, and research directions*. arXiv preprint arXiv:1909.11315.

Dang, S., Amin, O., Shihada, B., & Alouini, M. S. (2020). What should 6G be? *Nature Electronics*, *3*(1), 20–29. doi:10.103841928-019-0355-6

Elmeadawy, S., & Shubair, R. M. (2020). *Enabling Technologies for 6G FutureWireless Communications: Opportunities and Challenges*. arXiv:2002.06068.

Goian, A., AlHajri, M. I., Shubair, R. M., Weruaga, L., Kulaib, A. R., AlMemari, R., & Darweesh, M. (2015). Fast detection of coherent signals using pre-conditioned root-music based on toeplitz matrix reconstruction. In *2015 IEEE 11th International Conference on Wireless and Mobile Computing, Networking and Communications (WiMob)*. IEEE. 10.1109/WiMOB.2015.7347957

Gui, G., Liu, M., Tang, F., Kato, N., & Adachi, F. (2020). 6G: Opening new horizons for integration of comfort, security and intelligence. *IEEE Wireless Communications*, *27*(5), 126–132. doi:10.1109/MWC.001.1900516

Gupta, R., Shukla, A., & Tanwar, S. (2020, December). BATS: A blockchain and AI-empowered drone-assisted telesurgery system towards 6G. *IEEE Transactions on Network Science and Engineering*.

Jiang & Liu. (2017). An overview of 5G requirements. In *5G Mobile Communications*. Springer.

Lu, Y., & Zheng, X. (2020). 6G: A survey on technologies, scenarios, challenges, and the related issues. *Journal of Industrial Information Integration*, *19*, 100158. doi:10.1016/j.jii.2020.100158

Lua, Y., & Zheng, X. (2020). 6G: A survey on technologies, scenarios, challenges, and the related issues. *Journal of Industrial Information Integration*, *19*, 100158. doi:10.1016/j.jii.2020.100158

Marco, G. (2020). Toward 6G networks: Use cases and technologies. *IEEE Communications Magazine*, *58*(3), 55–61. doi:10.1109/MCOM.001.1900411

Nagatsuma, T. (2019). Advances in Terahertz Communications Accelerated by Photonics Technologies. *Proceedings of the 2019 24th International Conference on OptoElectronics and Communications Conference (OECC)*, 1–3.

Nawaz, S. J., Sharma, S. K., Wyne, S., Patwary, M. N., & Asaduzzaman, M. (2019). Quantum machine learning for 6G communication networks: State-of-the-art and vision for the future. *IEEE Access: Practical Innovations, Open Solutions*, *7*, 46317–46350. doi:10.1109/ACCESS.2019.2909490

Ng, Duong, Zhong, & Schober. (2019). *Wireless information and power transfer: Theory and practice*. John Wiley & Sons.

Pon, Leow, Rahim, Eteng, & Kamarudin. (2019). Printed spiral resonator for displacement-tolerant near-field wireless energy transfer. *IEEE Access, 7*.

Sengupta, K., Nagatsuma, T., & Mittleman, D. M. (2018). Terahertz integrated electronic and hybrid electronic–photonic systems. *Nature Electronics*, *1*(12), 622–635. doi:10.103841928-018-0173-2

Shahraki, A., Taherkordi, A., & Haugen, Ø. (2021). Tonta: Trend-based online network traffic analysis in ad-hoc iot networks. *Computer Networks*, *194*, 108125. doi:10.1016/j.comnet.2021.108125

Tang, W., Chen, M. Z., Dai, J. Y., Zeng, Y., Zhao, X., Jin, S., Cheng, Q., & Cui, T. J. (2019). *Wireless communications with programmable metasurface: New paradigms, opportunities, and challenges on transceiver design*. arXiv:1907.01956.

Tariq, F., Khandaker, M., Wong, K.-K., Imran, M., Bennis, M., & Debbah, M. (2019). *A speculative study on 6g*. arXiv:1902.06700.

Tsiropoulos, G. I., Yadav, A., Zeng, M., & Dobre, O. A. (2017). Cooperation in 5G HetNets: Advanced spectrum access and D2D assisted communications. *IEEE Wireless Communications*, *24*(5), 110–117. doi:10.1109/MWC.2017.1700082

Yang, P., Xiao, Y., Xiao, M., & Li, S. (2019). 6g wireless communications: Vision and potential techniques. *IEEE Network*, *33*(4), 70–75. doi:10.1109/MNET.2019.1800418

Zaidi, A. A., Baldemair, R., Moles-Cases, V., He, N., Werner, K., & Cedergren, A. (2018, June). OFDM Numerology Design for 5G New Radio to Support IoT, eMBB, and MBSFN. *IEEE Communication Standards Magazine*, *2*(2), 78–83. doi:10.1109/MCOMSTD.2018.1700021

Zhang, L., Liang, Y.-C., & Niyato, D. (2019, August). 6G visions: Mobile ultrabroadband, super internet-of-things, and artificial intelligence. *China Communications*, *16*(8), 1–14. doi:10.23919/JCC.2019.08.001

Zhang, S., Liu, J., Guo, H., Qi, M., & Kato, N. (2020). Envisioning device-to-device communications in 6G. *IEEE Network*, *34*(3), 86–91. doi:10.1109/MNET.001.1900652

Zhang, W., Ding, D.-S., Sheng, Y.-B., Zhou, L., Shi, B.-S., & Guo, G.-C. (2017). Quantum secure direct communication with quantum memory. *Physical Review Letters*, *118*(22), 220501. doi:10.1103/PhysRevLett.118.220501 PMID:28621985

Zhang, Z., Xiao, Y., Ma, Z., Xiao, M., Ding, Z., Lei, X., Karagiannidis, G. K., & Fan, P. (2019). 6G wireless networks: Vision, requirements, architecture, and key technologies. *IEEE Vehicular Technology Magazine*, *14*(3), 28–41. doi:10.1109/MVT.2019.2921208

Chapter 12
Defending IoT Security Infrastructure with the 6G Network, and Blockchain and Intelligent Learning Models for the Future Research Roadmap

Janani K.
SRM Institute of Science and Technology, India

Ramamoorthy S.
SRM Institute of Science and Technology, India

ABSTRACT

The battle for wireless networks' 6G has started. Many review articles expressed their views and critically evaluated the overall state of the art in order to give readers a clear picture of current advances. Most of the studies, though, are focused on broad views and a big-picture perspective, and do not address the real issues arising from internet of things (IoT) use cases. The authors give a sample of IoT use cases that are indicative of a wide range of IoT implementations, which is a novel approach in the review. The use cases chosen represent the most research-intensive industries that potentially profit from 6G and its enabling technologies. Healthcare, smart grid, transportation, and blockchain learning techniques are among these sectors. They also identified some of the actual hurdles and lessons encountered in putting these use cases into action. The evaluation focuses on the major requirements of the scenarios and how they intersect with the primary drivers for the next generation of wireless networks.

DOI: 10.4018/978-1-6684-3804-6.ch012

INTRODUCTION

Smart infrastructure focusing to provide effective solutions. Many Smart city improve this both government and private organization emerge to deployment of ICT (information communication technology) to find rational effective solutions to the highest list of challenges facing in the smart city (Syed A. S, 2021). IoT device has more increasing the count of heterogeneous devices interconnect with the Internet. Main challenges is to given safe and security to these devices, which work with lower energy, constraint resource, communication protocol, geographical devices. IOT devices are free network access, very easy to pull any hackers. The combination of these devices with build and flexible network can easily intrusion in the IoT network.

Internet of Things paradigm will entail billions of smart devices with computing, monitoring, and actuation abilities that can be hooked up to the internet in the long term. IoT privacy concerns are significant challenges, and they're also key enablers in the creation of a "trust and accessible ecosystem." The dearth of protocols built specifically for resource constrained devices and heterogeneous capabilities is stressed. Wireless sensor networks include Wi-Fi, Zigbee, LoRaWAN, Bluetooth, MQTT, CoAP, Z-Wave, and 6LoWPAN connect IoT devices (Salman,2019). These techniques have limited security because anybody can monitor the signal as it transmits. As a result, in IoT systems, secured data transfer is critical.

IoT is the extreme complexity and huge security vulnerabilities of IoT devices to the internet, the number of attacks and vulnerabilities is rapidly expanding. From the other hand, due to the limits of Devices, implementing preventive measures is becoming more challenging. Restricted memory, low computing power, and a low battery life are among the drawbacks. At this time, attackers are focusing their efforts on IoT devices with nefarious intent. Hackers, computer hackers, and fraudsters could all be involved in the attack (Husamuddin, 2017). The control measure against acts including such security breaches, unauthorized access, physical manipulation, falsification, and internet attacks is classified as security in IoT. As a result, security is critical for the IoT business to function properly. The network is only as strong as its most vulnerable connection. As a result, expanding the number of components in the network increases the network's susceptibility. Because of the complex and heterogeneous architecture, implementing security in IoT infrastructure is tough. As a result, it becomes difficult for developers to keep track of many hardware and software standards. However, many IoT frameworks and standards have recently been developed to assist developers in designing devices for various consumer needs (Abosata, 2021).

A blockchain is a secured and decentralized storage system. This is finding success in different of fields. There have been a few important topics which are still being researched when blockchain system is introduced inside the Internet of Things (IoT) field to share and exchange network traffic, records, confirmation, and secret services, with a special focus just on security of internet systems inside the IoT industry. It gets its derived from the fact that it's made up of valid transactions, every one containing a hash value of the one before it. To safely replicate a programmer across several entities, a blockchain uses encryption, smart contracts, and consensus methods (Wang, 2020). PoW- Proof of Stability PoS, and other consensus techniques. The Byzantine Fault Tolerance Algorithm and also the Byzantine Fault Tolerance Protocol ensure each entity records a certain transaction in same order. Smart contracts are pre-defined processes that are kept inside the blockchain. When group form transactions, they execute them immediately.

Figure 1. IoT smart infrastrcture network with blockchain security

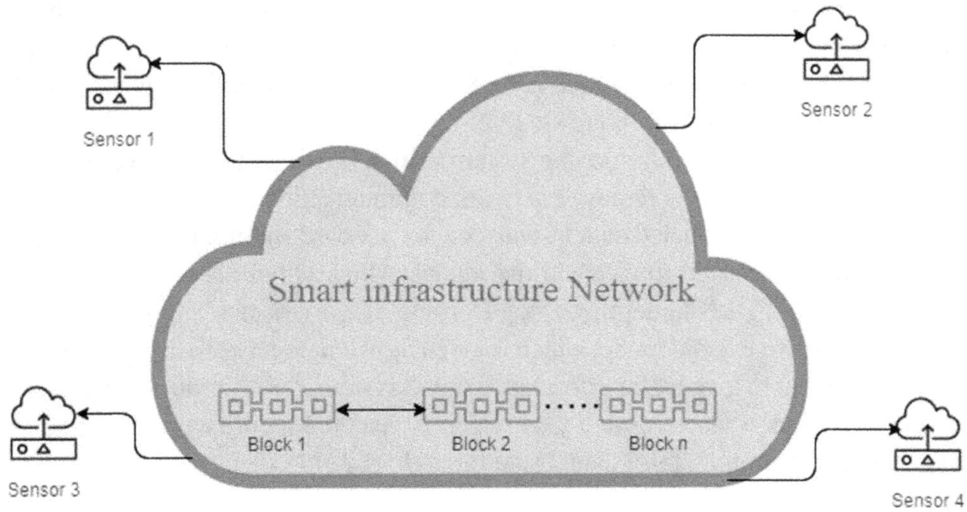

Blockchain is named after how transactions are store blocks of data's, which is encrypted data by hashing part of previous (n-1) block. Blockchain is unchangeable storage blocks can only be included. A P2P(peer-peer) network communication for the blockchain of records, permitting request to access the data contained in each record by transmit the entire structure to all the nodes used on a blockchain consensus algorithm to give the solution to more secure IoT communication devices (Lei, K, 2018). Other than more secure to apply consensus with hash encryption to given safety interconnected with blocks together.

The block is group of valid transaction. sensor validate the communication with old transaction once the verification is completed next initiate insert to the existing BC. Every transaction can occurred to that time will stored into the block according to the transaction. Which is able to stow data without relying on a third party. The mean of "BCT" begin from its structure blockchain, IoT is interlinked of smart devices to collect information and change decisions (Khacef, 2019). BCT is essentially a best complement to internet of things with the improving privacy and security and interoperability, flexibility, scalability.

Xie, J (2019) The blockchain followed by hash function i is hash of block i-1, The initial block is called genesis block having there is no parent block a BC including following steps: 1) The block validation version rules are followed 2) Parent block in the hash function 3) timestamp (TS) storing the current time 4) nonce begin from 0 and increasing for each block hash calculation.5) transaction number calculated 6) Merkleroot is whole transaction hash value in the block. 7) each block having a hierarchy from the previous transaction block Figure 1.

Deep learning is most promising approach in current years. Structuring an efficient deep learning model, Blockchain Technology (BCT) is a various distributed encrypted ledger on the peer to peer (P2P) process model and digital signature encryption key that allows verify users' transaction next that added to a block of the system, there is no involve any intermediates and third party, BCT can support B2B, B2C, and C2C communication (Zheng, Z, 2017).

They can design on end-to-end security for IoT devices.RNN together a temporal layer to bring sequential things and then take to learn multi-faceted different with the hidden the recurrent cell, which is classified in high network collision with high accuracy in reducing malicious action which is potentially improving IoT security system.AE's which is very important of feature extraction which is increasing knowledge various way, AE hidden layer this is have a two part h=f(x) encoder function and decoder function this is given the output r=g(h) which is detecting malware this is combination of DBN to improve a malware removing method (Kimbugwe, 2021).

RBMs using for anomaly network removing system, which is extract from RBM more vital features via an unsupervised model. DBNs is removed network communication malicious attack, DBNs are the unsupervised learning model unlabeled data in-between for standard significant features representation. GANs produced by only one need to pass via the model, which is built securing an architecture the cyberspace of IoT system. EDLN's merging Deep learning classifiers can help give successful model higher performance and give expand model which is given light heterogeneous and homogeneous classifiers is shared environment to give high performance of the system. DRL permit to learn agent to adjust policy given optimal solution via error and trial which is detected security problem like DoS/DDoS, data injection, spoofing, jamming, false data, brute force (Lin, 2017).

CLASSIFICATION OF ATTACKS RELATED TO IoT INFRASTRUCTURE

Smart Infrastructure IoT attacks

As shown Figure 2 In smart infrastructure worldwide, billions of interconnected 'things' are implemented. The growth of internet of Things (IoT) has exposed a slew of vulnerabilities which malicious hackers and other hostile entities can attack. Smart infrastructure is intended to boost production and efficiency, but if information technology is ignored, it can pose major hazards to citizens and government (Aversano, 2017). There's an unknown number of possible vulnerabilities and approaches; nonetheless, the following are some of the most popular threats:

Figure 2. Smart infrastrcture attacks and countermeasures

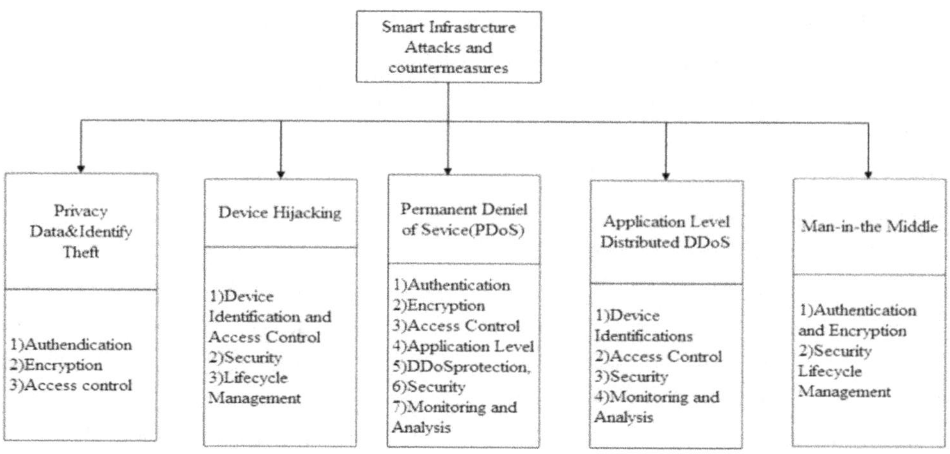

DOS/DDOS Attack

DoS is functional related attack, The device node is alive and active but there is no response to the end nodes, this attacker's exchange data monitor stream to the affected keeps network collision high, and eventually the permitted nodes can't communicate with one another. Entire node keep node busy with high traffic make delay by the responding (Gavra, 2020). Here some authorized node tries to communicate each other due to huge traffic cause node request is denied. DoS attack shutdown system permanently not allow to access to anyone.

Replay Attack

This attack is launched to hack on perception layer in internet of things, when one or many networks are using similar identities in one system. This threat mainly affected system privacy, the IoT device is spoofed without having trust access can be successful (Zhang 2014). The consequences of the attack are interlinked but system confidentiality information has more compromised which is not taken easily. The solution sates are two possible ways:1) prevention using nonce or timestamp or by protected using nonce, this is detected spoofed nodes in IoT.the trust.

Sybil Attack

Sybil nodes launched the attack, and these routers generated a copy with the intention of communicating data with another network. This duplicate node physically there at one node at same time virtually there have one to multiple location (Haseeb 2020), Sybil is act which look like authorized user but it's not. The basic network station only votes once, however this Sybil threat node has the ability to vote several times.

Malicious Code Injection

The computer malicious code attacks involve altering lines of code and change it with programming code may damage the end user's computer. IoT wearable devices affected by injected malicious program code allowing hackers to exploit the computer vulnerability. The attackers are threatening (Liang, 2016).

Proposed solution this attack RWFS- Randomized Watermarking Filtering Scheme for IoT is a technology to overcome this attack, very early technology used for en-route filtering find eliminate the malicious data. Homomorphic techniques are help to reduce size of pocket so that good security can be achieved using less power (Wara, 2020).

Malware

Malware is most harmful and adjust the confidentiality, availability and integrity of data in the computer, the programming code created purposefully to undertake without power activity which chance of benefits to its propagator. And malware recovery is processing now, but still, which is very complexity to detecting malware.

Ex (Virus, Trojans, Worms, Rootkits, Spyware, Botnet)

Timing Attack

This is a different form of spying attack in which hackers attempt to steal specific data from the system. The information on the whole encryption key, with a focus on the key encryption time. The attackers initiate this timing assault on the presentation layer with the goal of stealing information from the device.

Solution for a timing attack that has been proposed Ring LWE has been increased to bit checking to reduce branches in additive modules; by removing branches, there is a lower probability that the key will be altered, making this technology much more secure against select plain text attacks.

Sinkhole Attack

This sinkhole is adjusting node attack, making this node more appealing and accessible to its neighbors. By enabling the collision of all neighbor nodes to go to it, the duplicate node network is illegally sending data to neighbor routes. Every data sender believes that data has been transferred successfully if they receive the discards. This output saves energy and reduces network collisions, but it does not reach the end point destination in real time (Groves, 2019).

Man in the Middle Attacks

The Man in the Middle attack is one of the most widely used WSN attacks, with the primary purpose of stealing critical information (data) from IoT devices. MIM attack is when a hacker manipulates all of the information to send data from one device or sensor to another. At first, attackers just sniffed the wireless network, but as they learn more about the system, they may mimic an authorized user. The MIM starts on the computer's WSN, which is where the data is stored. The proposed solution necessitates the use of a valid digital signature that authorizes all nodes in the system, as well as an external continuous monitoring mechanism that will protect our devices from these attacks (Ngo, Q. D.2020)

Botnet Attack

This attack acting a connection of network communication devices defected by malware which allow attackers to control the systems, these attacks included all kind of virus activity in this botnet. This is continue looping method, which can easily steal the data, DDOS attack can be execute (Prates, 2020) N. System in the IP address exchanged into bot, this is found more virus and vulnerability nodes and change into bot and group of botnets at the end system will loss of confidentiality during this attack Figure 3.

IoT WITH BLOCKCHAIN WITH 6G BASED APPLICATIONS

In 6G-enabled IoT, a classification of distributed ledgers is provided. Smart factory, supply chain management, farming industry, power home, healthcare, media and internet freedom managerial staff, farming, and smart cars and unmanned aircraft were among the traditional areas of interest for various objectives. With low operating and capital expense, blockchain and 6G heightened security and throughput. Figure 4 illustrates the blockchain use for 6G-enabled IoT. These subsequent sections provide a more extensive explanation.

Figure 3. IoT system architecture

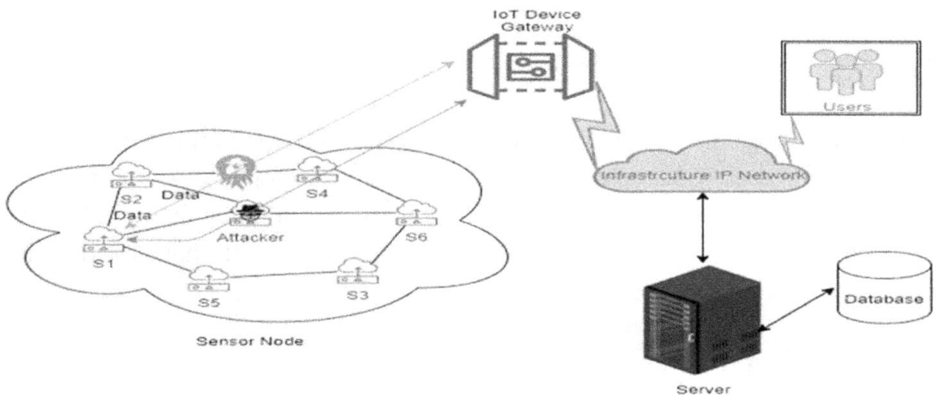

Figure 4. IoT application based implementation 2021 Survey

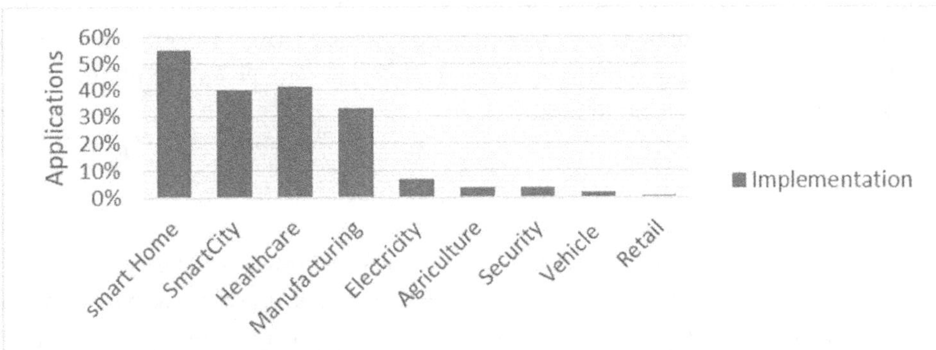

Smart Maritime

This paper by installing a private Blockchain network that is linked to a control center in the earthly region, we include an authentication mechanism blockchain based to provide secure communications for communication systems aided UAV sensor module for maritime IoT critical applications. Because of their ability to detect data and improve coverage area, UAVs have the ability to be extensively used in important IoT systems. Securing UAV connection with the relevant network is regarded as among the most difficult solutions to fix. To protect the data transfer of a communication systems aided UAV sensor module for marine IoT systems, we introduce an authentication technique based on Blockchain..

Smart Home

Chronic illness develops over time and is extremely important since it can have a greater effect on the health of persons with disabilities. Since they do not have signs, most people believe they are not afflicted or afflicted by the disease. However, this does not rule out the possibility of a chronic condition in someone who has no symptoms. The design phase execution of a blockchain system for detecting different patient medical records are presented in this paper.

Smart City

The goal of smart city development is to improve the standard of living for those who live there. Cloud - based IoT solutions were used to accomplish this goal. Blockchain technology is another viable idea that can provide a plethora of useful services to its customers. It's an immutable, digital electronic register that's used to keep track of things. Virtual goods with some worth were developed mainly for digital currencies like bitcoin.

Smart Manufacturing

The demand for predictive machine delivery and safety is driving the growth of smart manufacturing systems. To that goal, a variety of machine learning methods are being investigated. Data protection and monitoring is also another issue that is becoming increasingly significant in the sector. To address the issues raised above, we used a combination of blockchain and machine learning techniques to protect system operations and manage a dataset to combat the false dataset. Big data approaches were employed to organize and evaluate the obtained dataset. The private Hyperledger Fabric platform has been used to create a blockchain network (Kocakulak, 2017).

Smart Grid

The environment is fast turning away from fossil fuels and toward renewable energy (e.g., smart grid). The smart grid's successful implementation necessitates attack- and leak-proofing, as well as decentralization of the energy distribution network to provide openness in the smart grid system. This modernized energy system, though, poses a myriad of obstacles. Single-point failure owing to a central approach, sluggish payment services, and technologies concerns are all factors to consider. We developed a revolutionary blockchain-based safety strategy to address every one of these smart grid concerns.

Smart Parking

In congested cities, finding adequate parking places is a serious issue, generating traffic congestion and pollution, as well as squandering motorists' time. Drivers can use parking guidance processes to get real-time parking spaces and reserve parking areas. Drivers must now divulge sensitive data, such as their preferred destinations, in order to use smart parking systems. Furthermore, present methods are centralized, making them prone to bottlenecks, single points of failure, and service provider privacy breaches. We propose a confidentiality and integrity parking system based on blockchain and personal data in this study. Various parking lot owners form a coalition blockchain to secure the privacy, openness, and available parking options. Then, in order to protect the drivers' security, we use a personal information extraction approach to collect parking offers via blockchain nodes secretly. Furthermore, a short mutually interdependent signature is employed to allow users to verify for the purpose of anonymously booking available parking from parking proprietors. Our tests showed that the suggested system protects drivers' privacy while requiring minimal transmission and processing.

Smart Agriculture

Current privacy and security technologies for sustainable IoT-based agriculture. We presented an incorporating green IoT-based agriculture architectural design. We were able to identify the attack scenarios for green IoT-based farming into five groups after doing thorough research and analysis, covering attacks on security, authentication, confidentiality, scalability, and reliability attributes. We also looked at privacy-focused blockchain-based technologies and consensus mechanisms for sustainable IoT-based agriculture. ML approaches, datasets for vulnerability scanning, scalability evaluation of blockchain-based systems, and how to choose the optimum consensus protocol remain tough study fields. And the development of workable and interoperable cryptographic protocols, both of these should be researched more in the coming years.

IoT BLOCKCHAIN PLATFORM DESIGN SUGGESTIONS

IoT Layered Design

Architectural pattern is an input impedance where each layer is isolated from the others, allowing new modules to be fully removed without disrupting the entire system.

Physical Layer

The IoT physical layer is made up of many connected devices that can communicate, compute, and store data.

Connectivity Layer

The connectivity layer's principal role is routing control, which is required since physical devices do not have access to international web services Figure 2.

Service Layer

This service layer contains all components that structure basic services in order to deliver different blockchain technology functionalities such as the distributed ledger is a consensus of copied, shared, and synchronized digital data that's also distributed over the blockchain system and owned by network members. It creates a safe storage capacity for recording the device's design as well as receiving information from sensor sensors. Within minutes, any modifications to the ledger are mirrored in all instances of the blockchain network. If anybody and only authorized users can run a peer to authenticate users, the ledger can be allowed to access or permission less.

Big Data Analytics

The blockchain is supported by the big data module as just an effective mode of online storing data. Many different parties' transaction data is stored in variable stiffness of ledger accounts, making it a suitable source for more research. All of these entities can be granted access to every single connection and have easy access to the data.

Smart Contract

The smart contract is a business process that is called via an outside application server to handle ledger entry and modifications. Every networking peer is downloaded and configured with it.

Event Management

The event planning sends out events when a project is created to a ledger or if the smart contract's specified criterion is fulfilled.

API Interface

The API interface presents the blockchain channel's capabilities as functions that can be used by any client application that has access to and manages the system.

Figure 5. IoT with blockchain layer based secure

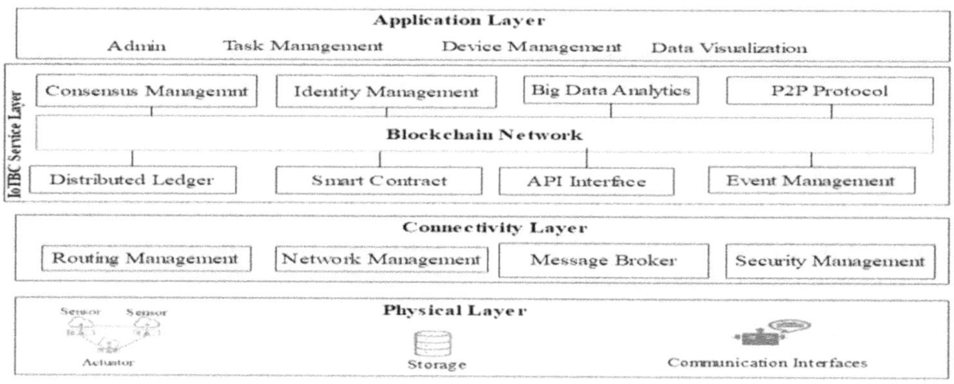

Application Layer

The application layer, which requires multiple connections to display information from sensor devices, as well as control and manipulate them Figure 5.

BLOCKCHAIN CONSENSUS ALGORITHM

This Consensus main core solution of the blockchain technology. This is help repeat node failure ensuring give the best solution to the distributed system. Failure node will divide into two sections: byzantine fault nodes (BF) and crash fault nodes (CF). The BF nodes act arbitrarily. They can able to send fake messages to other nodes or deliver different data to different nodes node fail just by halting; that mean, which can only stop job schedule. There are no malicious attacks right now. This situation data only be lost or delayed .

Proof of work is decentralization system has good scalability, computing energy of nodes decide the attribution of the holder rights of proof of work, Bitcoin, miners' transaction system first time noticed this application computationally-intensive simply verify task in order to create a new blocks using Proof of Work based BlockChain. The Problem is solved the answer is published and automatically new block is created in the chain.

Figure 6. Blockchain structure

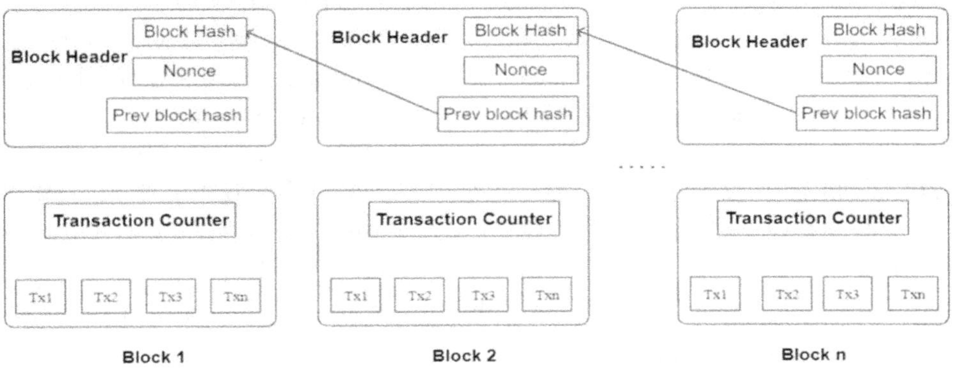

It's crucial that Block 1 and Block 2 are linked by their hash numbers. The importance of this 'hash number' derives from the fact that it links the new block to last valid block on the blockchain. If, on the other hand, the Block 1 Hash number on Block 2 differed from Block 1, the hash numbers will not equal, and Block 2 will not be confirmed. The Block Header is the first block on the network, and it has no previous block hash function. Updating a block necessitates reproducing all successors and performing the work they involve which really is nearly impossible. This prevents tampering with the blockchain Figure 6.

Proof of stack the proof of stake (PoS) consensus protocol was developed as an alternative to the PoW protocol in terms of addressing scalability and sustainable environmental problems. authorized to evaluate the truthful of a block present the users with many stakes are high trustfulness then who with minimum cryptocurrencies. If they not properly validated the users automatically proof of stack for intestinal collusion happened.

As the name implies, nodes on a system stake a certain number of cryptocurrencies in order to become candidates for validating new blocks and earning the charge associated with them. The node that will validate the new block is then chosen by an algorithm from a pool of candidates. To keep the selection fair to everybody on the system, this algorithm mixes the number of stake (amount of bitcoin) with other considerations

Proof of location the consensus in Proof of Location is totally about being able to find a given location and generate a various hash key that can be accepted by other parties once done available publically.

PoL is a blockchain technique that allows a device's specific geographical data to be broadcast so that other machines can trust on the information without blindly trusting the transmitting device. Distribution network, adaptive insurance, KYC verification, and even provided to the customers are examples of leadership role applications. Today's spatial solutions are wonderful for determining a device's position, but they can't provide reliable and trustworthy location verification. Because they are bidirectional and unencrypted, relying on GPS for smart contracts which need to execute when spatial criteria are met is troublesome. Dynamic PoL enables a permission-free and autonomous network of radio transmitters to provide safe location confirmation services via decentralised time synchronization.

Practical Byzantine Fault Tolerance

BFT is a fully distributed characteristic that allows it to attain consensus, even when some nodes refuse to communicate or respond with inaccurate information. The goal of a BFT method is to protect against system problems by utilizing group decision (both right and faulty nodes) with the target of decreasing the influence of node failures. Byzantine Generals' Issue is the source of BFT. pBFT aims to create a viable Byzantine finite state replication that really can work even if the system contains malevolent nodes. In a pBFT-enabled distributed network, nodes are sorted progressively, with one serving as the main (or leader node) or others as secondary (Sun, 2018). It's worth noting that by migrating from secondary to main, any qualified node in the system could become the main. The goal is for all number of nodes to contribute to a majority rule consensus on the status of the system Figure 7.

Figure 7. Practical Byzantine fault tolerance transaction model

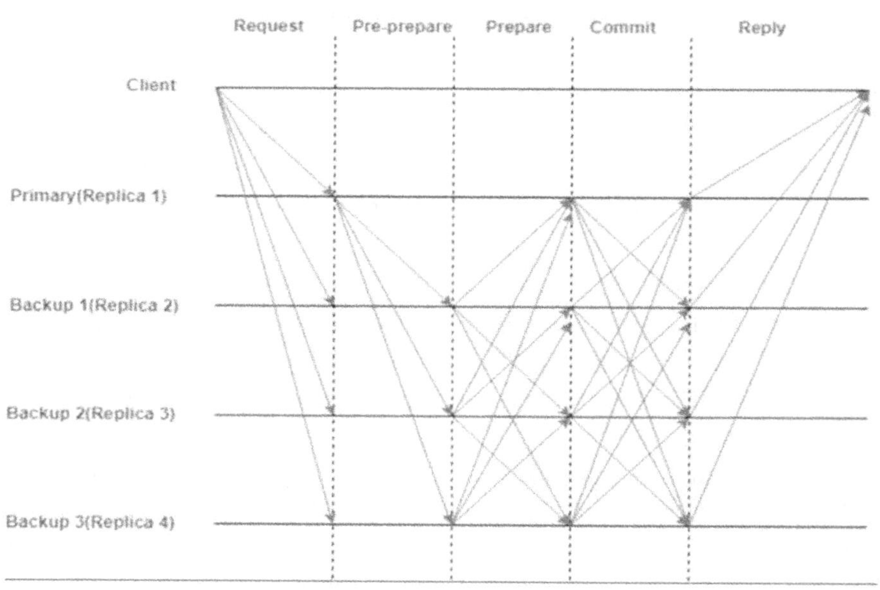

Every view pBFT consensus round changes the primary node, which can be replaced by a perspective shift protocol if a predetermined time has elapsed without such leading node sending a request to the backup systems. If necessary, a plurality of the honest miners can vote just on current leading datatype validity and substitute it with next in order.

Constraints of pBFT

Due to the large communication cost that increases dramatically with each additional node in a network, the pBFT consensus system only works efficiently whenever the number of sensor nodes in the distributed system is minimal. Sybil attacks, in which one entity controls several identities, are vulnerable to the pBFT techniques. Sybil attacks get harder to carry out as the number of nodes in a network grows. However, because pBFT techniques have scalability issues, they are employed in conjunction with the other methods. To its communication cost, pBFT does not scale effectively. The time it takes to react to a request grows as the number of nodes in a network grows where n represents the size of communications and k is the number of nodes.

Proof of Burn (PoB)

Instead, then investing heavily infrastructure, validators in PoB 'burn' bitcoins by transferring them to a location in which they are permanently lost. Auditors acquire the right to mined on the network based on an arbitrary selection procedure by transferring the bitcoins to an inaccessible location. As a result, auditors have a long-term commitment in return for a quick loss when they burn tokens.

Proof of Capacity (PoC)

Validators are meant to contribute their disk space instead than investing heavily gear or burning coins inside the PoC consensus. Auditors with more disk space have a higher possibility of being selected to mine the next transaction and winning the block incentive.

Proof of Elapsed Time (PoET)

PoET is a consensus method that uses only fair techniques to determine the next block. In permissioned Public blockchains, it's commonly employed. Every validator on the network has an equal opportunity to construct their unique block using this process. All networks do this by waiting for an unspecified period of time and then providing a proof of their waiting to the block. The blocks that have been formed are broadcasted to the entire network for others to evaluate. The verifier with the smallest duration number in the proof section is declared the winner. The winner's validator node's block is added to the Blockchain. Additional tests are built into the algorithm to prevent nodes from ever winning an election either from spawning a large number of nodes.

Proof of Activity

Combine a Proof of Work and a Proof of Stake component. The classic method of mining starts with miners competing to be the one to solve a riddle and collect their prize. The difference would be that the chains currently mined don't have any actions in them. They're just forms with packet headers and the address again for mining incentive. The system converts to a proof of stake process as once practically blank transaction is mined. To verify the transaction, the transmitting data is used to choose a different group of consensus mechanism. Validators are selected once more, and so on, till all the validators have signed a successful block. The successful miner as well as the approvers who validated the block divide the system fees.

Proof of Weight (PoWeight)

This consensus model is important for the Proof-of-Weight classification of consensus algorithms. In a PoWeight system, instead of your proportion of coins owned in the inter - relations your probability of "uncovering" a next block, some other comparatively balanced number is utilized. Consider the following scenario: The Solid evidence algorithm in Filecoin is weighted based as to how much Shared data you're keeping. Weight values for items like Proof-of-Reputation could be included in other frameworks.

Proof of Zero

PoZ is a development of PoS, with key distinction being the ability to perform Zero-Knowledge transactions. PoZ, like traditional Proof of Stake, allows all customers the opportunity to make interest on your coins. Last but just not distant, in terms of power and CPU consumption, PoZ is significantly less expensive than PoW. Zcrypt is a decentralised cryptocurrency that employs zero-knowledge proofs. It employs a Proof of Work/Proof of Stake forms of communication on the LyraZ algorithm.

Raft

Raft is a consensus method that was created as a replacement for Paxos. By separating logic, it was intended to be more intelligible than Paxos, but is also legally deemed safe and provides some additional features. Raft presents a general technique to spread a device across a cluster of computers, guaranteeing so each node relies on same set of transfer functions (Jennath, H. S, 2019).

Proof of Signature

PoSIGN consensus is established through exchanges between such a network of physical and virtual organization. accordingly, that are always in contact, ensuring precise, fast, and signature-verified payments. The ZOLT method also saves a great deal of energy because it doesn't need nodes to tackle computationally difficult tasks in order to earn trading fees or develop new tokens. It makes use of a network of STATIC nodes that have been registered. To join in consensus, STATIC nodes must be enrolled and operational. Because all nodes are registered and understood, the licensing validity of STATIC nodes is confirmed via Proof-of-Signature (Jahid, 2021).

Direct Acyclic Graph Tangle (DAG)

IoT a uses the Tangle DAG consensus protocol. To transmit an Iota payment, you must first verify two previously received contracts (Awan S, 2021). As more transaction are uploaded to the Tangle, the two-for-one, compensation consensus improves the authenticity of transaction. Because transactions build consensus, someone that can fabricate 1/3 of an event may possibly persuade the entire network that their incorrect actions are genuine. Iota is kind of "double-checking" all the channel's activities on a central infrastructure called "The Coordination" until there is some sufficient activity traffic that producing 1/3rd of it gets impossible. The Coordination, according to Iota, acts as a set of wheel chocks for the network, and will be eliminated that once Tangle has grown large sufficient Table 1.

Table 1. Blockchain algorithm for IoT security

S. No	Algorithm Name	Attack	key feature
1	Byzantine Fault Tolerance	Man-in- Middle	BFT is decentralized distribution system to reach consensus
2	Proof of Work	DDOS	POW used for confirm transaction and create new blocks
3	Proof of Stake	Fake Stake	POS is used in the cryptocurrency consensus mechanism
4	proof of location	Physical attack	POL is coordinate the location to broadcast
5	Time based Consensus algorithm	DDOS	Minimize each block process waiting time each
6	Lightweight Scalable Blockchain	Sybil attack	To reduce the blockchain's packet overhead and memory requirements, IoT system data is stored off-chain
7	POET	IP Spoofing	Timely force to transfer hardware system
8	DTC	Timing Attack	Time based distributed ledger monitoring
9	Repucoin	Sinkhole Attack	The Renown
10	AlgoRand	Botnet attack	pooling based on the selection
11	FBA	Identity Theft	pooling based on the plenum intersection

IoT SECURITY SOLUTION USING BLOCKCHAIN PLATFORMS

Smart Contract

Traditional Internet and IoT security systems focus on a single trusted centralized power, leaving them vulnerable to a broad range of assaults, from data loss to Denial of Service (Shurman,2020). We discovered that smart contract-driven security measures need not necessitate a major change to the existing network architecture, but it takes advantage of Blockchain technology exceptional characteristics and the smart contract's comprehensive programming features. Almost everything key studies rely on Blockchain technology's trust - free, decentralized, interfere, and verifiable properties. Some of the research papers also used the configurable flexibility of smart contracts in combination with Blockchain capabilities to provide identification (Varfolomeev, 2021). On top of Blockchain for IoT, authorization and authentication solutions are available. The decentralized and trustless character of Blockchain, together with the software development features of smart contracts, has been solely utilized in primary research that offered Web smart contract-driven information security. Moreover, primary researchers have assessed advantage of the interfere nature of Ethereum smart contracts to protect data, IoT applications, and proof validity. Smart contracts, like Blockchain, do not even have a positive leaning toward resolving security flaws and vulnerabilities in the Web and IoT (Melamed, 2018).

Figure 8. Smart contract IoT security solution model

Blockchain grows in size for each connectivity request, which may limit its scaling in meeting the demands of specific IoT systems Figure 8.

Lightweight Scalable Blockchain (LSB)

This LSB, a cooperation with Data61 and UNSW, has created a distributed lightweight scalable blockchain design for IoT which essentially removes the overhead costs of traditional BC while retaining the majority of its own privacy and security features (Abbas, 2021). To optimize power consumption, IoT systems benefit from a private distributed ledger that works similarly to BC but which is administered remotely. To design a publicly accessible distributed BlockChain that assures point-to-point security and privacy, high-resource devices offer a network infrastructure (Shah, 2020).

Ethereum

Ethereum has its own digital money, Ether. The Digital currency blockchain network is a decentralised platform composed of thousands of Ethereum virtual machines. The Ethereum Blockchain is used for the construction of Smart Contracts (Al Hayajneh, 2020). The Free Software project is currently listed as among the most disrupting tools, and it serves as the foundation for the operation of other projects. Ethereum is only second behind Bitcoin in terms of cryptocurrency popularity. Ethereum Smart Contracts are Turing Complete and may be designed and implemented in languages such as Solidity. It is calculated and stored as code, and this is executed on an Ethereum Network (Czachorowski, K, 2019).

Merkle Tree

The Merkle tree is used by the Blockchain to ensure transactions integrity. This data structure is simply a binary tree of hashing algorithms in its entirety. Every non-leaf node is indeed a hash of the prior hashes, while each leaf node is a hash of the preceding hash value. Merkle proofs are used to determine whether a piece of data fits to the branch and is compatible with such a data set without revealing it (Mbarek, B, 2020).

SDN Blockchain for IoT Security

SDN is a framework to assess that secures inter-controller interaction by combining SDN with blockchain technologies. Each domain is given a single master controller and a number of backup controllers via SDN. The managers are blockchain users; the controller unit produces blocks, and the duplicate controllers verify its activity. Every SDN domain is administered by a single control system who interacts with the master of the other domains via blockchain. The controller unit generates blocks of dynamic network changes, which are then validated by the duplicate regulators using a suggested reputational method (Tzagkarakis, 2019).

Machine Learning/Deep Learning for IoT Security

Because of their peculiar characteristics of solving problems, learning algorithms are widely utilized in many practical uses. These methods are used to make machines that develop continuously as they gain more experience. Deep learning has recently become widely used in practice (Elsisi, 2021). The creation of new algorithms as well as the accessibility of massive data, as well as the rise of low-computation-cost techniques, have fueled the present development of machine learning. Over the last few years, ML and DL have progressed significantly, beginning as experimental curiosity and advancing to actual machinery with a wide range of applications.

Despite the fact that DL is a sub-field of ML, it refers to new improvements in learning approaches that use numerous non-linear handling layers for racist and discriminatory or generative showcasing abstraction and transition for pattern recognition. In this study, ML methods relate to standard ML methods with engineering features, whereas DL methods refer to new advances in learning methods that leverage numerous non-linear computation layers for discriminative or formative feature abstract ideas and transformation for pattern recognition(Oyinloye, 2021). The purpose of dividing ML and DL into two sections is to provide visitors with a thorough understanding of each.

Learning algorithms, in theory, strive to enhance performance in completing a task through teaching and developing from experience. In vulnerability scanning, for example, the aim is to classify behavior of the system as normal and anomalous. Improved classification performance can lead to better performance, and indeed the experiences wherein the algorithms learn are a repository of typical dynamic systems. There are three ways of learning algorithms: unsupervised, supervised, and reinforcement learning. On the basis of a learned projection, classifiers create a classification or prediction model that is produced by monitoring the input data. To put it another way, these strategies capture the connections between both the input variables and the desired output.

As a result, learning samples are required at the start of supervised methods to train dataset the algorithms, that are used to make predictions or classify fresh input. Deep networks have recently made significant progress in supervised learning. These systems can be thought of as multilayer systems with minimum units, which each evaluates the input data. Even though many practical implementations of DL have come from supervised learning approaches for learning representations, current research has made progress in developing DL systems that learn crucial input portrayals without the need for pre-labeled training data. These ML algorithms are unsupervised approaches that are used to analyze the data that has not been categorized. An unsupervised machine learning algorithm's goal is to classify the input data into distinct groups by looking at how comparable they are. RL is the third most popular form of ML. The data from an environment is used to train RL algorithms. The goal of RL is to comprehend an environment and determine the best strategies to a given ingredient in various ecosystems Figure 9.

The training data in RL are in the middle of the unsupervised and supervised spectrum. Instead of providing the correct output for a given input, the machine learning model in RL are considered to signal whether an action is correct or incorrect; if an action is incorrect, the problem persists until the correct action is determined. As a result, RL is a form of trial-and-error training. Different metrics could be used to categories and compare ML/DL for IoT security. The classification for Information devices utilizing ML/DL is shown in Figure 9. IoT protection may be classified as follows: IoT system, IoT security risks, IoT security learning techniques, ML/DL for layer security, and lastly problems, challenges, and future directions Below are detailed explanations for every category.

Figure 9. ML/DL security solution model

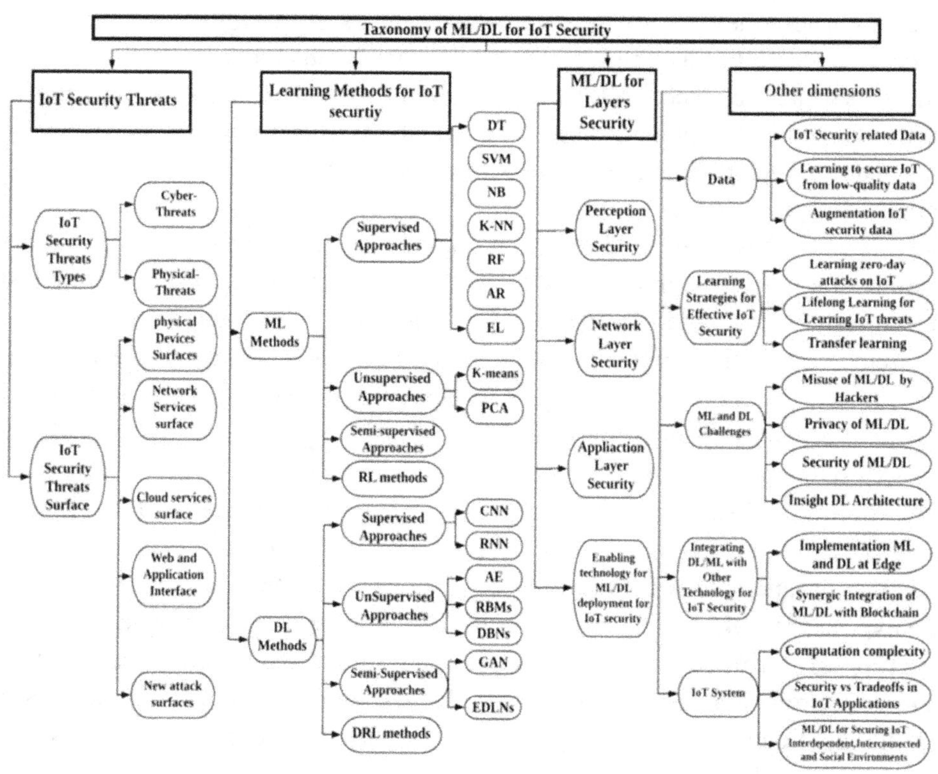

MACHINE LEARNING ALGORITHM FOR IoT SECURITY

We give a quick summary of various ML techniques and related applications in IoT in this part

Supervised Learning

Supervised learning methods employ tagged datasets to identify frequency, calculate the bandwidth, apply adaptive filters, and identify locations in Distributed systems. There are two sorts of processes in this team: regression and categorization. Among the most common classification techniques are SVM, naive Bayes, random forest, and decision tree. Polynomial and logistic regression are two often used regression algorithms. These algorithms are sometimes called as "instance-based" techniques since they anticipate result for each new data point using the learnt model. For Information devices, trained to learn techniques such as SVM, DT, and naïve Bayes (NB) have been frequently used. The non-linear requirements for a solution model, for example, are stored in SVMs. SVM, on the other hand, is ineffective for large amounts of data. In comparison to SVM, applying random forest techniques and conforming to the vast dataset available is easier. It has a higher degree of precision; hence it takes less time to anticipate. Training, on the other hand, takes longer than SVM and NB. Regression analysis and neighboring methods necessitate a large amount of memory and the processing of the data with a large number of features. Inside the communication layer as well as the cloud, supervised learning approaches have been utilized to detect intrusion and DDoS threats for IoT environments Table 2.

Unsupervised Learning

Unsupervised training algorithms deal with large datasets and rely on heuristics to learn patterns in data. Unsupervised techniques, in instance, detect abnormalities, trends, anomalies, and aggregate classes. To categorize the data in unlabeled data, classification methods are used. Unsupervised techniques can be utilized in IoT without knowing what the preferred result. K-means and hierarchical clustering are two standard grouping techniques based on unsupervised ML methods. Because it is a simple approach that produces clusters based on patterns observed in the pieces of data, K-means grouping is the most common. Cluster/edge groups form all around cluster centers, resulting in the establishment of clusters of same size. However, the number of nodes must be specified during network lifetime, that isn't always done precisely and effectively. In the communication protocol, unsupervised machine learning approaches are commonly employed to detect anomalies and Sybil assaults .

Reinforcement Learning

The RL approaches learn by experimenting with different behaviors in a given environment and determining the best set of actions to maximize reward. This incentive scheme is helpful in solving a variety of IoT security concerns. RL does not require any prior understanding of the world and instead learns from history by engaging with it to identify the best action in a given situation. Although RL methods are straightforward, identifying the best policy takes a very long time. The main difficulties in dynamic IoT network systems are sluggish convergence and an appropriate transition probability mechanism or policy.

Deep Learning for IoT Activity Design and Prediction

Deep learning is regarded as the cornerstone of modern artificial intelligence. DL is broadly applied in domains such as machine learning, voice recognition, automation, and many others. Deep learning does have a number of advantages over typical ML algorithms. (1) Deep learning can match intricate complex interrelationships among characteristics due to the usage of numerous hidden layers inside a neural network topology. (2) Rather of depending on palm statistical characteristics, famous designs such as convolution neural networks CNNs and LSTM networks can extract and detect useful features directly from raw data as classical computer vision does. (3) Deep learning is especially well adapted to working with "big data" issues.

IoT systems can produce a massive amount of data since they link billions of devices globally to sense and share data. DL offers a lot of potential for analyzing user activity in complex IoT systems. Deep learning may also be able to help IoT systems learn complicated response occurs more quickly than classical learning methods. The Internet of Things (IoT) is a full ecosystem with a wide range of devices and interconnections, a large number of visitors, and a massive amount of data. Rather than focusing on specific devices or layers, it is vital to look at the entire IoT. and the behaviors displayed to detect potential risks within an IoT system. We are concentrating our efforts on the three following issues: (1) to categories, train, and recover the physical fingerprint of an IoT device; (2) to explore network behaviors in IoT; and (3) to model information misuse in an IoT context.

As Figure 10 To intelligently provide such a remedy to emerging or zero-day threats, IoT devices can indeed be monitored. ML/DL are strong data analysis methods for learning about 'usual' and 'unusual' behavior inside the Distributed device based on the how IoT parts and sensors interact with each other. Each component of the IoT platform's input information can be gathered and examined in order to determine normal patterns of communication, allowing for early detecting malicious behavior. Since they can effectively forecast future unknown threats by learning from clarify the issue, ML/DL algorithms could be useful in anticipating new attacks, that are typically variants of earlier attempts. As a result, for safe and effective networks, IoT devices should migrate from just supporting communication security between sensors to threat protection intelligence provided by DL/ML approaches. Machine learning algorithms are those that necessitate designed features, whereas DL algorithms are significant developments in learning methods that employ a few non-linear processing layers for exclusionary or formative feature abstract concept and conversion for pattern recognition in IoT security, which can provide researchers and practitioners with a manual guidance to developing a successful and end-to-end secure system smartness (Istiaque, 2021).

Machine Learning Algorithm with 6G Communication

(1) Overall network distribution that is self-optimized, with active solutions based on the network knowledge and prediction being evaluated. (2) development of intelligent 6G mobile applications that understand from user activity and act as a situationally virtual .in addition, running on smart phones or remotely via integer arithmetic methods. (3) development of contextual inference algorithms and conceptual communication styles for incorporating sources of knowledge into interaction This will be extremely useful in properly implementing holographic telecommunications. Because future 6G networks will be more dynamic, system awareness of anticipated problems and environmental changes will be required to do is provide particular intervention, which will be hampered by three tasks. To start, the interaction

Figure 10. Role of ML/DL in IoT attacks

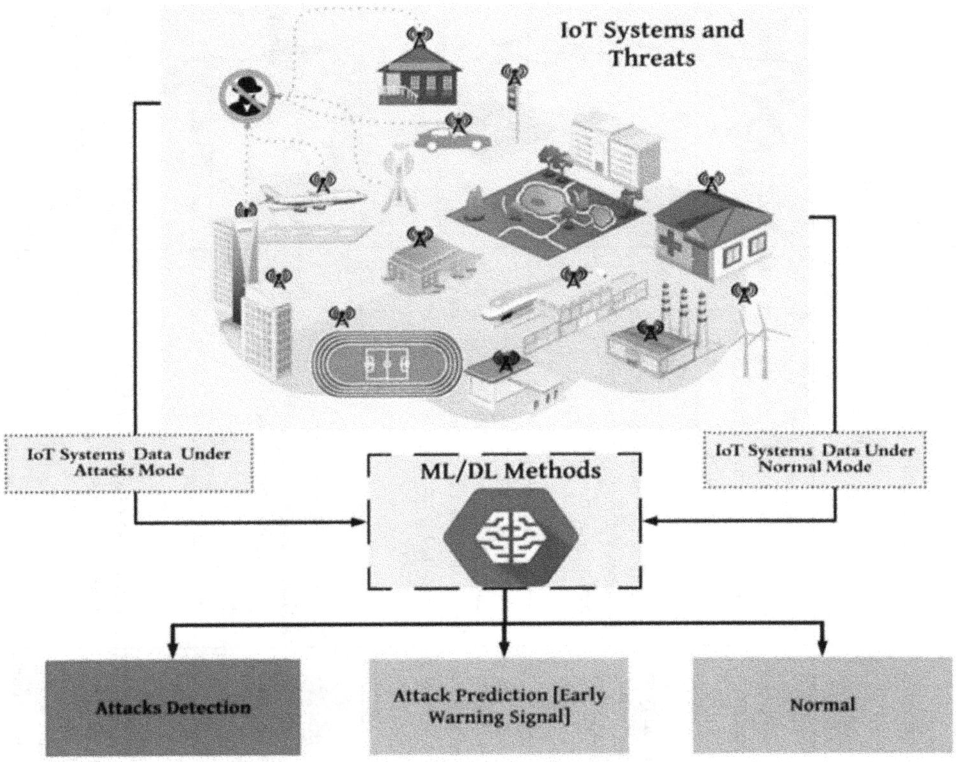

system must be able to fully integrate learning in ability to answer quickly to changes service needs of mobile customers. Second, for adapting to new conditions, regular and unplanned system modifications are required. Third, the challenges of heterogeneity in hardware and software components should be solved in order to fully realize the promise of ML (Hussain, 2020).

In order to fulfil labour in diverse industries, 6G's goal of intelligent connection of all things will replace people with a big number of AI machines. ML-enabled computers are prone to just being deceived by examples to the contrary, despite their great efficiency [40]. High-precision ML models are vulnerable to tiny perturbations, hence ML-enabled computers are susceptible to being tricked by examples to the contrary.

Machine vision can be used by 6G in a variety of settings, and the malicious use of counterexamples poses a significant threat to 6G's privacy. Adversarial attacks, on the one hand, can compromise privacy by providing the ML system the illusion of power over the machine. A smart car, for example, may travel wrongly if a street sign is misidentified (Chen, 2021)). As a result of these attacks, any industrial sector dispersed over 6G may issue the incorrect command. Confrontational attacks, on either hand, can make privacy protection methods useless. Similarly, harmful detection based on machine learning (ML) used by 6G smart network management, face recognition used by smart terminals, and other systems could be deceived, leaving attackers and criminals unnoticed. AI is the most cutting-edge technology accessible today, but the threat of attacks on machine learning (ML) must be addressed if AI is to better serve the public in the 6G era.

Table 2. Learning algorithms for IoT security

Algorithm	Methods	Pros	Cons	Effective area in IoT Security
DT	By describing training samples as leaves and branches, the DT-based technique establishes a model to learn from data. The pre-trained algorithm is again utilised to forecast the fresh sample's classification.	DT is a straightforward, straightforward, and clear strategy.	Due to its building nature, DT necessitates a lot of storage. Only when only a few DTs are present is it straightforward to comprehend DT-based approaches.	Anomaly detection and questionable traffic sources are detected.
KNN	KNN identifies a sample data based on votes of a certain number of its closest neighbours.	For malware detection, KNN is a popular and effective machine learning algorithm.	Because the ideal k values ranged through one dataset to the next, finding the optimal k value can be a difficult and time-consuming procedure.	Attack detection and anomaly detection
SVM	In the feature dimensions of multiple classes, SVMs generate a splitting hyperplane such that the length between the higher dimensional space and the closest neighboring measurements made of each category is greatest.	SVMs are well-known for their ability to generalise and suitability for data with a large set of feature variables but a limited number of data points.	It is difficult to choose the best kernel. SVM-based models are challenging to completely comprehend.	Intrusion virus and threats in power grid monitoring
NB	The posterior calculation is performed using NB. With the premise of feature independence, it use Bayes' theorem to forecast the likelihood that a given feature set of unlabelled data would fit a specified labeling.	The simplicity, ease of operation, cheap training sample required, and resilience to unnecessary features make NB a popular choice.	Because NB processes features independently, it is unable to extract useful information from feature correlations and interactions.	Network intrusion detection
k-Means clustering	k-Means clustering is an unsupervised method for identifying data groups based on feature similarity. The number of nodes formed by the method is denoted by the letter k.	When creating labelled data is challenging, unsupervised algorithms are usually a viable alternative. Since it does not require data samples, k-Means grouping can be utilised for private data anonymization in an IoT device.	In terms of detecting known threats, k-Means clustering algorithm is less successful than supervised learning methods.	In an IoT system, Sybil detection in commercial WSNs and personal data anonymization
CNN	CNNs primarily try to reduce the number of data parameters needed by employing minimal interactions, parameter sharing, and product is formed models, resulting in fewer interconnections across layers over ANNs.	CNNs are reliable supervised deep learning methods that outperform the competition. When compared to ANNs, CNNs have more scalability and training time complexity due to their new features. CNNs could be useful in IoT security since they can automatically learn features from the raw sensitive information.	Because CNNs have a significant computational cost, using them to support onboard surveillance systems on source of energy devices is difficult.	Anti-malware protection CNNs will learn aspects of raw security necessary data, allowing them to build a final security framework for IoT devices.

continues on following page

Table 2. Continued

Algorithm	Methods	Pros	Cons	Effective area in IoT Security
RNNs	RNNs use a temporal layer to collect sequence information and then use the secret component of the recurring cell to train multi-faceted differences.	RNNs and their derivatives have excelled in a wide range of applications involving sequence information. In some circumstances, RNNs could be useful in IoT security because IoT security data consists of sequential data.	The issue of disappearing or expanding gradients is the fundamental disadvantage of RNNs.	When it comes to identifying malicious activity, RNNs can classify network activity with high precision. RNNs and their variations have a great deal of potential for increasing IoT security system, especially when it comes to time series-based threats.
RBM	RBMs are unsupervised learning models that use fully convolutional models.There are no links at all between the pair of nodes in the very same layer, hence they are totally undirected systems.	Using a feedback system on RBMs enables again for unsupervised collection of a number of important characteristics.	RBMs have a significant computational cost, making them difficult to install on source of energy IoT devices that support internal surveillance systems.	RBMs could be used to identify network anomalies.
GANs	The GAN framework uses an adversarial method to train two models at the same time. The generating model learns underlying distribution of the data and produces new data. The racist and discriminatory model suggests that a sampling will come from of the training data instead of the produces better results when it comes to data sampling.	Unlike RBMs and DBNs, when require an unknown amount of iterations of a Markovchain to generate a sample, GANs only need one run through the generating a sample.	GAN training is insecure and challenging. It's challenging to learn how to use GAN to create data samples.	GANs could be utilised to create an infrastructure for safeguarding IoT devices' internet.
EDLNs	EDLNs are created by combining generative, exclusionary, and hybrid models.	Modeling diversity, performance was evaluated, and model generalisation can all be improved by combining DL classifications.	The system's time complexity can be considerably raised.	Further research into the application of GANs in protecting IoT systems is needed, specifically into the feasibility of using light homogeneous classifiers in a distributed context to increase system performance and accuracy.

CONCLUSION

Internet of Things Application privacy issues and threats interfaces are explored in this survey. A thorough examination of the possible applications of blockchain and ML and DL approaches in IoT security and privacy is offered. Because numerous technologies, ranging from Wi-Fi and physical devices communication to AWS based infrastructures, must be safety and secured and coupled with the different technologies, which needs for protecting IoT devices are becoming complicated. The growth of machine learning and deep learning has enabled the creation of a number of strong analytical approaches that can be used to improve IoT security. The merits, limitations, and applications of these strategies in IoT security are then contrasted at the end of every subsection. Following that, the application of blockchain with 6G Network as well as Machine Learning, and also Deep Learning methods with 6G Network for protecting the primary IoT layers will be discussed. are analyzed Finally, a detailed list of concerns, obstacles, and research perspectives linked to the usage of blockchain, Machine Learning, and Deep Learning in efficiently protecting IoT systems is given and classed based on data; Blockchain based communication with learning methodologies; and future directions. In the interconnected, networked, and reactive ecosystems of IoT systems, machine learning and deep learning for IoT security; varied security choices in IoT systems; and synergic combination of intelligent learning technology with blockchain for IoT Infrastrcture. The goal of the objective of the research work is to build a helpful guidebook. that will motivate academics to progress IoT privacy and security beyond simply providing safe interaction among System elements to building intelligent end-to-end IoT protection methodologies.

REFERENCES

Abbas, K., Tawalbeh, L. A. A., Rafiq, A., Muthanna, A., Elgendy, I. A., El-Latif, A., & Ahmed, A. (2021). Convergence of Blockchain and IoT for Secure Transportation Systems in Smart Cities. *Security and Communication Networks*.

Abosata, N., Al-Rubaye, S., Inalhan, G., & Emmanouilidis, C. (2021). Internet of Things for System Integrity: A Comprehensive Survey on Security, Attacks and Countermeasures for Industrial Applications. *Sensors (Basel)*, *21*(11), 3654. doi:10.339021113654 PMID:34073975

Al Hayajneh, A., Bhuiyan, M. Z. A., & McAndrew, I. (2020). Improving Internet of Things (IoT) security with software-defined networking (SDN). *Computers*, *9*(1), 8.

Aversano, L., Bernardi, M. L., Cimitile, M., & Pecori, R. (2021). A systematic review on Deep Learning approaches for IoT security. *Computer Science Review*, *40*, 100389.

Awan, S., Ahmed, S., Ullah, F., Nawaz, A., Khan, A., Uddin, M. I., ... Alyami, H. (2021). IoT with BlockChain: A Futuristic Approach in Agriculture and Food Supply Chain. *Wireless Communications and Mobile Computing*.

Chen, M. Y., Fan, M. H., & Huang, L. X. (2021). AI-based vehicular network toward 6G and IoT: Deep learning approaches. *ACM Transactions on Management Information Systems*, *13*(1), 1–12.

Czachorowski, K., Solesvik, M., & Kondratenko, Y. (2019). The application of blockchain technology in the maritime industry. In *Green IT Engineering: Social, Business and Industrial Applications* (pp. 561–577). Springer. doi:10.1007/978-3-030-00253-4_24

Elsisi, M., Mahmoud, K., Lehtonen, M., & Darwish, M. M. (2021). Reliable industry 4.0 based on machine learning and IOT for analyzing, monitoring, and securing smart meters. *Sensors (Basel)*, *21*(2), 487.

Gavra, V. D., & Pop, O. A. (2020, October). Usage of ZigBee and LoRa wireless technologies in IoT systems. In *2020 IEEE 26th International Symposium for Design and Technology in Electronic Packaging (SIITME)* (pp. 221-224). IEEE.

Groves, B., & Pu, C. (2019, November). A Gini index-based countermeasure against sybil attack in the internet of things. In *MILCOM 2019-2019 IEEE Military Communications Conference (MILCOM)* (pp. 1-6). IEEE. doi:10.1109/ICESS49830.2020.9301593

Haseeb, J., Mansoori, M., Al-Sahaf, H., & Welch, I. (2020, December). IoT Attacks: Features Identification and Clustering. In *2020 IEEE 19th International Conference on Trust, Security and Privacy in Computing and Communications (TrustCom)* (pp. 353-360). IEEE.

Husamuddin, M., & Qayyum, M. (2017, March). Internet of Things: A study on security and privacy threats. In *2017 2nd International Conference on Anti-Cyber Crimes (ICACC)* (pp. 93-97). IEEE.

Hussain, F., Hussain, R., Hassan, S. A., & Hossain, E. (2020). Machine learning in IoT security: Current solutions and future challenges. *IEEE Communications Surveys and Tutorials*, *22*(3), 1686–1721.

Istiaque Ahmed, K., Tahir, M., Hadi Habaebi, M., Lun Lau, S., & Ahad, A. (2021). Machine Learning for Authentication and Authorization in IoT: Taxonomy, Challenges and Future Research Direction. *Sensors (Basel)*, *21*(15), 5122.

Jahid, A., Alsharif, M. H., & Hall, T. J. (2021). *The Convergence of Blockchain, IoT and 6G: Potential, Opportunities, Challenges and Research Roadmap*. arXiv preprint arXiv:2109.03184.

Jennath, H. S., Adarsh, S., Chandran, N. V., Ananthan, R., Sabir, A., & Asharaf, S. (2019). Parkchain: A blockchain powered parking solution for smart cities. *Frontiers in Blockchain*, *2*, 6.

Khacef, K., & Pujolle, G. (2019, March). Secure Peer-to-Peer communication based on Blockchain. In *Workshops of the International Conference on Advanced Information Networking and Applications* (pp. 662-672). Springer.

Kimbugwe, N., Pei, T., & Kyebambe, M. N. (2021). Application of Deep Learning for Quality of Service Enhancement in Internet of Things: A Review. *Energies*, *14*(19), 6384. doi:10.3390/en14196384

Kocakulak, M., & Butun, I. (2017, January). An overview of Wireless Sensor Networks towards internet of things. In *2017 IEEE 7th annual computing and communication workshop and conference (CCWC)* (pp. 1-6). IEEE.

Lei, K., Zhang, Q., Xu, L., & Qi, Z. (2018, December). Reputation-based byzantine fault-tolerance for consortium blockchain. In *2018 IEEE 24th International Conference on Parallel and Distributed Systems (ICPADS)* (pp. 604-611). IEEE.

Liang, L., Zheng, K., Sheng, Q., & Huang, X. (2016, December). A denial-of-service attack method for an iot system. In *2016 8th international conference on Information Technology in Medicine and Education (ITME)* (pp. 360-364). IEEE. 10.1109/ITME.2016.0087

Lin, T. (2020, November). Deep Learning for IoT. In *2020 IEEE 39th International Performance Computing and Communications Conference (IPCCC)* (pp. 1-4). IEEE.

Mafamane, R., Ouadou, M., Hassani, A. T. J., & Minaoui, K. (2021, April). Study of the heterogeneity problem in the Internet of Things and Cloud Computing integration. In *2020 10th International Symposium on Signal, Image, Video and Communications (ISIVC)* (pp. 1-6). IEEE. 10.1109/ISIVC49222.2021.9487539

Mbarek, B., Ge, M., & Pitner, T. (2020, September). Blockchain-Based Access Control for IoT in Smart Home Systems. In *International Conference on Database and Expert Systems Applications* (pp. 17-32). Springer. 10.1007/978-3-030-59051-2_2

Melamed, T. (n.d.). *An Active Man-In-The-Middle Attack on Bluetooth Smart Devices.* doi:10.1109/GLOBECOM42002.2020.9322070

Ngo, Q. D., Nguyen, H. T., Le, V. H., & Nguyen, D. H. (2020). A survey of IoT malware and detection methods based on static features. *ICT Express*, *6*(4), 280–286. doi:10.1016/j.icte.2020.04.005

Oyinloye, D. P., Teh, J. S., Jamil, N., & Alawida, M. (2021). Blockchain Consensus: An Overview of Alternative Protocols. *Symmetry*, *13*(8), 1363.

Prates, N., Vergütz, A., Macedo, R. T., Santos, A., & Nogueira, M. (2020, December). A Defense Mechanism for Timing-based Side-Channel Attacks on IoT Traffic. In *GLOBECOM 2020-2020 IEEE Global Communications Conference* (pp. 1-6). IEEE.

Salman, T., & Jain, R. (2019). *A survey of protocols and standards for internet of things.* arXiv preprint arXiv:1903.11549.

Shah, P., Arora, M., & Adhvaryu, K. (2020, October). Lightweight Cryptography Algorithms in IoT-A Study. In *2020 Fourth International Conference on I-SMAC (IoT in Social, Mobile, Analytics and Cloud) (I-SMAC)* (pp. 332-336). IEEE.

Shurman, M., Obeidat, A. A. R., & Al-Shurman, S. A. D. (2020, April). Blockchain and smart contract for IoT. In *2020 11th International Conference on Information and Communication Systems (ICICS)* (pp. 361-366). IEEE.

Sun, H., Hua, S., Zhou, E., Pi, B., Sun, J., & Yamashita, K. (2018, June). Using ethereum blockchain in Internet of Things: A solution for electric vehicle battery refueling. In *International Conference on Blockchain* (pp. 3-17). Springer.

Syed, A. S., Sierra-Sosa, D., Kumar, A., & Elmaghraby, A. (2021). IoT in Smart Cities: A Survey of Technologies, Practices and Challenges. *Smart Cities*, *4*(2), 429–475. doi:10.3390martcities4020024

Tzagkarakis, C., Petroulakis, N., & Ioannidis, S. (2019, June). *Botnet attack detection at the IoT edge based on sparse representation. In 2019 Global IoT Summit (GIoTS)*. IEEE.

Varfolomeev, A. A., Alfarhani, L. H., & Oleiwi, Z. C. (2021). Secure-reliable smart contract applications based blockchain technology in smart cities environment. *Procedia Computer Science*, *186*, 669–676.

Wang, Q., Zhu, X., Ni, Y., Gu, L., & Zhu, H. (2020). Blockchain for the IoT and industrial IoT: A review. *Internet of Things*, *10*, 100081.

Wara, M. S., & Yu, Q. (2020, December). New Replay Attacks on ZigBee Devices for Internet-of-Things (IoT) Applications. In *2020 IEEE International Conference on Embedded Software and Systems (ICESS)* (pp. 1-6). IEEE.

Xie, J., Yu, F. R., Huang, T., Xie, R., Liu, J., & Liu, Y. (2019). A survey on the scalability of blockchain systems. *IEEE Network*, *33*(5), 166–173. doi:10.1109/MNET.001.1800290

Zhang, Z. K., Cho, M. C. Y., Wang, C. W., Hsu, C. W., Chen, C. K., & Shieh, S. (2014, November). IoT security: Ongoing challenges and research opportunities. In *2014 IEEE 7th international conference on service-oriented computing and applications* (pp. 230-234). IEEE.

Zheng, Z., Xie, S., Dai, H., Chen, X., & Wang, H. (2017, June). An overview of blockchain technology: Architecture, consensus, and future trends. In *2017 IEEE international congress on big data (BigData congress)* (pp. 557-564). IEEE.

Chapter 13
Wireless Brain-Computer Interface (WBCI) and 6G Technology Security Issues, Safety Mechanisms

Saravana Kumar Ganesan
Karpagam College of Engineering, India

Parthasarathy V.
Karpagam Academy of Higher Education, India

Arunachalam M.
Sri Krishna College of Engineering and Technology, India

Viswa Bharathy A. M.
GITAM School of Technology, GITAM University, India

ABSTRACT

A brain-computer interface (BCI) is a communication system that does not depend upon the brain's normal output pathways of peripheral nerves and muscles. Wireless brain-computer interface (WBCI) systems are a branch of BCI systems with an exclusive method to acquire the electrical activities of the brain, that is, electroencephalogram (EEG) using an effective non-invasive, implantable electrode scheme and employment of wireless communication schemes to transfer the acquired EEG for further processing. The five paramount security and privacy issues are authentication, access control, malicious behavior, encryption, and communication. With the appropriate implementation of the wireless BCI in the context of 6G technology, this chapter presents a comprehensive overview of WBCI and 6G technology and outlines artificial intelligence-based schemes' utility to address security and privacy issues arising out of 6G network deployment to the contexts revolving around WBCIs.

DOI: 10.4018/978-1-6684-3804-6.ch013

INTRODUCTION

Wireless Brain Computer Interface (WBCI) acquires and manipulates the electrical signals of human nervous system and it is uniquely characterized by wireless transmission of acquired signals to the processing units. The WBCIs provide an edge over conventional BCIs in terms of rendering expediency to the end users through sophisticated signal processing unit. The replacement of wired signal transfer between the functional blocks of conventional BCIs empowers the WBCIs to offer compactness and durability. Modern day computing systems houses the signal processing blocks of WBCIs to facilitate unrestricted physical movements and smooth functional enactment (Chen, M et al, 2019)

Contemporary Wired BCI systems employ non-active electrodes to acquire the electrical signals of the brain. Owing to the massive size, error prone structure, large number of electrodes, the conventional electrode placement scheme eludes provision of comfort to the user. This book chapter highlights the benefits of maneuver-friendly, wireless, sub-micron thickness EEG acquisition system. The system can be placed on scalp targeting occipital lobe and it comprises of three flexible conductive polymer electrodes and a paper thin skin electrode made through aerosol jet print. This offers highly desirable comfort to the end user in terms highly miniaturized size exhibiting high fidelity acquisition. This system also curtails the onset of artifacts caused by end user movement, wiring units and an ineffective electrode placement scheme and provides better resolution with seated subjects (Strukov, D et al, 2019)

A CNN is a feed-forward neural network and enhanced structural version of multiple layered Artificial Neural Network (ANN) and yields ML-based outcomes. It encompasses convolutional layer, fully connected layer and pooling layer. The layers perform the convolution operation and computations on the given data sets and yield an scaled down datasets in terms of dimension. For accomplishing the expected result, the neurons are connected in cascade structure by the fully connected layer (Chen, M et al, 2019). The structure exhibits sophisticated capability in reducing the dimensions of the input data and also captures the exclusive inter-dependent spatial orientation of the EEG patterns.

The fore most publication of white paper in 6G technology was presented by the leading global communication technocrats at World's First 6G summit convened in Finland. This innovative initiative paved the way to laid a foundation for 6G networking. The 6G network is perceived as a network characterized and driven by Artificial Intelligence (AI) technologies. The 5G technology also employs the AI techniques in its architecture whereas the 6G networking is hypothesized to integrate the evolving AI tools for enhancing the functional levels of the networking (Yang, P, 2019). The extenuation of risk associated with security and privacy aspects of the networking has to be carried out with utmost priority. The 6G technology finds a unique place in various emerging technologies that includes AI-based software, molecular communications, quantum communications, Block Chain, Tera Hertz (THz) technology, and Visible Light Communication (VLC) technology are among the technologies used in this study. All of these technologies have a lot of potential for usage in 6G network applications including multi-sensory X Reality (multi-sensory XR), networked robotics and autonomous systems, wireless brain-computer connections, and blockchain and distributed ledger technologies, to name a few (Letaief, K.B et al, 2019)

On the 5G based Internet systems, there is a high intensity malicious attacks, susceptibility of the system to attacks, intrusive attacks, and protection issues. This will no longer sufficient in 6G, as physical safety will become increasingly reliant on safety algorithms as well as the firewall structures of the communication networks. As a result, reliability of the 6G network holds a paramount role. The terms "trust," "security," and "privacy" are all related and they refer to various aspects of the identical technologies.

The Modeling, policies, and procedures for trust must all be defined for the networks to cater a gamut of applications. 6G connects the physical and digital worlds, making the critical loopholes in information security to safety (Gindraux, S, 2002). A set of examination measures on Personal Information (PI) on usable and identified data sets, that takes into account personal characteristics, time and space aspects of data, and the context in which data is captured and analyzed shall be employed to provide smart services, as well as can act as benchmark assessment for Personally Identifiable Information (PII). These strategies and outlines can be utilised to enable effective management schemes that fulfill supervisory necessities for citizen data defense in different economies that implement technology beyond 5G. To build a trustworthy 6G network and due concern to regulatory aspects, impacts on the larger society in different spheres are found to be challenges (Letaief, K.B et al, 2019). The present situation maintains a status quo in terms of regulatory mechanisms, economic incentives, and technology. This chapter provides a complete review of WBCI and 6G technologies, as well as the applicability of AI-based techniques in addressing security and privacy challenges originating from 6G network deployments in WBCI contexts.

The objectives of this book chapter are,

1. Identification of effective EEG signal acquisition scheme
2. Utilizing AI based 6G technology for EEG transmission
3. Optimization of EEG transmission through an appropriate antenna design scheme and CNN.
4. Highlighting security threats in interlinking WBCI and 6G
5. Scheme for building a trustworthy 6G network with WBCI

Background

BCIs interacts with neural sensory systems, and acts as an interface with the external environment, offers acquisition and interpretation of the signals generated by the brain for persons suffering from Amyotrophic Lateral Sclerosis (ALS) and similar set of neuronal debilitative conditions. The acquisition element of conventional BCI systems typically consists of less precise, less efficacy amplification systems and has the potential to introduce various types of artifacts and in turn reduces the signal quality (Wolpaw, 2002). With the presence of the cables between the signal acquiring electrodes and pre-processing blocks, acquisition scheme is found to be encumbering. As a result, the time spent preparing to measure EEG signals is usually quite lengthy. Because of these limitations, the conventional BCI systems restrict the mobility of the end user and this reduces manoeuvring scape of the complete system.

Neurophysiological signal collection and processing systems that wirelessly communicate acquires physiological signals to the translation unit are known as Wireless Brain Computer Interface (BCI) systems. With the use of a wireless Communication modes which involves wireless modules such as Zigbee and Bluetooth. WBCI systems can remove the signal acquisition scheme of conventional BCIs at various strategic operating points and improves the fidelity of the signal. The mobility of BCI systems is considerably increased by removing wire connections. The positions and activities of individuals utilizing remains unadulterated. These appealing features of wireless BCI devices encourage researchers to move beyond small-scale trials and explore real-world applications (Wolpaw, 2002).

Despite WBCI systems offers a wide variety of benefits, there exists a set of limitations to be addressed, such as signal quality, compactness and sleek design, and the discovery of viable applications. For a more precise classification of user intentions, the fidelity of the original EEG signals has to enriched optimally (Lebedev, M. A et al, 2017). A wide variety of artifacts introducing sources, such as the intermixing of co-existing bodily electrical activities and supply lines, can straightforwardly disrupt the recorded EEG signals. These impediments wreak havoc on signal measurements involving large leakage currents and contact impedances. EEG signals are easily distorted as a result of these challenges, and the quality of measured signals is frequently poor. As a result of these considerations, application accuracy suffers and for daily use and long fashionable, compact, and low-weight WBCI devices are required.

To address these difficulties, research organisations are focusing on the following aspects:

1. Cutting-edge electrode placement scheme, implantable and user friendly EEG acquisition schemes
2. Optimal Power Saving and sinking schemes, compact, convenient and BCI systems
3. Massive scale production of BCI systems and high amount of feasibility for end users.

Artificial Intelligence (AI) can be defined as, an ensemble of technologies that employ a computing unit to imitate discretionary behavior and minimal humanoid transaction and leading to matching or surpassing mortal efficiency in various learning and or application oriented activities. Internal characteristics such as duration of the signal, peak-to-peak voltage levels, frequency bands, power usage, recording schemas, and brain's neural activities are continually fed to the AI based feature extraction and feature classification algorithms for improved BCI efficacy (Lebedev, M. A et al, 2017). The extraction and classification activities of the algorithms are found to be superior in comparison with the conventional operative procedures then produce the needed functional consequences at the same time. Despite the fact that these investigations are mostly in the preclinical research stage, their continuous progress may highlight therapeutically actionable alterations in BCIs.

The architecture of the Convolutional Neural Networks can emulate the human brain's composite functional structure. It employs only a hefty datasets for training and derives a sophisticated model that learns features using back propagation and gradient descent optimization algorithms and extracts features using filtering, normalization, and nonlinear activation procedures.

For feature extraction, each and every layer of the network architecture comprises of numerous computational cores possessing equal weightage. Each computational core is exhibiting multiple dimensional weightage structures. Each neuron weightage in the computational layer is calculated by carrying out multiplication operation on two values, previous values, present offset values. The kernel searches the higher layer's input data sequentially in accordance with functional steps and carries out feature extraction operations with specific operational contexts. To obtain the convolution sub graph, the computational structures and the values from the preceding layer are subjected to multiplication operations. Two fundamental aspects of connectivity scheme and weightage allocation scheme plays a larger role in the operation of the network architectures (Zappone A, 2020). The local receptive area is analogous to the local connection. Its primary purpose is to carry out feature extraction with suitable details through adopting a scheme with appropriate network attributes. The distribution of weights indicates that each and every neuron of the network architecture has equal weight, appropriate network attributes, reduced computation time and minimal over fitting.

This chapter provides a comprehensive summary of the development of security and privacy aspects of Wireless Communication networks from the 1G to 5G as illustrated in the following Figure 1.

In 6G, end-to-end network security measures based on machine learning must be considered. Machine Learning (ML) is currently employed in a variety of services, network components, and the entities of the network. The utility of multiple artificial intelligence algorithms in 6G has to be synchronized and this has to be ensured across the space as adhered in the spectrum sharing schemes which uses machine algorithms. The frequency utility information must be safely communicated across peers competing for the same frequency slot in order for intelligent spectrum sharing to work (Latva-aho, M et al 2019). The first scenario would be to ensure information sharing between competing peers through scheduling. The learning algorithms can be used to determine the credibility of competing nodes, permitting for appropriate and secured information distribution. Conversely, security will still be required after regulating the layer architecture and subsequently moving to the accessible frequency slots as approved by the two mechanisms, providing and requesting. A change in routing processes in the network layer with intent to modify physical layer can be considered as a primitive step. Such systems that use machine learning at each tier, at the primitive layers, physical and network layers, the appropriate initiatives has to be suitable frequency slot distribution, route scheming with adequate safety measures, able to endure payload will necessitate synchronized ML methods. The exponential growth of data usage by the public worldwide is illustrated in the Figure 2.

Figure 1. Evolution of communication technologies

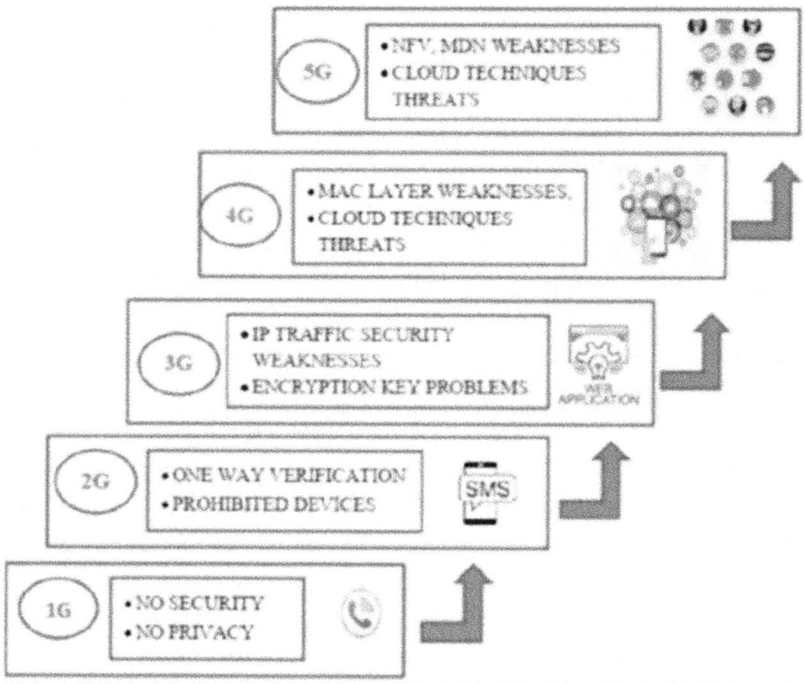

The wireless medium's intrinsic openness makes it vulnerable to interference. Intentional or inadvertent interference are both possible. Proximity Devices shall cause unintentional interference (Zhang, L., et al, 2017). The limitations of the current literature were recognized as the following research problems.

1. The application of learning algorithm as firewall structures for 6G technologies.
2. Addressing Layer features through Machine Learning
3. The scope of machine learning in security with respective to WBCI

Figure 2.

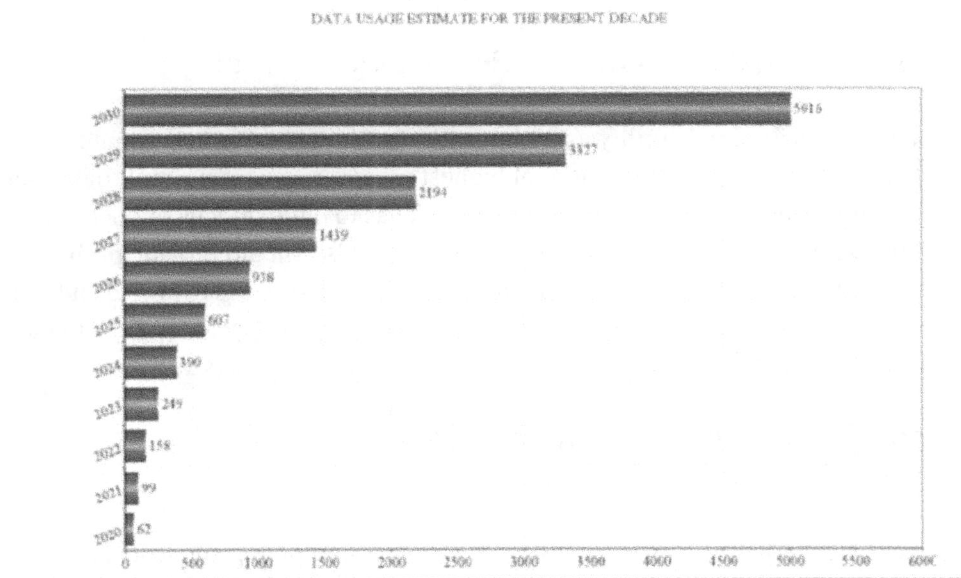

MAIN FOCUS OF THE CHAPTER

Issues, Controversies, Problems

The functions of trust, safety, and confidentiality are all intertwined in evolution of upgraded network technologies and found to be distinct. The fundamental challenges and problems in two main areas, trust, security, and privacy, are addressed in this book chapter.

Trustworthy 6G

The technical developments, monetary benefits, regulatory framework defines the level of malicious attacks, trust deficit, maintenance of confidentiality and protection algorithms for all online transactions (6G Research Visions [6GRV], 2020). A Trustworthy 6G is defined as a network which protects the various aspects of its physical layer from malevolent intrusions and possessing the characteristic exhibit reliability amidst multiple on slaughtering scenarios.

Research Issue 1: Innate and Unique Threats in 6G Technologies

The advancement from 5G to 6G systems, the variety and large groups of new nodes connected through internet and their associated control circuitry will last to present major safety and confidentiality threats, as well their exists an array of new threat vectors also. The advancement of AI will blur the distinction between actual and fraudulent information, allowing for the development of increasingly sophisticated assaults. From the ten billion sizes of present day networks to the hundred billion sizes of upcoming 6G networks, the amount of new nodes connected through internet will increase tenfold. The economic and social reliance on IT and networks will deepen as a consequence of this implementation and utility of evolved technologies will localize IT and network security as crucial.

Research Issue 2: Trust Level Across Point-to-Point Trust

The cloud structure finds its simple utility as for trust services for all users, according to existing "open Internet" legislation. The present day intrusions of highest order varying from simple data contamination to complete hijacking of the systems. The evolved network has to build it locus of operational context in and around the malicious attacks irrespective of their depth. It also should provide probable results and concurrently when hostile actors challenge to disrupt it. The 6G network has to provide inbuilt implanted trust, ensuing in a substantially higher level of information protective mechanisms in 6G and ensuring error free data packets transfer across innumerous nodes to which 6G offers services.

Research Issue 3: Quantum Computing based Ciphering and 6G Protection Architecture

The contemporary standard norm flouts the utility of quantum computing and instead depends on classical ciphering. In 6G networks, the trend toward cloud computing architectures is projected to evolve. The large scale quantum computing are in the verge extending their applications to real time environment and this juncture requires the utility of Artificial Intelligence based algorithms to hold Centre stage in invoking innovative ciphering techniques.

Research Issue 4: The Trade-Off Issues of Machine Learning in AI driven 6G

The upward migration of operational frequencies in to Tera Hertz requires larger bandwidth caters to a global space with is forecasted as a space with a presence of one million devices per square kilometer positioned across innumerous global points, oceanic, networks with space and ground elements, on-line protective mechanisms and the space is characterized by softwarization based security and utility of the physical resources in virtual mode and has closely interspersed and cloud oriented structure which also uses the principle of machine learning algorithms. Conversely, protective algorithms can able to perceive and deploy the proper network functionalities to provide operative and reinforced attacks. Along with machine learning, deep learning methodologies with evolutionary mechanisms at the fundamental level has to be applied such that they recognize, restrict, alleviate, and thwart malicious entries or dynamic intrusions, continuous deep learning will be required on a packet/byte level.

Research Issue 5: Physical Layer Security in 6G

The safety approaches for fundamental layers of the network may be effective in safeguarding the maximum number of crucial and understudied sections of the network, such as those connecting implantable electrodes / sensors in a human body to all the master nodes. The supreme and appropriate aspects of the Physical layer, that has to be manipulated is for the definition of safety algorithms in a 6G background are earmarked by expansive characteristics of the network, diverse nodes and numerous types of malicious attacks. It also depends upon the functioning of physical layer as a separate protective scheme that can interact with upper layers is a critical research issue to be addressed by Physical layer security in 6G technologies.

Research Issue 6: Privacy as Highly Vulnerable Facet in 6G.

The significance of the for 6G network in comparison with its predecessor in terms of core functional devices, is that it, envisages a futuristic immersive environment and this characteristic holds a large scale of scrutiny in the contemporary research circles. The present juncture fails to highlight the utility of connectivity schedule, deidentified datasets to preserve individual identifiably. Intended for numerous technological applications in digital domain, that includes smart healthcare, industrial automation, and smart transportation, this is a big, unsolved issue. Companies are looking for new methods to maneuver private data to generate new business income, while courts around the world are deciding whether their individual private rights are violated stemming from extrication of adherence to non-intrusion policies. Block chain, distributed ledger technology, and differential privacy techniques may be considered as possible solutions.

Problems and Controversies

6G is projected as an exhaustive interconnecting space with the functional nodes ranging across all domains. This closely interspersed dissimilar network elements transmits massive volumes of data, which is typically highly vulnerable to attacks (Chowdhury, M.Z et al, 2020). Contextually, it leads to define a comprehensive security policy to ensure that user's interests are entirely protected. The less sophisticated network elements can be safe guarded by proper implementation of establishing firewall structures for the physical layer.

Solutions for the Fundamental Lower Layers of the 6G Network

Continuous Evolution of Communication Technologies from 5G to 6G will enhance the primary services provided by the 5G technology and supporting state of the art applications that include, Holographic Projection and Augmented Reality. A comprehensive approach is required to handle the security of the different systems. The present day sensor networks design acquires and captures assorted types of data heaps and requires cutting-edge security techniques. This requirement introduces additional restraints on proficiency of devices used, interconnectivity schemas and network structures.

Physical layer security, or modifying the physical layer's security policy, could be one of the 6G connectivity's confidentiality enablers. Securing Physical Layer or restructuring the protocol of the Physical Layers Security enriches the confidentiality in 6G networks. The appropriate mixture of State-of-the-art AI procedures and Distributed Computing Architecture shall be utilized to enrich the conventional ciphering techniques and supporting safety aspects of systems without cryptographic capabilities. These include IoT devices and nano-devices, as well as the Internet of Body Area Networks, in which individual nodes will serve as future Internet nodes.

By characterizing the scenarios in which the transmission and reception of data occurs a context-aware strategy may be defined, which reduces the power consumption requirements of all nodes and overall computational complexity of the algorithms. The dynamic environment that can be anticipated is characterized by hand-off conditions, density of nodes, utility of spectrum and presence of multiple technologies (Gyongyosi, L et al, 2019)

To provide adequate safety features, the strategical aspects of security algorithm has to be designed in accordance with, the response initiatives required at any given time to various levels of malicious attacks and a comprehensive study of the channel characteristics used for communication. The AI algorithms for catering security can be designed to counter time-variant situations and the specifications for design include functional aspects of operational contexts. The energy requirements for each and every node of the network and the authenticity level of the operational context shall also be considered as design factors [6GRV, 2020]. The security facilitation measures should conforms to the fundamental 6G core criteria requirements that includes, extended power availability, low energy utilization, exhaustive customization and architectures to cater diverse platforms.

The acquisition of brain signals, processing which includes feature extraction and feature classification by utilizing Brain Computer Interface (BCI) irrespective of wired or wireless less exchange and which is facilitated by a wireless mobile communication, largely categorized under Personal Assistive System category and this operation is mentioned as Brain To Thing Communication (BTC). This technique carries out an extensive experience on a various platforms in the context of BTC collaboration with previous, present generations'(Bernstein, D et al 2017) In comparison to present systems, which are primarily mediated by cellphones, this technique would permit for processing functional blocks and implantable nodes. The system's performance assessment requires to be significantly different than present fifth generation systems The efficacy of the WBCI is measured through a performance metric, Quality of Physical Experience (QoPE) which assesses novel assistive provisions provided by the system namely, as interaction with the external world through electrical activities generated by muscle movements, evoked potentials through stimulus paradigms, or electrical activity of the brain. The futuristic systems includes WBCI automobile systems, BAN based networks, Replication of sensors and neurons to constitute "artificial brain," and augmented reality, Virtual networks are all possibilities.

6G and the Brain:

The use of brain's electrical activities in BMI poses significant issues for communication systems (Kabir H et al, 2020), particularly those that are meant to handle transmissions involving the responses evoked by end users in response to the stimulus applied rather than brains as shown in Figure 3.

Artificial Intelligence and its application in diverse domains are evolving at an exponential pace owing to incredible novel expansions in data processing, data transfer operations. The studies on brain and it's neuronal circuitry is continuously marveling the researchers and presently existence of a plethora of paradigms on brain's functionality are available for the engineering community to explore new avenues for developing applications which can render well-being to the various social strata (Kaddoum, G, 2016).

As a contemporary development, the sharp progress in technical improvements in acquisition devices, processing schemes, and units for carrying out complex computational activities have continuously strengthened the research in neurological sciences and the expertise on functional aspects of neurological system had reinforced the expansion of schemas of computation. Human-Type Communication (HTC) refers to the group of applications that had connected people in multiple ways that includes 2G services to 5G services. In addition to this, with the advent of Internet of Things (IoT), the entry of wireless communication schemes exclusively designed for interconnecting computing nodes had found a prominent spot (Gyongyosi, L et al, 2019). Undeniably, the rise of Machine-Type Communication (MTC) systems has transformed the wireless network technologies and acts as the primary propelling factor in molding the enduring disposition of 6G wireless systems (Kantola, R et al, 2015).

This chapter forecasts that BTC shall progress as ensuing cutting edge wireless transfer modes. BTC connection schemes essentially to be developed in a manner to impeccably link the electrical activity of the brain to a wireless network and hypothetically deliver full duplex exchange amongst the end user's implantable acquisition nodes and the numerous processing units associated. An exclusive facet of BTC links is that they necessitate the linkage to play a counterpart role to the competencies of the human brain. Obviously, providing BTC will henceforth offers an entire innovative group of perplexing research issues.

Figure 3.

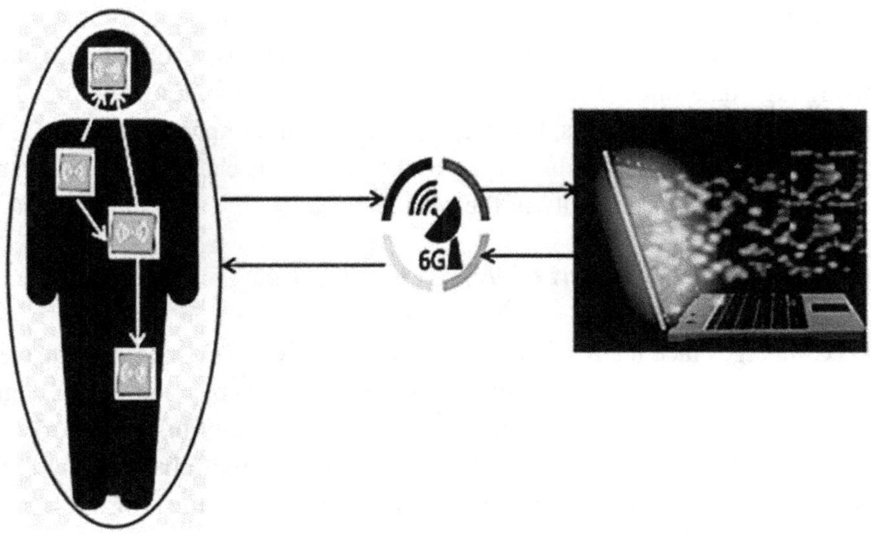

Research Issue 1: Establishment of Uplink and Downlinks

The existence of conventional hurdles that prevails in ensuring appropriate HTC and MTC services prominently lies establishment of downlink communication in the HTC services and uplink communication in MTC services. Conversely, in BTC, both the uplink and downlink schemes shall pose a logjam for effective data transfer. In the given context, substantial data has to be received in the downlink from the implanted electrode schemes to craft an augmented and virtual experiences. In the meantime, the data from the brain has to be trasnferred from the implanted electrodes to the processing units. This will ensure that responses evoked in line with stimulus paradigms by the brain will be properly transferred to the processing blocks for further operations. As a paramount concern, presence of data streams in pipeline intended for uplink transmission can pose a challenge that has to be dealt in the context of latency. The size of the data generated by the brain to establish cognitive services is found to be in the order of tera bytes and a subsequent result, the challenge can be construed as design of a communication scheme to offer very high speed transfer and minimal errors.

Research Issue 2: Cognitive Limitations and Computational Complexity

The human brain, despite its tremendous processing capabilities, has explicit cognitive and perceptional schemas. Numerous neurons firing and complex interconnections can drive various parameters namely, environment, focus, energy cycles, cognitive restrictions. These factors can effectuate a group of cognitive limitations and these limitations pose a sizeable amount of computational complexities. The depth of cognitive limitations determines the efficacy of the system in terms of QoPE.

Research Issue 3: Data Rates, Latency as Design Issues

The 6G brain communication scheme implementation finds limitation in delivering three types of services, providing improved EEG data transfer in broadband context, steadfast and low processing time, unwavering communication links between computational units and brain and a scope to accommodate the presence of internet driven parallel devices to establish a network. The limitation associated with latency does not involve in promising high data rates. These restricts thrusts the BCI based 6G systems to be designed to concurrently provide high steadfastness, low processing time, and improved data rates . Owing to the precarious characteristic of data flow in the Owing to the precarious characteristic of data flow in the, WBCI systems also adds up the complexity of the issue.

Research Issue 4: Establishment of WBCI Networks

Although B2B BTC linkages face many of the same issues as B2C BTC links, they also add a new dimension including the interactions between human brains with varying physiology and cognitive ability. To address this problem, researchers need to have a deeper knowledge of brain networks and how they interact. B2B communications, by definition, introduces a slew of multidisciplinary bottlenecks that demands an intense analysis of the prevailing scenarious associated with the transmission and reception characteristics of networks and also includes exchanges amidst the neuronal systems of parallel users with disparate contexts, demographics, and characteristics.

This chapter gives an exhaustive look at projected multidisciplinary issues that combines neurology, wireless communications, processing of signals, machine learning based algorithms. With the effective schemes for wireless and effective data transmission, it's also worth noting that this conjunction will happen in the 6G, and will facilitate WBCIs to effectively address various issues associated with of neural signals acquisition and communication with computational blocks. This chapter grouped the projected paybacks of this cooperative exploration as, Neurosciences for Wireless Communication, examines how advances in neuroscience may empower novel applications in 6G based WBCI, and novel signal acquisition and transmission schemes and aspires to replicate the human brain's functionalities. The book chapter presents the mechanics of 6G to establish new neuroscience research and development paradigms, such as novel 6G-empowered WBCIs.

SOLUTIONS AND RECOMMENDATIONS

This chapter finds a wide range of solutions for the research issues highlighted in Section 3. In the context of WBCI and 6G, this chapter presents the following solutions.

Facilitating Confidentiality in 6G Networks through Securing Physical Layer

The utility of 6G technologies for establishing communication between computational blocks and body area networks is ensured by facilitation of confidentiality through providing a gamut of security provisions at Physical layer. In this context, scientific research is increasingly focused on Wireless Body Area Networks, particularly patchable and implantable ultra-miniaturized devices, as well as data transactions between nodes of BAN. The futuristic trend that shall evolve is the human-being's neurological system as a node in WBAN. The nodes shall exhibit better acquisition characteristics to collect various biological signals for a variety of applications that includes monitoring, data collection and drugs development. The algorithms that facilitates safety provisions turns out effective gateways in protecting the nodes with high level of vulnerability to malevolent attacks and prone to connectivity disruption among the BAN. In ensuring the cohesiveness of the structure the algorithms should also ensure the communication hubs occupy minimal space and lower energy consumption.

The successful implementation of this scheme paves way to two enthralling and promising applications, Molecular Communication (MC) and Human Bond Communication (HBC). HBC is an innovative notion that integrates sensory sources of humans, sensory datas, data conversion, transmission, duplication at the receiving side to establish holistic transfer of among humans, and BCI. In addition to application development that includes, medical, gaming, augmented reality and virtual reality. MC involves the application of tenets of information theory to characterize the information transfer between the cells in the human physiological systems. This technology can derive immense benefits from this scheme to ensure security among the intra cellular communications and evolve into establishing Internet of Bio Nano things

Visible Light Communication Based Safety Mechanism for Physical Layer in 6G Networks:

The high response time and the complexity of the contemporary safety methods had catalyzed to formulate Physical Layer Security (PLS) scheme. Specifically for a 6G technology network, high data rate services is the foremost factor addressed by this scheme. The Visible Light Communications (VLC), one of the variants of optical wireless communications has several advantages, accommodation to large data pools, effective insulation mechanisms against electromagnetic fields, innate safety aspects, and open spectrum. This technology provides feasible solutions to the challenges endured by 6G. This technology employs optical frequencies by white LEDs for data encoding and the preliminary studies had concluded that the data rate for 6G is accurately matched by white LEDs. The desired applications includes, VLC underwater sensor network, Wireless Body Area Network (WBAN), Personal Area Network (PAN), Wireless Local Area Network (WLAN). VLC will also be beneficial in situations where typical low efficiency RF transmission fails to deliver.

New Era Sensor Networks

The 6G technologies can able to deliver it's optimal services for systems constructed by Brain simulating designs (Ahmad, I et al, 2019). An in-depth study of brain's structure and functional modalities guides the designer to design sensors which consumes less power as that of a neuron and an development of an interconnection scheme resembling that of the brain to develop comprehensive Intelligent sensor networks.

1. Sensor Networks with Intelligence:

The primitive sensor networks relied on exclusive sensors and non-reconfigurable connectivity scheme to the computational units, and this restricted their strategic positioning. Wireless Sensor Networks (WSNs) addresses this limitation and paved to a scape with large maneuvering potential. Specifically, the WSNs with low power consumption characterized by nodes with larger power life and exhibiting ability to utilize power reaping methods are highly useful in Internet of Things (IoT) era. Energy management in WSN has been the subject of extensive research, highlighting energy charging mechanisms and dissipating modes (Zhang, X et al 2017). The data acquisition, transfer of data and data reception aspects of these WSNs devour the major part of stored energy and extensive research has been carried out on channel compensation techniques, processor designs, and interconnection protocols. A moderate research in progress on development of decision schemes to fix data acquisition levels for different applications.

The parametric methods that includes Auto Regressive Moving Average (ARMA) or Moving Average (MA) method for time-series forecasting are straightforward and requires minimal computations and a potential to produce good yields for unpretentious observations. The complexity of the processing has to appropriately scale to cater multifaceted scenarios that includes high baud rate, presence of multiple sensor platforms and all-in-one nodes ((Zhang, L et al 2017), . The schemes should also ensure the quality of the data to be broadcasted for a given wireless communication channel. Several forms of microprocessor optimizations for the IoT have resulted from kernel schemes review and applied for specific data conversions (e.g., Transform tools for converting time domain description to frequency domain description). All of these methodologies, however, are propelled by data and the objective remains as effective data transmission from sensors to a computational unit for further operations.

2. Ultra low power inference:

Depending on the application, the power consumed by the upgraded versions of Spiking Neural Networks (SNN) for interpreting the behavior of the human brain has to be reduced by 10% to 20%. (Latva-aho, M et al, 2019). The sophistical level of the operation for any given SNN categorizes it into three types, low, medium and high and the power budget of the SNN varies respectively.

Numerous favorable guidelines are being framed, comprising of SNN derivational schemes from ANN s that has a temporal coding schemes which employs miniscule number of spikes outdating the contemporary coding schemas. The schemes with a capacity to tune itself to a specific power consumption level are highly desirable, that is, low power consumption and exhibits high level of sensitivity to occurrence of spikes ultra-low power wake-up nets for a smaller event, larger SNN for larger event (Kantola, R et al 2015) . Novel neuro scientific insights into various coding schemas of the brain, and interconnectivity methods shall serve as stimulus for solutions with high efficacy, swift response time and low power consumption.

3. Learning at Acquisition nodes:

The interpretation mechanism found to be extremely useful for detecting comprehensive characteristics, it is limited by the fact that does not the exhibit the efficacy of the active model, and the training is carried out remote cloud points on normalized data sets and the training yields are then installed on acquisition nodes. A gamut of programmes must carry out their operations in a variable environment conditions or adapting the user's unique qualities as in the case of BCI. The ability to acquire feasibility traits on the acquisition nodes, for any given bounded data, and is a desirable characteristic for real time applications (Zappone, A et al 2020). The computational costs, a sizeable latency and huge data heaps restrict the usage of deep learning techniques for learning at acquisition nodes. By imitating the processing schemes of the human brain, few techniques, not limited to, Hierarchial Temporal Memory (HTM), Spike-Timing Dependent Plasticity (STDP) are found to exhibit encouraging outcomes in the functional contexts but they still fall short of the precision of the more classic ANN solution. Further research activities on brain learning processes will facilitate development of learning schemes for acquisition nodes with low power.

This chapter aims to converge the predicted future applications and their problems in two threads, 6G Wireless Communication Technologies and neurophysiological sciences. The first is about how existing and new scientific/technological breakthroughs in neurosciences can be integrated into wireless systems. This includes everything from Implantable and wireless acquisition nodes, Performance metrics and successful brain imitation initiatives. The subsequent theme concerns intertwining wireless communication technologies (namely, 6G) and personal assistive systems for people debilitated by Systemic Sclerosis. These endeavors unearth the innate capabilities and constraints associated with the utility of the communication technologies for the neurological applications.

In the upcoming decade, the techniques employing DL can be used to minimize the time constraints associated with the wireless transmission of the biological signals. This shall be achieved by modeling the differences between the ideal wireless channel characteristics for 6G communications and practical wireless channel characteristics. The process can be carried out in online mode.

This chapter, in particular, forecasts the impact and an uncertainty surrounding deep learning-driven PHY layer technology and concludes that, performance deterioration shall be owed to off-site training and inappropriate channel compensation schemes. The impact of communication channels to introduce noise signals is decreased noise because channel coding assumes radio channel compensation. Furthermore, synchronization is done to achieve a high level of similarity for a targeted neurological signal length in the background of given channel, which is minimally prone to irregularities. The positioning technique is constructed on the near ideal channel characteristics which are less susceptible to channel discrepancies.

The most appropriate physical layer properties is to support the development of safety protocols for the 6G technologies personified by wide coverage and wide varieties of malevolent attacks.

- Using AI approaches for fine-tuning the safety algorithms of the physical layer in real time.
- Development of authorization protocols that take advantage of newly unknown PHY -layer characteristics while retaining ULL quality of service (QoS)
- Enabling transfer learning and non-intrusion of of acquisition node networks over diverse communication interfaces by designing suitable, exclusive processing schemes.
- Techniques for guaranteeing platform security when electrodes are dispersed throughout the platform.

In summary, Realization of establishing an active wireless communication scheme for neurological signal based systems is a burgeoning study field with a slew of unsolved issues arising from the described hurdles. One of the first unsolved issues in this field is the necessity for new methodologies that interlaces neurology and wireless communication schemes in order to accomplish high QoPE metrics. These techniques can also support to develop a solution based on data acquisition and processing schemes which arebased on DL machine learning approaches. These solutions can dynamically construct QoPE measurements by learning from network users' brain behavior. The key drawback of this technique highlights the requirement of high volume of datasets and sustained monitoring. This also creates a scope for QoPE improvement as in 6G based WBCIs.

As an alternative, development of laborious analytical scheme to model the various traits of brain's in addition to the data-driven approach. To understand the brain as a commanding center with a feedback scheme, one can employ current techniques from control theory and neuroscience, prospect and cognitive hierarchy theory and this insight can also be used to develop a statistical behavior model for a wide varieties of input (from the wireless network). The analysis the brain control system's transfer functions and comprehends its behaviour in response to various input excitations from a wireless network. We may be able to analyse how QoS metrics are transformed into QoPE using this method. Existing studies on generation of the brain's control messages can be used for the improvement. The behavioral investigations and outcomes can be integrated with behavioral structures that presents a decision making model of human brain.

CONCLUSION

With the research phase of 5G networks coming to a conclusion and deployment looming, 6G technologies has risen to top of innumerable academics' minds. The 6G technologies will surely improve the provision of services in comparison with previous generations. This chapter delves into the concerns associated with safety and non-intrusion of individual's private space surrounding 6G technologies, as well as providing a summary of the pinnacles scaled by different communication technologies ranging from 5G to 1G, setting the groundwork for the 6G network's development. This chapter also looks at four major aspects of the 6G network in order to reveal security concerns about future technologies. The inquiry comes to a close with a consideration of the possible uses for WBCI that the 6G network will enable.

These PADs will improve self-reliance functionalities and independence for people suffering from various disorders arising from system spinal cord damages obstructing the communication channel between the brain and physiological subsystems with the appropriate implementation of wireless BCI in the context of 6G technology. This chapter provides a complete overview of WBCI and 6G technologies, as well as the applicability of AI-based strategies in addressing security and privacy challenges originating from the deployment of 6G networks in WBCI contexts

REFERENCES

Ahmad, I., Shahabuddin, S., Kumar, T., Okwuibe, J., Gurtov, A., & Ylianttila, M. (2019). Security for 5G and Beyond. *IEEE Communications Surveys and Tutorials*, *21*, 4.

Bernstein, D., & Lange, T. (2017). Post-quantum cryptography. *Nature*, *549*, 7671.

Chen, M., Challita, U., Saad, W. C., Yin, C., & Debbah, M. (2019). Artificial neural networks based machine learning for wireless networks: A tutorial. *IEEE Comm. Tutorials and Surveys*, *21*, 3039–3071.

Gindraux, S. (2002). From 2G to 3G: a guide to mobile security. *Third International Conference on 3G Mobile Communication Technologies.*

Gyongyosi, L., & Imre, S. (2019). A survey on quantum computing technology. *Computer Science Review*, *31*, 51–71.

Kabir, H., Mohsin, H., & Kantola, R. (2020). Implementing Security Policy Management for 5G Customer Edge Nodes. *NOMS 2020 - 2020 IEEE/IFIP Network Operations and Management Symposium,* 1-8.

Kaddoum, G. (2016). Wireless chaos-based communication systems: A comprehensive survey. *IEEE Access: Practical Innovations, Open Solutions*, *4*, 2621–2648.

Kantola, R., Llorente, J., Santos, N., & Beijar. (2015). Policy Based Communications for 5G Mobile with Customer Edge Switching. *Wiley Security and Communication Networks.*

Latva-aho, M., & Leppänen, M. (2019). *Key Drivers and Research Challenges for 6G Ubiquitous Wireless Intelligence. In 6G Flagship.* University of Oulu.

Letaief, K. B., Chen, W., Shi, Y., Zhang, J., & Zhang, Y. J. A. (2019). The Roadmap to 6G: AI Empowered Wireless Networks. *IEEE Communications Magazine*, *57*(8), 84–90. doi:10.1109/MCOM.2019.1900271

Strukov, D., Indiveri, G., Grollier, J. S., & Fusi, S. (2019). Building brain inspired computing. *Nature Communications*, 10.

Wolpaw, J. R., Birbaumer, N. D., McFarland, J., Pfurtscheller, G., & Vaughan, T. M. (2002). Brain–Computer Interfaces for Communication and Control. *Clinical Neurophysiology*, *113*(6), 767–791.

Yang, P., Xiao, Y., Xiao, M., & Li, S. (2019). 6G Wireless Communications: Vision and Potential Techniques. *IEEE Network*, *33*(4), 70–75. doi:10.1109/MNET.2019.1800418

Zhang, L., Yan, Z., & Kantola, R. (2017). *Privacy-preserving trust management for unwanted traffic control.* Elsevier Future Generation Computer Systems.

Zhang, X., Kunz, A. S., & Schroder, S. (2017). Overview of 5G security in 3GPP. *2017 IEEE Conference on Standards for Communications and Networking (CSCN)*, 181-186.

Chapter 14
6G:
Transformation of Smart Cities With Blockchain and AI

Sunil Tekale
Malla Reddy College of Engineering, India

Rajagopal R.
Narsimha Reddy Engineering College, India

Bhoopathy V.
Malla Reddy College of Engineering, India

ABSTRACT

It is really a very challenging task for the private sector, public sector, and the government to make use of this blockchain technology for various applications in smart cities. The task is not going to be easy for the system for providing better and the best services for the people in the country. The solution for all the policies to be implemented is to be by making use of blockchain technology. This technology will create trust on the policymakers and will also help in reducing the cost of data storing along with security for the data stored. The technology provides security by decentralization process and will allow anyone to verify the records. As we all know, trust is going to be a big question when it comes to finance and any other kind of dealing, but blockchain will help to regain and maintain the factor of trust as it is a transparent system allowing the people to verify and check what is happening behind the policy. This banking concept is considered as it plays a vital role in every person's life. 6G will change remote correspondence by utilizing AI.

DOI: 10.4018/978-1-6684-3804-6.ch014

INTRODUCTION

In order to address to the problems in banking sector and in order to provide security for the data stored by reducing the cost and also by building the trust the blockchain technology is used. This technology provides a very transparent and amicable solution for the problem described as it will create an impact on every single individual. This technology in order to maintain privacy and security will make use of complex mathematical algorithms and maintains transparency in the system by distributing the data across many systems with a timestamp, by making the data stored as immutable. This will help to see that the data is not hacked and it is near to impossible to do so. All this can be achieved by reducing the cost of the project. As the importance of data is increasing day by day and the survival of the business is also based on possession of data. It is the time to convert existing system and transform the same on to blockchain technology. The various benefits of this technology are it helps in reducing the conversation between the consumer and the provider, it also helps in building trust among the consumers, it provides network resilient to frauds and cybercrimes. This system also helps in reducing the middle men in the system and provides tamper proof solution.

Today we do not have a single largest database which addresses the issues and provides services to our citizens in the country, which in this case is referred as Nitizens. The paper deals right from creation of the database, so as to cater to the needs of the nitizens in terms of services and issues faced. Blockchain technology is used in order to store the details of the nitizens as the same will help to provide security and privacy as it will not allow any record to be modified once it is recorded. The database created with blockchain will be finest and simplest in terms of handling challenges pertaining to security. Blockchain makes use of various mathematical algorithms which are complicated in nature to provide huge amount of security to the data stored.The cost to store the data using this technology will be very less compared to various other methods with huge and high security features provided in terms of various layers. The data stored here is not centralized but will be stored in decentralized fashion with perfect cryptography. To make the application more smarter we can use Artificial Intelligence in order to see that the nitizen do not face any kind of problems in terms of getting timely and right kind of service. The role of AI is to make the nitizen alert about various duties and responsibilities which the nitizen has to follow and implement.

Figure.1 shown as Block chain network the database will get the information about the nitizen as soon as the nitizen is born. The initial entry of this will happen with the help of hospital administrator. The administrator will feed the necessary information as per the format of the database along with the security code available. Like the initial name along with the details of the parents (Mother and Father) including details like place of birth, time of birth, caste, religion (which can be optional) along with security id which can be the image of the retinal of father+mother+the child. After the initial information is provided the software will generate a unique id which is further referred as nitizens id-NID.As the initial update happens with the security code of the hospital administrator the record of the newly born will be made public, so that when the nitizen visits any other hospital then the in charge over there can access the data with the help of the security code or the nitizen id provided. The nitizen profile can be viewed by any organization as and when the nitizen requires the service from that organization by feeding the NID including the scanning of the retina. At every point of time when the data has to be entered the in charge has to do so by his personnel identification only. This will not only help the organization and the nitizen but will also improve the efficiency of the overall system in terms of avoiding duplicate entries and avoiding feeding of the data for multiple times at multiple places. The government as and when requires can take a glance of the data available and can plan for the schemes which are beneficial for the nitizens at large.

Figure.2 shown as clouds in all these technologies hold great promise for use in various6G network applications, such as multi-sensory X Reality (multi-sensory XR) applications, connected robotics and autonomous systems, wireless brain-computer interactions, and blockchain and distributed ledger technologies.

Figure 1. Block chain network

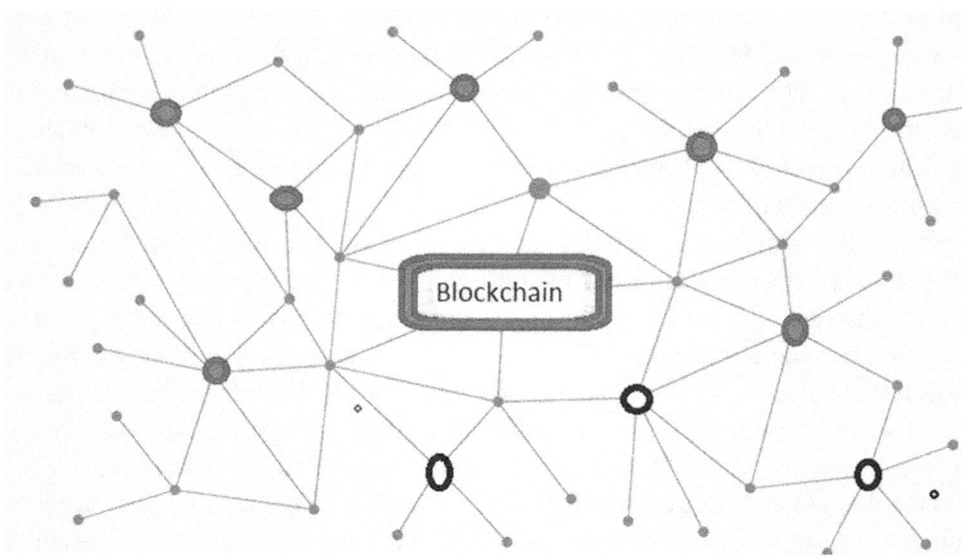

Figure 2. Security and privacy issues in the 6G network.

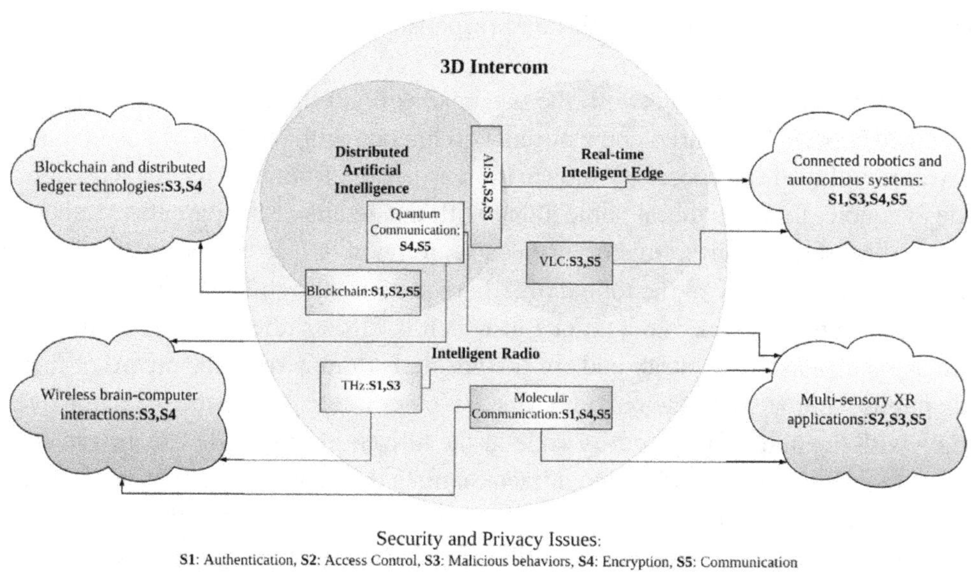

The important aspect is to safe guard the data in terms of theft and also loss of data by means of complex authentication process and backup mechanism. The overall concept is to see that not even a single time the details pertaining to the nitizens are to be entered for the second time.

There will be lot of challenges in terms of safeguarding the data available as the same will be very important for the country at large(Carley et al., 2006). The paper also deals with the algorithm designed to generate the nitizens id, which takes care of the duplicate values in the database.

Limitations: The paper deals with creation and updating of the records pertaining to the nitizens in the country and showcasing the data in terms of graphs so as to make comfort in understanding the data available. As the data is voluminous it is very important to view with various angles by generating different graphs. The concept related to generation of security id is taken care along with the method of using the data by various organizations is mentioned. The challenging aspect of security is taken care of and issues pertaining are addressed. The overall objective is to see that the data stored pertaining to the nitizen is no time entered again by any organization at large.

Literature review: The concept covered are limited to an extent of a particular organization and not across all the organizations, which basically will result in duplicate data stored and will not serve the purpose of the nitizen database(Verbeke, 2001). It literature do not address the security issues in maintaining the database as well the backup mechanism for the same. The existing data base which is used is created by every individual organization and will definitely have duplicate values when integrated. So in order to avoid duplicate data and in order to provide a unique identification number the concept of having nitizen id is evolved. This will help us to avoid the problem we face if the existing individual organization databases are integrated.

Some components of the 5G network have already been considered, and some components have already deployed AI as their backbone, e.g., channel coding and estimation in the physical layer, multiple access in the MAC layer, and various applications in the network layer (Huang et al., 2019). However, AI applications are not common, and the, the support for AI-driven technologies in 5G networks is limited by the constraints of the traditional architecture that was available in the early stages of its conception.

Accordingly, there is no support for the distributed AI or the intelligent radio, as these two areas are fully AI-based. Moreover, although the real-time intelligent edge, such as vehicle networks, have already been implemented for 5G networks, emergency conditions cannot be handled in "real-time" due to latency issues. However, 6G networks can. For example, the radio latency of 6G networks is 0.1 ms, which is one-tenth of that of 5G networks (Kantola & Pouttu, 2019). Moreover, 5G coverage is still only at the ground level; space and underwater communications at some levels of 3D intercoms are not possible.

Compared with all other technologies expected to be used in 6G networks, AI is widely considered to be one of the key parts of the future network infrastructure. It is an understatement to say that artificial intelligence has attracted a lot of attention in the field of network. Although AI in the 5G network is ostensibly operated in isolated areas where massive amounts of training data and powerful but private computing hubs are available, AI will become more of a core component of the 6G network. The physical layers, which includes devices such as data links and network infrastructure; and the computing layers, which include software-defined networks, network function virtualization and cloud/edge/fog computing, etc.

Emerging smart city and industry applications such as smart vehicles, autonomous driving, delivery drones, smart manufacturing, and smart robots generally require precision control, strict guarantee of ultra-low latency (typically below 1 ms) and reliability of 99.9999% [13]. These applications are well-investigated in 5G under the URLLC use case. On the other hand, industrial and massive IoT are

Table 1. Comparison between the Key Performance Indicator of 5G and 6G

Celluar network	5g	6g
Data rate	1 Gb/s	1 Tb/s
Bandwidth(MHz)	100	>=300
Latency(ms)	1	0.001-0.1
Reliablility(%)	99.999	99.99999
Spectral efficiency	3xover 4G	10-100X Over 5G
Energy efficiency	10X100X Over 4G	10-100x Over 5G
Connection density (devices/Km2)	10^6	$>=10^7$
Traffic Capacity	10 $Mb/s/m^2$	1-10 $Gb/s/m^3$(volumetric)
Mobility(Km/h)	500	1000
Network Trait	Transmission-centric	Transmission-and computing centric

well-explored under the 5G mMTC use case, which in general corresponds to low-power applications. However, in the era of 6G, it is foreseen that all the aforementioned applications will be widespread. Extreme network densification give rise to a large number of wirelessly connected sensors, cameras, caches, computing elements, and controllers (ranging from low-power ones to high-power ones) concentrated within every small area. Consequently, a huge volume of data will be generated and will need to be transmitted across the 6G network. Furthermore, given the advancement in the area of XR (e.g., augmented reality), it will be introduced into smart factory for remote control and inspection, which will require a large bandwidth. Besides, future smart applications will be empowered by AI, which consist of computational-intensive operations. Furthermore, energy consumption will be enormous with the massive amounts of communications, processing, and sensing [14], [15]. Hence, a new use case category on massive autonomous systems - concerning latency, reliability, bandwidth, data density, connection density, computational load and energy efficiency - need to be defined for 6G, and both mMTC and URLLC would just be special use cases under this new category. Besides that, intercity travel with high-speed transportation such as high-speed trains will become more prevalent across the nation, and this will foster the need for extremely high reliable connectivity.

Algorithm: The algorithm uses almost the same concept of luhn algo[5], which is used for generating the number for financial cards like the American, Visa, Master card issued by various financial institutions.

The formula verifies a number against its included check digit(Renn, Ortwin; Klinke, 2001). Counting from the check digit, which is the rightmost and moving left, double the value of every second digit. If for all the products p < 10 where p is equal to either n or 2n, then calculate

$$(\sum_{i=0}^{k} 2n_{2i} + \sum_{i=0}^{k} n_{2i+1})$$

Modulus 10, where n ∈ Z10 and k ∈ N but if ∃ p ≥ 10 add all the digits of the product generated to obtain a single digit, that is ∀a, b, c ∈ Z10, p = ab = a+b = c before calculation of the checksum modulo 10. If the total modulo10 is equal to 0 (if the total ends with zero) then the number is valid according to the Luhn formula otherwise it is invalid (Gandhi, 2015).

Let a1,a2...and be an account number that will have a check digit added making it of the form a1,a2... anx. from the rightmost double every even positioned digit to obtain v1v2...vnx. If ∃ vi ≥ 10, sum the digits of the product (e.g. 10 = 1+0 = 1,18 = 1+8 = 9) to obtain a single digit before computing the check digit x = 10−(n Σ i=1 vi) mod 10 but if ∀ vi, vi < 10 calculate the check digit x directly as x = 10−(n Σ i=1 vi) mod 10.

The check digit x will only be zero if the checksum is a multiple of 10 or has 10 as a factor otherwise the check digit is in the set {1, 2, ..., 9}. Example The check digit x of (888x) using the Luhn formula is

Table 2. Check Digit Calculation in Luhn Formula

Number under consideration	8	8	8	X
Multiply every even position digit from the right most by 2 to get product p	16	8	16	x
If p>=10 that is p=ab a,b,n add digits together a+b=n	7	8	7	x

Checksums =4ai=0 n2i+ 4a i=0 n2i+1 = 22.If ∀ p, p < 10 calculate the checksum = aki=1 pi where k, i N.Check digit x = 10−checksum(s)(mod 10) = 8. Therefore the number that passes the Luhn formula is 8888 (Gandhi, 2015). Using the above algorithm and added to that we can have a prefix of 4 digits to indicate the name of the state where the record is created first time, this will be the nitizen id and then the same can be used for all further operations. With this we can have no's up to 1000 crores.

The concept is to see that the population that exists in our country and which will be for coming 100 years we must able to accommodate(Epstein, M., 2002). So that there will be a unique number which is referred as NID will be useful for any further actions.

Proposed Methodology: In the proposed methodology a huge, large and voluminous database will emerge where the data pertaining to every born is maintained, right from the day the person comes on the earth till the day the person remains on the earth. All entries pertaining to every nitizen will be maintained in this database and will be tracked using nitizen identification number. During the course of making entries care is to be taken so that every entry made is done with the help of personnel identification of the authorized person in charge. So as to have responsibility and clarity in terms of any ambiguity that arises at any given point of time(Chiaraviglio, 2016). The proposed method will take care of duplicate data with the help of nitizen identification number provided. The flow of the data will be from one organization to other and continues at every place where the nitizen wishes to take the required services. This concept will help to reduce reentry of any part of nitizen information. Figure.3 shown nitizen database the proposed method takes care of the various measures in order to ensure safety and privacy of the data that is generated and entered. The proposed method also help in taking care of backup of the data and availability of the same all the time.

So using the above number and then adding up a prefix of 4 chars by taking the name of the state in to account and unique number is generated. For example for the state of Telangana the number will be TELG5113 1235 5698 5623, similarly for MAHA5113 1235 5698 5623 and for GUJT5113 1235 5698 5623 and so (Ali et al., 2017) .

The database has to undergo process of normalization

Table 3. Results: The method used will help to generate a unique Id for the nitizen so as to use for tracking of information which is created as

Random number	5113 1235 5698 5623
Check using Luhn algorithm Step:1 Reverse	3265 8965 5321 3115
Step 2: Multiply even position no with 2. If no>9 then subtract 9	3461 8961 5622 3211
Add all the digits	60
Then divide by 10 is the remainder is zero then it I a valid number	60 mod 10 Remainder 0

Figure 3. Nitizen database

The database architecture along with the scaling aspect of the database. Figure 4 shown the architecture used is a three tier architecture where the process of data entry happens at client side and the validation will happen at server side along with storage at the database side, which is the backend side .Different software are used in order to

Incremental backup procedure is used for talking the backup on a regular basis. It is also important to perform scaling of the data stored. In order to improve the performance of the database.

6G smart city applications involve vast amounts of communications, data exchange, sensing, storage, controls, and computing in supporting the upcoming massive interconnectivity in dense 6G networks. Indeed, supporting such connectivity requires sufficient capacity and efficient spectrum usage. Network densification is a viable way to support massive interconnectivity. By densely deploying low-power BSs, network capacity increases and more users/devices can be supported. The capacity can be fur-

Figure 4. Design and develop the three different layers

ther enhanced by exploiting the spectrum beyond sub-6 GHz bands such as mmWave and THz, which provide a substantial amount of bandwidth, and with appropriate access mechanisms, the network can effectively provide support for data-intensive use cases such as: (i) the delivery of large-size sensory data in vehicular communications, (ii) very-low latency and data-rate intensive sensory data delivery in XR, holographic communications and UAV applications in smart city, and (iii) connected autonomous robotics in Industry 4.0. However, network densification alone cannot keep up with the rapid growth in the number of wireless devices. In addition, inter-cell interference can severely limit the achievable capacity when BSs are in proximity. Although exploiting the mmWave or THz band potentially provides extremely-high data rates and capacity, it suffers from high channel attenuation and is susceptible to blockages. Another approach for supporting massive access is to reshape the traffic data arrival distribution into random access attempt distribution which can reduce burstiness of data arrivals. However, this may only be suitable for low-rate MTC devices(David & Berndt, 2018).

Conclusion: The proposed method works in accordance to the procedure adapted right from the inception phase of capturing the data till the process of providing information as and when required by making using of the nitizen identification number which is generated using luhn algorithm followed by adding up a prefix in terms of name of the state. The process of maintaining the data for specified number of years is also maintained including the process of backup mechanism. AI will be crucial to effectively implementing the key technologies and intelligently optimize the 6G performance under vastly dynamic conditions and environments. Therefore, it is hard to guarantee reliable communications and is problematic, because enabling future super-smart city require ultra-high reliable communications providing data rates of up to 1 Tbps under dynamic and dense scenarios with a discussion of the potential applications that the 6G network will support(Gursu et al., 2017).

REFERENCES

Ali, M. S., Hossain, E., & Kim, D. I. (2017). LTE/LTE-A Random Access for Massive Machine-Type Communications in Smart Cities. *IEEE Communications Magazine*, *55*(1), 76–83. doi:10.1109/MCOM.2017.1600215CM

Carley, K. M., Fridsma, D. B., Casman, E., Yahja, A., Altman, N., Chen, L. C., Kaminsky, B., & Nave, D. (2006). BioWar: Scalable agent-based model of bioattacks. *IEEE Transactions on Systems, Man, and Cybernetics. Part A, Systems and Humans*, *36*(2), 252–265. doi:10.1109/TSMCA.2005.851291

Chiaraviglio, L. (2016). 5G in rural and low-income areas: Are we ready? *ITU Kaleidoscope, 2016.*

David, K., & Berndt, H. (2018). 6G Vision and requirements: Is there any need for beyond 5g? *IEEE Vehicular Technology Magazine*, *13*(3), 72–80. doi:10.1109/MVT.2018.2848498

Epstein, M. J. (2002). Modeling Civil Violence: An Agent-Based Computational Approach on JSTOR. *Proceedings of the National Academy of Sciences of the United States of America*, *99*(10), 7243–7250. http://www.jstor.org/stable/3057848?origin=JSTOR-pdf&seq=1#page_scan_tab_contents

Gursu, H. M., Vilgelm, M., Kellerer, W., & Reisslein, M. (2017). Hybrid Collision Avoidance-Tree Resolution for M2M Random Access. *IEEE Transactions on Aerospace and Electronic Systems*, *53*(4), 1974–1987. doi:10.1109/TAES.2017.2677839

Huang, T., Yang, W., Wu, J., Ma, J., Zhang, X., & Zhang, D. (2019). A Survey on Green 6G Network: Architecture and Technologies. *IEEE Access: Practical Innovations, Open Solutions*, *7*, 175758–175768. doi:10.1109/ACCESS.2019.2957648

Kantola, R., & Pouttu, A. (2019). *Key drivers and research challenges for 6G ubiquitous wireless intelligence*. Issue September.

Renn, O., & Klinke, A. (2001). Environmental Risks : Perception, Evaluation and Management Article information. Environmental Risks: Perception, Evaluation and Management.

Verbeke, W. (2001). Beliefs, attitude and behaviour towards fresh meat revisited after the Belgian dioxin crisis. *Food Quality and Preference*, *12*(8), 489–498. doi:10.1016/S0950-3293(01)00042-8

Compilation of References

(2016). Qiu, Wu, Ding, Xu, & Feng. (2016). A survey of machine learning for big data processing. *EURASIP Journal on Advances in Signal Processing*, (1), 67.

Elmeadawy, S., & Shubair, R. M. (2020). *Enabling Technologies for 6G FutureWireless Communications: Opportunities and Challenges*. arXiv:2002.06068.

Sengupta, K., Nagatsuma, T., & Mittleman, D. M. (2018). Terahertz integrated electronic and hybrid electronic–photonic systems. *Nature Electronics, 1*(12), 622–635. doi:10.103841928-018-0173-2

Nagatsuma, T. (2019). Advances in Terahertz Communications Accelerated by Photonics Technologies. *Proceedings of the 2019 24th International Conference on OptoElectronics and Communications Conference (OECC)*, 1–3.

Chowdhury, M. Z., Shahjalal, M., Ahmed, S., & Jang, Y. M. (2019). *6g wireless communication systems: Applications, requirements, technologies, challenges, and research directions*. arXiv preprint arXiv:1909.11315.

Pon, Leow, Rahim, Eteng, & Kamarudin. (2019). Printed spiral resonator for displacement-tolerant near-field wireless energy transfer. *IEEE Access, 7*.

Yang, P., Xiao, Y., Xiao, M., & Li, S. (2019). 6g wireless communications: Vision and potential techniques. *IEEE Network, 33*(4), 70–75. doi:10.1109/MNET.2019.1800418

Goian, A., AlHajri, M. I., Shubair, R. M., Weruaga, L., Kulaib, A. R., AlMemari, R., & Darweesh, M. (2015). Fast detection of coherent signals using pre-conditioned root-music based on toeplitz matrix reconstruction. In *2015 IEEE 11th International Conference on Wireless and Mobile Computing, Networking and Communications (WiMob)*. IEEE. 10.1109/WiMOB.2015.7347957

AlHajri, M., Goian, A., Darweesh, M., AlMemari, R., Shubair, R., Weruaga, L., & Kulaib, A. (2015). Hybrid rss-doa technique for enhanced wsn localization in a correlated environment. In *2015 International Conference on Information and Communication Technology Research (ICTRC)*. IEEE. 10.1109/ICTRC.2015.7156466

Tsiropoulos, G. I., Yadav, A., Zeng, M., & Dobre, O. A. (2017). Cooperation in 5G HetNets: Advanced spectrum access and D2D assisted communications. *IEEE Wireless Communications, 24*(5), 110–117. doi:10.1109/MWC.2017.1700082

Tang, W., Chen, M. Z., Dai, J. Y., Zeng, Y., Zhao, X., Jin, S., Cheng, Q., & Cui, T. J. (2019). *Wireless communications with programmable metasurface: New paradigms, opportunities, and challenges on transceiver design*. arXiv:1907.01956.

Alsharif, M. H., Kelechi, A. H., Yahya, K., & Chaudhry, S. A. (2020). Machine Learning Algorithms for Smart Data Analysis in Internet of Things Environment. *Taxonomies and Research Trends. Symmetry, 12*, 88.

Albreem, M., Alsharif, M. H., & Kim, S. A. (2020). Low Complexity Near-Optimal Iterative Linear Detector for Massive MIMO in Realistic Radio Channels of 5G Communication Systems. *Entropy (Basel, Switzerland)*, *22*(4), 388. doi:10.3390/e22040388 PMID:33286162

Ng, Duong, Zhong, & Schober. (2019). *Wireless information and power transfer: Theory and practice*. John Wiley & Sons.

Shahraki, A., Taherkordi, A., & Haugen, Ø. (2021). Tonta: Trend-based online network traffic analysis in ad-hoc iot networks. *Computer Networks*, *194*, 108125. doi:10.1016/j.comnet.2021.108125

Zhang, S., Liu, J., Guo, H., Qi, M., & Kato, N. (2020). Envisioning device-to-device communications in 6G. *IEEE Network*, *34*(3), 86–91. doi:10.1109/MNET.001.1900652

Zaidi, A. A., Baldemair, R., Moles-Cases, V., He, N., Werner, K., & Cedergren, A. (2018, June). OFDM Numerology Design for 5G New Radio to Support IoT, eMBB, and MBSFN. *IEEE Communication Standards Magazine*, *2*(2), 78–83. doi:10.1109/MCOMSTD.2018.1700021

Tariq, F., Khandaker, M., Wong, K.-K., Imran, M., Bennis, M., & Debbah, M. (2019). *A speculative study on 6g*. arXiv:1902.06700.

Jiang & Liu. (2017). An overview of 5G requirements. In *5G Mobile Communications*. Springer.

Gupta, R., Shukla, A., & Tanwar, S. (2020, December). BATS: A blockchain and AI-empowered drone-assisted telesurgery system towards 6G. *IEEE Transactions on Network Science and Engineering*.

Zhang, L., Liang, Y.-C., & Niyato, D. (2019, August). 6G visions: Mobile ultrabroadband, super internet-of-things, and artificial intelligence. *China Communications*, *16*(8), 1–14. doi:10.23919/JCC.2019.08.001

Lu, Y., & Zheng, X. (2020). 6G: A survey on technologies, scenarios, challenges, and the related issues. *Journal of Industrial Information Integration*, *19*, 100158. doi:10.1016/j.jii.2020.100158

Marco, G. (2020). Toward 6G networks: Use cases and technologies. *IEEE Communications Magazine*, *58*(3), 55–61. doi:10.1109/MCOM.001.1900411

Aazam, M. E.-N., & Huh. (2015). Fog computing micro datacenter based dynamic resource estimation and pricing model for iot. *IEEE 29th International Conference on Advanced Information Networking and Applications*, 687–694.

Aazhang, B., Ahokangas, P., Alves, H., Alouini, M. S., Beek, J., Benn, H., . . . Zhu, P. (2019). *Key drivers and research challenges for 6G ubiquitous wireless intelligence (white paper)*. Oulu, Finland: 6G Flagship, University of Oulu.

Abadi, M., Chu, A., Goodfellow, I., McMahan, H. B., Mironov, I., Talwar, K., & Zhang, L. (2016). Deep learning with differential privacy. *Proceedings of the 2016 ACM SIGSAC Conference on Computer and Communications Security*, 308–318.

Abbas, K., Tawalbeh, L. A. A., Rafiq, A., Muthanna, A., Elgendy, I. A., El-Latif, A., & Ahmed, A. (2021). Convergence of Blockchain and IoT for Secure Transportation Systems in Smart Cities. *Security and Communication Networks*.

Abosata, N., Al-Rubaye, S., Inalhan, G., & Emmanouilidis, C. (2021). Internet of Things for System Integrity: A Comprehensive Survey on Security, Attacks and Countermeasures for Industrial Applications. *Sensors (Basel)*, *21*(11), 3654. doi:10.339021113654 PMID:34073975

Ahmad, I., Shahabuddin, S., Kumar, T., Okwuibe, J., Gurtov, A., & Ylianttila, M. (2019). Security for 5G and Beyond. *IEEE Communications Surveys and Tutorials*, *21*, 4.

Ailaja, Seetharamulu, & Jabbar. (2018). Machine learning in healthcare: A review. *2018 Second International Conference on Electronics Communication and Aerospace Technology (ICECA)*, 910-914.

Akan, O. B., Ramezani, H., Khan, T., Abbasi, N. A., & Kuscu, M. (2016). Fundamentals of molecular information and communication science. *Proceedings of the IEEE, 105*(2), 306–318. doi:10.1109/JPROC.2016.2537306

Akhtar, M. W., Hassan, S. A., Ghaffar, R., Jung, H., Garg, S., & Shamim Hossain, M. (2020). The shift to 6G communications: Vision and requirements. *Human-centric Computing and Information Sciences, 10*(1), 1–27.

Akyildiz, I. F., Jornet, J. M., & Han, C. (2014). Terahertz band: Next frontier for wireless communications. *Physical Communication, 12*, 16–32. doi:10.1016/j.phycom.2014.01.006

Al Hayajneh, A., Bhuiyan, M. Z. A., & McAndrew, I. (2020). Improving Internet of Things (IoT) security with software-defined networking (SDN). *Computers, 9*(1), 8.

Al-Ansi, A., Al-Ansi, A. M., Muthanna, A., Elgendy, I. A., & Koucheryavy, A. (2021). Survey on Intelligence Edge Computing in 6G: Characteristics, Challenges, Potential Use Cases, and Market Drivers. *Future Internet, 13*(5), 118. doi:10.3390/fi13050118

Al-Eryani, Y., & Hossain, E. (2019). The D-OMA method for massive multiple access in 6G: Performance, security, and challenges. *IEEE Vehicular Technology Magazine, 14*(3), 92–99. doi:10.1109/MVT.2019.2919279

Ali, S., Saad, W., Rajatheva, N., Chang, K., Steinbach, D., Sliwa, B., Wietfeld, C., Mei, K., Shiri, H., Zepernick, H.-J., Chu, T. M. C., Ahmad, I., Huusko, J., Suutala, J., Bhadauria, S., Bhatia, V., Mitra, R., Amuru, S., Abbas, R., & Malik, H. (2020). *6G White Paper on Machine Learning in Wireless Communication Networks*. Academic Press.

Ali, M. S., Hossain, E., & Kim, D. I. (2017). LTE/LTE-A Random Access for Massive Machine-Type Communications in Smart Cities. *IEEE Communications Magazine, 55*(1), 76–83. doi:10.1109/MCOM.2017.1600215CM

Alsharif, M. H., & Nordin, R. (2017). Evolution towards fifth generation (5G) wireless networks: Current trends and challenges in the deployment of millimetre wave, massive MIMO, and small cells. *Telecommunication Systems, 64*(4), 617–637. doi:10.100711235-016-0195-x

Analytics, S. (n.d.). *Smartphones: Global Artificial Intelligence Technologies Forecast: 2010 to 2023*. Retrieved September 26, 2021, https://www.strategyanalytics.com/access-services/devices/mobile-phones/emerging-device-technologies/market-data/report-detail/smartphones-global-artificial-intelligence-technologies-forecast-2010-to-2023

Arel, I., Rose, D. C., & Karnowski, T. P. (2010). Deep machine learning-a new frontier in artificial intelligence research. *IEEE Computational Intelligence Magazine, 5*(4), 13–18.

Arias, Salado, Medina, & Zambrano. (2021). *5G technology deployment in Latin America: An analysis of public policy and regulation environment*. Academic Press.

Aversano, L., Bernardi, M. L., Cimitile, M., & Pecori, R. (2021). A systematic review on Deep Learning approaches for IoT security. *Computer Science Review, 40*, 100389.

Awan, S., Ahmed, S., Ullah, F., Nawaz, A., Khan, A., Uddin, M. I., ... Alyami, H. (2021). IoT with BlockChain: A Futuristic Approach in Agriculture and Food Supply Chain. *Wireless Communications and Mobile Computing*.

Babukarthik, R. G. (2021). 15 Lethal Vulnerability of Robotics in Industrial Sectors. In *Big Data Management in Sensing: Applications in AI and IoT* (pp. 227–238). River Publishers.

Banumathi, J., Sangeetha, S. K. B., & Dhaya, R. (2022). Robust Cooperative Spectrum Sensing Techniques for a Practical Framework Employing Cognitive Radios in 5G Networks. *Artificial Intelligent Techniques for Wireless Communication and Networking*, 121-138.

Barber, D. (2012). *Bayesian Reasoning and Machine Learning*. Cambridge University Press.

Basir, R., Qaisar, S., Ali, M., Aldwairi, M., Ashraf, M., Mahmood, A., & Gidlund, M. (2019). Fog Computing Enabling Industrial Internet of Things: State-of-the-Art and Research Challenges. *Sensors (Basel), 19*, 4807. doi:10.339019214807

Bergano, N. S. (2012). Undersea fber-optic cables make the web worldwide. *2012 17th Opto-Electronics and Communications Conference.*

Bernstein, D., & Lange, T. (2017). Post-quantum cryptography. *Nature, 549*, 7671.

Bibri, S. E., & Krogstie, J. (2017). Smart sustainable cities of the future: An extensive interdisciplinary literature review. *Sustainable Cities and Society, 31*, 183–212.

Bin Ahammed & Patgiri. (2020). 6G and AI: The Emergence of Future Forefront Technology. *Advanced Communication Technologies and Signal Processing (ACTS),* 1-6. . doi:10.1109/ACTS49415.2020.9350396

Biral, A., Centenaro, M., Zanella, A., Vangelista, L., & Zorzi, M. (2015). The challenges of M2M massive access in wireless cellular networks. *Digital Communications and Networks, 1*(1), 1–19. doi:10.1016/j.dcan.2015.02.001

Bird, E., Fox-Skelly, J., Jenner, N., Larbey, R., Weitkamp, E., & Winfield, A. (2020). *The Ethics of Artificial Intelligence: Issues and Initiatives*. European Parliamentary Research Service. Available: https://www.europarl.europa.eu/RegData/etudes/STUD/2020/634452/EPRS_STU(2020)634452_EN.pdf

Bouzinis, P. S., Diamantoulakis, P. D., & Karagiannidis, G. K. (2021) *Wireless Federated Learning (WFL) for 6G Networks — Part I: Research Challenges and Future Trends.* arXiv:2105.00842.

Brito, A., Sebastião, P., & Souto, N. (2019, May). Jamming for Unauthorized UAV Operations-Communications Link. In *2019 International Young Engineers Forum (YEF-ECE)* (pp. 94-98). IEEE.

Brown, S. (2021). *Machine learning, explained.* https://mitsloan.mit.edu/ideas-made-to-matter/machine-learning-explained

Calvanese Strinati, E., Barbarossa, S., Gonzalez-Jimenez, J. L., Ktenas, D., Cassiau, N., Maret, L., & Dehos, C. (2019). 6G: The Next Frontier: From Holographic Messaging to Artificial Intelligence Using Subterahertz and Visible Light Communication. *IEEE Vehicular Technology Magazine, 14*(3), 42–50. doi:10.1109/MVT.2019.2921162

Carley, K. M., Fridsma, D. B., Casman, E., Yahja, A., Altman, N., Chen, L. C., Kaminsky, B., & Nave, D. (2006). BioWar: Scalable agent-based model of bioattacks. *IEEE Transactions on Systems, Man, and Cybernetics. Part A, Systems and Humans, 36*(2), 252–265. doi:10.1109/TSMCA.2005.851291

Carroll, N. (2016). Key success factors for smart and connected health software solutions. *Computer, 49*(11), 22–28.

Chassignol, M., Khoroshavin, A., Klimova, A., & Bilyatdinova, A. (2018, January). Artificial intelligence trends in education: A narrative overview. *Procedia Computer Science, 136*, 16–24.

Chen, C., Bian, R., & Haas, H. (2018, May). Omnidirectional transmitter and receiver design for wireless infrared uplink transmission in LiFi. In *2018 IEEE International Conference on Communications Workshops (ICC Workshops)* (pp. 1-6). IEEE. 10.1109/ICCW.2018.8403727

Chen, M. Y., Fan, M. H., & Huang, L. X. (2021). AI-based vehicular network toward 6G and IoT: Deep learning approaches. *ACM Transactions on Management Information Systems, 13*(1), 1–12.

Chen, M., Challita, U., Saad, W. C., Yin, C., & Debbah, M. (2019). Artificial neural networks based machine learning for wireless networks: A tutorial. *IEEE Comm. Tutorials and Surveys, 21*, 3039–3071.

Chen, R., Li, C., Yan, S., Malaney, R., & Yuan, J. (2019). Physical layer security for ultra-reliable and low-latency communications. *IEEE Wireless Communications, 26*(5), 6–11. doi:10.1109/MWC.001.1900051

Chiaraviglio, L. (2016). 5G in rural and low-income areas: Are we ready? *ITU Kaleidoscope, 2016.*

Chiaraviglio, L., Cacciapuoti, A. S., Martino, G. D., Fiore, M., Montesano, M., Trucchi, D., & Melazzi, N. B. (2018). Planning 5G networks under EMF constraints: State of the art and vision. *IEEE Access: Practical Innovations, Open Solutions*, *6*, 51021–51037. doi:10.1109/ACCESS.2018.2868347

Chi, N., Zhou, Y., Wei, Y., & Hu, F. (2020). Visible Light Communication in 6G: Advances, Challenges, and Prospects. *IEEE Vehicular Technology Magazine*, *15*(4), 93–102. doi:10.1109/MVT.2020.3017153

Cho, S., Chen, G., & Coon, J. P. (2019). Enhancement of physical layer security with simultaneous beamforming and jamming for visible light communication systems. *IEEE Transactions on Information Forensics and Security*, *14*(10), 2633–2648. doi:10.1109/TIFS.2019.2904440

Christidis, K., & Devetsikiotis, M. (2016). Blockchains and smart contracts for the internet of things. *IEEE Access: Practical Innovations, Open Solutions*, *4*, 2292–2303. doi:10.1109/ACCESS.2016.2566339

Cortes, C., & Vapnik, V. (1995). Support-vector networks. *Machine Learning*, *20*(3), 273–297. doi:10.1007/BF00994018

Cover, T., & Hart, P. (1967). Nearest neighbor pattern classification. *IEEE Transactions on Information Theory*, *13*(1), 21–27. doi:10.1109/TIT.1967.1053964

Czachorowski, K., Solesvik, M., & Kondratenko, Y. (2019). The application of blockchain technology in the maritime industry. In *Green IT Engineering: Social, Business and Industrial Applications* (pp. 561–577). Springer. doi:10.1007/978-3-030-00253-4_24

Dang, S., Amin, O., Shihada, B., & Alouini, M. S. (2020). What should 6G be? *Nature Electronics*, *3*(1), 20–29. doi:10.103841928-019-0355-6

David Neumann, W. U., & Wiese, T. (2017), Deep channel estimation. *21th International ITG Workshop on Smart Antennas*, 1–6.

David, K., & Berndt, H. (2018). 6G vision and requirements: Is there any need for beyond 5G? *IEEE Vehicular Technology Magazine*, *13*(3), 72–80.

David, K., & Berndt, H. (2018). 6G Vision and requirements: Is there any need for beyond 5g? *IEEE Vehicular Technology Magazine*, *13*(3), 72–80. doi:10.1109/MVT.2018.2848498

Deng, Zhao, Fang, Yin, Dustdar, & Zomaya. (2020). Edge Intelligence: The Confluence of Edge Computing and Artificial Intelligence. *IEEE Internet of Things Journal, 7*(8).

Devedžic, V. (2004). Web intelligence and artificial intelligence in education. *Educ. Technol. Soc.*, *7*(4), 29–39.

Devi, M., Dhaya, R., Kanthavel, R., Algarni, F., & Dixikha, P. (2019, May). Data Science for Internet of Things (IoT). In *International Conference on Computer Networks and Inventive Communication Technologies* (pp. 60-70). Springer.

Dhaya, Kanthavel, Devi, & Algarni. (2019). Assessment on Recurrent Applications of Machine Learning and its Behaviors. *International Journal of Engineering and Advanced Technology, 8*(6S3).

Dhaya, R., & Kanthavel, R. (2022). Energy Efficient Resource Allocation Algorithm for Agriculture IoT. *Wireless Personal Communication.* doi:10.1007/s11277-022-09607-z

Dhaya, R., & Kanthavel, R. (2022). IoE based private multi-data center cloud architecture framework. *Computers and Electrical Engineering, 100.* j. compeleceng.2022.107933 doi:10.1016/

Dhaya, Ujwal, Sharma, Singh, R., Selvan, & Krah. (2022). Energy-Efficient Resource Allocation and Migration in Private Cloud Data Centre. *Wireless Communications and Mobile Computing*. doi:10.1155/2022/3174716

Dhaya, R., & Kanthavel, R. (2020). Accomplishment Of Intellectual Supply Managing Algorithm For Ubiquitous Cloud Computing. *Gedrag & Organisatie Review*, *33*(3), 236–246.

Dhaya, R., & Kanthavel, R. (2022). Dynamic automated infrastructure for efficient cloud data centre. *Cmc-Computers Materials & Continua*, *71*(1), 1625–1639.

Dhaya, R., Kanthavel, R., & Ahilan, A. (2021). Developing an energy-efficient ubiquitous agriculture mobile sensor network-based threshold built-in MAC routing protocol (TBMP). *Soft Computing*, *25*(18), 12333–12342.

Dhaya, R., Kanthavel, R., Algarni, F., Jayarajan, P., & Mahor, A. (2020). Reinforcement Learning Concepts Ministering Smart City Applications Using IoT. In *Internet of Things in Smart Technologies for Sustainable Urban Development* (pp. 19–41). Springer.

Dhaya, R., Kanthavel, R., & Mahalakshmi, M. (2021). Enriched recognition and monitoring algorithm for private cloud data centre. *Soft Computing*, 1–11.

Ding & Poor. (2019). Simple semi-grant-free transmission strategies assisted by non-orthogonal multiple access. *IEEE Transactions on Communications, 67*(6), 4464-4478.

Domingos, P. (2012). A few useful things to know about machine learning. *Communications of the ACM*, *55*(10), 78–87.

Dong, P., Zhang, H., Li, G. Y., Gaspar, I. S., & NaderiAlizadeh, N. (2019). Deep CNN-Based Channel Estimation for mmWave Massive MIMO Systems. *IEEE Journal of Selected Topics in Signal Processing*, *13*(5), 989–1000. doi:10.1109/JSTSP.2019.2925975

Du, Ai, Wang, & Zhang. (2017). Overview of deep learning. *Proceedings - 2016 31st Youth Academic Annual Conference of Chinese Association of Automation YAC 2016*.

Du, J., Jiang, C., Wang, J., Ren, Y., & Debbah, M. (2020, December). Machine Learning for 6G Wireless Networks: Carrying Forward Enhanced Bandwidth, Massive Access, and Ultrareliable/Low-Latency Service. *IEEE Vehicular Technology Magazine*, *15*(4), 122–134.

Efrati, V., Limongelli, C., & Sciarrone, F. (2014). *A data mining approach to the analysis of students' learning styles in an e-learning community: A case study. In Universal Access in Human-Computer Interaction. Universal Access to Information and Knowledge*. Springer International Publishing.

Elamassie, M., & Uysal, M. (2020). Vertical Underwater Visible Light Communication Links: Channel Modeling and Performance Analysis. In IEEE Transactions on Wireless Communications (Vol. 19). IEEE.

Elsayed, M., & Erol-Kantarci, M. (2019). AI-enabled future wireless networks: Challenges, opportunities, and open issues. *IEEE Vehicular Technology Magazine*, *14*(3), 70–77.

Elsisi, M., Mahmoud, K., Lehtonen, M., & Darwish, M. M. (2021). Reliable industry 4.0 based on machine learning and IOT for analyzing, monitoring, and securing smart meters. *Sensors (Basel)*, *21*(2), 487.

Epstein, M. J. (2002). Modeling Civil Violence: An Agent-Based Computational Approach on JSTOR. *Proceedings of the National Academy of Sciences of the United States of America*, *99*(10), 7243–7250. http://www.jstor.org/stable/3057848?origin=JSTOR-pdf&seq=1#page_scan_tab_contents

Fang, H., Wang, X., & Tomasin, X. (2019). Machine Learning for Intelligent Authentication in 5G and Beyond Wireless Networks. *IEEE Wireless Communications*, *26*(5), 55–61.

Farsad, N., & Goldsmith, A. (2018). Neural network detection of data sequences in communication systems. *IEEE Transactions on Signal Processing*, *66*(21), 5663–5678. doi:10.1109/TSP.2018.2868322

Farsad, N., Yilmaz, H. B., Eckford, A., Chae, C. B., & Guo, W. (2016). A comprehensive survey of recent advancements in molecular communication. *IEEE Communications Surveys and Tutorials*, *18*(3), 1887–1919. doi:10.1109/COMST.2016.2527741

Ferraro, P., King, C., & Shorten, R. (2018). Distributed ledger technology for smart cities, the sharing economy, and social compliance. *IEEE Access: Practical Innovations, Open Solutions*, *6*, 62728–62746. doi:10.1109/ACCESS.2018.2876766

Gai, K., Wu, Y., Zhu, L., Zhang, Z., & Qiu, M. (2019). Differential privacy-based blockchain for industrial internet-of-things. *IEEE Transactions on Industrial Informatics*, *16*(6), 4156–4165. doi:10.1109/TII.2019.2948094

Gavra, V. D., & Pop, O. A. (2020, October). Usage of ZigBee and LoRa wireless technologies in IoT systems. In *2020 IEEE 26th International Symposium for Design and Technology in Electronic Packaging (SIITME)* (pp. 221-224). IEEE.

Gindraux, S. (2002). From 2G to 3G: a guide to mobile security. *Third International Conference on 3G Mobile Communication Technologies.*

Giordani & Zorzi. (2019). Toward 6G networks: Use cases and technologies. *IEEE Communications Magazine, 58*(3), 55-61.

Gkikopouli, A., Nikolakopoulos, G., & Manesis, S. (2012). A survey on Underwater Wireless Sensor Networks and applications. *2012 20th Mediterranean Conference on Control & Automation (MED).*

Gong, X., & Cortese, C. (2017, September). A socialist market economy with Chinese characteristics: The accounting annual report of China Mobile. *Accounting Forum*, *41*(3), 206–220. doi:10.1016/j.accfor.2017.04.002

Groves, B., & Pu, C. (2019, November). A Gini index-based countermeasure against sybil attack in the internet of things. In *MILCOM 2019-2019 IEEE Military Communications Conference (MILCOM)* (pp. 1-6). IEEE. doi:10.1109/ICESS49830.2020.9301593

Gu. (2018). Network topology reconfiguration for FSO-based fronthaul/backhaul in 5G+ wireless networks. *IEEE Access, 6*, 69426-69437.

Gui, G., Liu, M., Tang, F., Kato, N., & Adachi, F. (2020). 6G: Opening New Horizons for Integration of Comfort, Security, and Intelligence. *IEEE Wireless Communications*, *27*(5), 126–132. doi:10.1109/MWC.001.1900516

Guo, W. (2020). Explainable Artificial Intelligence for 6G: Improving Trust between Human and Machine. *IEEE Communications Magazine*, *58*(6), 39–45. doi:10.1109/MCOM.001.2000050

Gupta, A., Fernando, X., & Das, O. (2021, June). Reliability and availability modeling techniques in 6g iot networks: A taxonomy and survey. In 2021 International Wireless Communications and Mobile Computing (IWCMC) (pp. 586-591). IEEE.

Gupta, R., Kumari, A., Tanwar, S., & Kumar, N. (2020). Blockchain-envisioned softwarized multi-swarming uavs to tackle covid-i9 situations. *IEEE Network*, *35*(2), 160–167.

Gursu, H. M., Vilgelm, M., Kellerer, W., & Reisslein, M. (2017). Hybrid Collision Avoidance-Tree Resolution for M2M Random Access. *IEEE Transactions on Aerospace and Electronic Systems*, *53*(4), 1974–1987. doi:10.1109/TAES.2017.2677839

Gyongyosi, L., & Imre, S. (2019). A survey on quantum computing technology. *Computer Science Review*, *31*, 51–71.

Gyongyosi, L., Imre, S., & Nguyen, H. V. (2018). A survey on quantum channel capacities. *IEEE Communications Surveys and Tutorials*, *20*(2), 1149–1205. doi:10.1109/COMST.2017.2786748

Hamamreh, J. M., Basar, E., & Arslan, H. (2017). OFDM-subcarrier index selection for enhancing security and reliability of 5G URLLC services. *IEEE Access: Practical Innovations, Open Solutions*, *5*, 25863–25875. doi:10.1109/ACCESS.2017.2768558

Han. (2018). Licensed-assisted access for LTE in unlicensed spectrum: A MAC protocol design. *IEEE Journal on Selected Areas in Communications, 34*(10), 2550-2561.

Han, G., Guizani, M., Bi, Y., Luan, T. H., Ota, K., Zhou, H., Guibene, W., & Rayes, A. (2017). Software-defined vehicular networks: Architecture, algorithms, and applications: Part 1. *IEEE Commun.*, *55*(7), 78–79. doi:10.1109/MCOM.2017.7981529

Han, S., Noh, Y., Liang, R., Chen, R., Cheng, Y.-J., & Gerla, M. (2014). Evaluation of underwater optical-acoustic hybrid network. *China Communications*, *11*(5), 49–59. doi:10.1109/CC.2014.6880460

Han, S., Xie, T., I, C.-L., Chai, L., Liu, Z., Yuan, Y., & Cui, C. (2020). Artificial-Intelligence-Enabled Air Interface for 6G. *Solutions, Challenges, and Standardization Impacts IEEE Communications Magazine*, *58*(10), 73–79. doi:10.1109/MCOM.001.2000218

Haseeb, J., Mansoori, M., Al-Sahaf, H., & Welch, I. (2020, December). IoT Attacks: Features Identification and Clustering. In *2020 IEEE 19th International Conference on Trust, Security and Privacy in Computing and Communications (TrustCom)* (pp. 353-360). IEEE.

He, H. T., Wen, C. K., Jin, S., & Li, G. Y. (2020). Model-driven deep learning for MIMO detection. *IEEE Transactions on Signal Processing*, *68*, 1702–1715. doi:10.1109/TSP.2020.2976585

Heuser & Zohner. (2012). Intelligent machine homicide. *Proc. Int. Workshop Constructive Side Channel Anal. Secure Design*, 249-264.

Hewa, T., Gür, G., Kalla, A., Ylianttila, M., Bracken, A., & Liyanage, M. (2020, March). *The role of blockchain in 6G: Challenges, opportunities and research directions. In 2020 2nd 6G Wireless Summit (6G SUMMIT)*. IEEE.

He, Y., Zhang, J., Wen, C., & Jin, S. (2019). Turbonet: A model-driven DNN decoder based on max-log-map algorithm for turbo code. *Proc. Of IEEE VTS Asia Pacific Wireless Comms. Symp. (APWCS)*. 10.1109/VTS-APWCS.2019.8851650

Hong, T., Liu, C., & Kadoch, M. (2019). *Machine learning based antenna design for physical layer security in ambient backscatter communications. Wireless Communications and Mobile Computing. How AI will Enable Network Intelligentizing?* Vision for Future Communications Summit, Lisbon, Portugal. Retrieved September 25, 2021, https://www.is-wireless.com/wp-content/uploads/2019/12/Network-Intelligentizing-presentation-on-NetWorld2020-summit.pdf

Hong, T., Liu, C., & Kadoch, M. (2019). Machine learning based antenna design for physical layer security in ambient backscatter communications. *Wireless Communications and Mobile Computing*, *2019*, 2019. doi:10.1155/2019/4870656

Hooper, M., Tian, Y., Zhou, R., Cao, B., Lauf, A. P., Watkins, L., & Alexis, W. (2016, November). Securing commercial wifi-based uavs from common security attacks. In *MILCOM 2016-2016 IEEE Military Communications Conference* (pp. 1213–1218). IEEE. doi:10.1109/MILCOM.2016.7795496

Huang, C. C. C., & Li, Q. (2017). A study of deep learning networks on mobile traffic forecasting. *IEEE 28th Annual International Symposium on Personal, Indoor, and Mobile Radio Communications (PIMRC),* 1–6. 10.1109/PIMRC.2017.8292737

Huang, H., Guo, S., Gui, G., Yang, Z., Zhang, J., Sari, H., & Adachi, F. (2020, February). Deep Learning for Physical-Layer 5G Wireless Techniques: Opportunities, Challenges and Solutions. *IEEE Wireless Communications*, *27*(1), 214–222. doi:10.1109/MWC.2019.1900027

Huang, T., Yang, W., Wu, J., Ma, J., Zhang, X., & Zhang, D. (2019). A survey on green 6G network: Architecture and technologies. *IEEE Access: Practical Innovations, Open Solutions*, *7*, 175758–175768. doi:10.1109/ACCESS.2019.2957648

Hu, J. Y., Yu, B., Jing, M. Y., Xiao, L. T., Jia, S. T., Qin, G. Q., & Long, G. L. (2016). Experimental quantum secure direct communication with single photons. *Light, Science & Applications*, *5*(9), e16144–e16144. doi:10.1038/lsa.2016.144 PMID:30167186

Hu, J., Chen, C., Cai, L., Khosravi, M. R., Pei, Q., & Wan, S. (2021). Uav-assisted vehicular edge computing for the 6g internet of vehicles: Architecture, intelligence, and challenges. *IEEE Communications Standards Magazine*, *5*(2), 12–18.

Huq & Busari. (2019). Terahertz-enabled wireless system for beyond-5G ultra-fast networks: A brief survey. *IEEE Network*, *33*(4), 89-95.

Husamuddin, M., & Qayyum, M. (2017, March). Internet of Things: A study on security and privacy threats. In *2017 2nd International Conference on Anti-Cyber Crimes (ICACC)* (pp. 93-97). IEEE.

Hussain, F., Hussain, R., Hassan, S. A., & Hossain, E. (2020). Machine learning in IoT security: Current solutions and future challenges. *IEEE Communications Surveys and Tutorials*, *22*(3), 1686–1721.

Islim, M. S., Ferreira, R. X., He, X., Xie, E., Videv, S., Viola, S., Watson, S., Bamiedakis, N., Penty, R. V., White, I. H., Kelly, A. E., Gu, E., Haas, H., & Dawson, M. D. (2017). Towards 10 Gb/s orthogonal frequency division multiplexing-based visible light communication using a GaN violet micro-LED. *Photonics Research*, *5*(2), A35–A43. doi:10.1364/PRJ.5.000A35

Istiaque Ahmed, K., Tahir, M., Hadi Habaebi, M., Lun Lau, S., & Ahad, A. (2021). Machine Learning for Authentication and Authorization in IoT: Taxonomy, Challenges and Future Research Direction. *Sensors (Basel)*, *21*(15), 5122.

Jaber & Tukmano. (2019). 5G backhaul challenges and emerging research directions: A survey. *IEEE Access*, *4*, 1743-1766.

Jagannath, A. J., & Melodia, J. (2021). Redefining Wireless Communication for 6G: Signal Processing Meets Deep Learning with Deep Unfolding. doi:10.36227/techrxiv.16434855.v2

Jahid, A., Alsharif, M. H., & Hall, T. J. (2021). *The Convergence of Blockchain, IoT and 6G: Potential, Opportunities, Challenges and Research Roadmap*. arXiv preprint arXiv:2109.03184.

Janjua, M. B., Duranay, A. E., & Arslan, H. (2020). Role of Wireless Communication in Healthcare System to Cater Disaster Situations under 6G Vision. *Frontiers in Communications and Networks*, *1*, 6.

Javad-Kalbasi, M., & Valaee, S. (2021, June). Re-configuration of UAV Relays in 6G Networks. In *2021 IEEE International Conference on Communications Workshops (ICC Workshops)* (pp. 1-6). IEEE.

Je, D., Jung, J., & Choi, S. (2021). Toward 6G Security: Technology Trends, Threats, and Solutions. *IEEE Communications Standards Magazine*, *5*(3), 64–71.

Jennath, H. S., Adarsh, S., Chandran, N. V., Ananthan, R., Sabir, A., & Asharaf, S. (2019). Parkchain: A blockchain powered parking solution for smart cities. *Frontiers in Blockchain*, *2*, 6.

Jiajia, J., Xianquan, W., Fajie, D., Xiao, F., Chunyue, L., & Zhongbo, S. (2020). A basic bio-inspired camouflage communication frame design and applications for secure underwater communication among military underwater platforms. *IEEE Access: Practical Innovations, Open Solutions*, *8*, 24927–24940. doi:10.1109/ACCESS.2020.2970746

Jiang, H., Qiu, H., He, N., Zhao, Z., Popoola, W., Ahmad, Z., & Rajbhandari, S. (2020). LDPC-coded CAP with spatial diversity for UVLC systems over generalized-gamma fading channel. *Sensors (Basel), 20*(12), 3378. doi:10.339020123378 PMID:32549316

Joakar. (2016). *A Methodology for Solving Problems with DataScience for Internet of Things.* Open Gardens.

Jordan, M. I., & Mitchell, T. M. (2015). Machine learning: Trends perspectives and prospects. *Science, 349*(6245), 255–260.

Jouhari, M., Ibrahimi, K., Tembine, H., & Ben-Othman, J. (2019). Underwater wireless sensor networks: A survey on enabling technologies, localization protocols, and internet of underwater things. *IEEE Access: Practical Innovations, Open Solutions*, *7*, 96879–96899. doi:10.1109/ACCESS.2019.2928876

Kabir, H., Mohsin, H., & Kantola, R. (2020). Implementing Security Policy Management for 5G Customer Edge Nodes. *NOMS 2020 - 2020 IEEE/IFIP Network Operations and Management Symposium,* 1-8.

Kaddoum, G. (2016). Wireless chaos-based communication systems: A comprehensive survey. *IEEE Access: Practical Innovations, Open Solutions*, *4*, 2621–2648.

Kanthavel, R., & Dhaya, R. (2013). Reliable Hierarchical Security Routing For Delay Tolerant Sensor Networks. International Information Institute (Tokyo).

Kanthavel, R., Dhaya, R., & Ahilan, A. (2022). AI-Based Efficient WUGS Network Channel Modeling and Clustered Cooperative Communication. ACM Transactions on Sensor Networks, 18(3), 1–14. doi:10.1145/3469034

Kanthavel, R., Indra Priyadharshini, S., Sudha, D., Sundara Velrani, K., & Dhaya, R. (2022). Multi-hoped cooperative communication-based wireless underground sensor network design. *International Journal of Communication Systems.* doi:10.1002/dac.5174

Kanthavel, R., & Dhaya, R. (2022). Prediction Model Using Reinforcement Deep Learning Technique for Osteoarthritis Disease Diagnosis. *Computer Systems Science and Engineering*, *42*(1), 257–269.

Kanthavel, R., & Dhaya, R. (2022). Wireless Underground Sensor Networks Channel Using Energy Efficient Clustered Communication. *Intelligent Automation And Soft Computing*, *31*(1), 649–659.

Kantola, R., Llorente, J., Santos, N., & Beijar. (2015). Policy Based Communications for 5G Mobile with Customer Edge Switching. *Wiley Security and Communication Networks.*

Kantola, R., & Pouttu, A. (2019). *Key drivers and research challenges for 6G ubiquitous wireless intelligence.* Issue September.

Kasgari, Saad, Mozaffari, & Poor. (2020). Experienced deep reinforcement learning with generative adversarial networks (GANs) for model-free ultra reliable low latency communication. *IEEE Trans. Commun.*

Katz, M., & Ahmed, I. (2020). Opportunities and Challenges for Visible Light Communications in 6G. *2020 2nd 6G Wireless Summit (6G SUMMIT).*

Kaur, J., Khan, M. A., Iftikhar, M., Imran, M., & Emad Ul Haq, Q. (2021). Machine Learning Techniques for 5G and Beyond. *IEEE Access: Practical Innovations, Open Solutions*, *9*, 23472–23488. doi:10.1109/ACCESS.2021.3051557

Kaushal, H., & Kaddoum, G. (2016). Underwater Optical Wireless Communication,&quot. *IEEE Access: Practical Innovations, Open Solutions*, *4*, 1518–1547. doi:10.1109/ACCESS.2016.2552538

Khacef, K., & Pujolle, G. (2019, March). Secure Peer-to-Peer communication based on Blockchain. In *Workshops of the International Conference on Advanced Information Networking and Applications* (pp. 662-672). Springer.

Khan, A. S., Balan, K., Javed, Y., Tarmizi, S., & Abdullah, J. (2019). Secure trust-based blockchain architecture to prevent attacks in VANET. *Sensors (Basel), 19*(22), 4954. doi:10.339019224954 PMID:31739437

Khan, L. U. (2017). Visible light communication: Applications, architecture, standardization and research challenges. *Digital Communications and Networks*, *3*(2), 78–88. doi:10.1016/j.dcan.2016.07.004

Kimbugwe, N., Pei, T., & Kyebambe, M. N. (2021). Application of Deep Learning for Quality of Service Enhancement in Internet of Things: A Review. *Energies*, *14*(19), 6384. doi:10.3390/en14196384

Kim, H., Kim, S. H., Hwang, J. Y., & Seo, C. (2019). Efficient privacy-preserving machine learning for blockchain network. *IEEE Access: Practical Innovations, Open Solutions*, *7*, 136481–136495. doi:10.1109/ACCESS.2019.2940052

Kiran, Kumar, & Kumar. (2020). *A Review of Machine Learning Algorithms on IoT Applications.* IEEE Xplore Part Number: CFP20V90-ART.

Kishk, M., Bader, A., & Alouini, M. S. (2020). Aerial base station deployment in 6G cellular networks using tethered drones: The mobility and endurance tradeoff. *IEEE Vehicular Technology Magazine*, *15*(4), 103–111.

Kiyomoto, S., Basu, A., Rahman, M. S., & Ruj, S. (2017, December). On blockchain-based authorization architecture for beyond-5G mobile services. In *2017 12th International Conference for Internet Technology and Secured Transactions (ICITST)* (pp. 136-141). IEEE.

Kocakulak, M., & Butun, I. (2017, January). An overview of Wireless Sensor Networks towards internet of things. In *2017 IEEE 7th annual computing and communication workshop and conference (CCWC)* (pp. 1-6). IEEE.

Kotobi, K., & Bilen, S. G. (2018). Secure blockchains for dynamic spectrum access: A decentralized database in moving cognitive radio networks enhances security and user access. *IEEE Vehicular Technology Magazine, 13*(1), 32-39.

Kotsiantis, S. B., Zaharakis, I., & Pintelas, P. (2007, July). Supervised machine learning: A review of classification techniques. *Emerg. Artif. Intell. Appl. Comput. Eng.*, *160*, 3–24.

Latva-aho, M., & Leppänen, M. (2019). *Key Drivers and Research Challenges for 6G Ubiquitous Wireless Intelligence. In 6G Flagship*. University of Oulu.

LeCun, Y., Bengio, Y., & Hinton, G. (2016). Deep learning. *Nature*, *521*(7553), 436.

Lei, K., Zhang, Q., Xu, L., & Qi, Z. (2018, December). Reputation-based byzantine fault-tolerance for consortium blockchain. In *2018 IEEE 24th International Conference on Parallel and Distributed Systems (ICPADS)* (pp. 604-611). IEEE.

Letaief & Zhang. (2019). The roadmap to 6G: AI empowered wireless networks. *IEEE Communications Magazine, 57*(8), 84-90,

Letaief, K. B., Chen, W., Shi, Y., Zhang, J., & Zhang, Y. A. (2019). The Roadmap to 6G. *AI Empowered Wireless Networks IEEE Communications Magazine*, *57*(8), 84–90. doi:10.1109/MCOM.2019.1900271

Letaief, K. B., Chen, W., Shi, Y., Zhang, J., & Zhang, Y.-J. A. (2019). The Roadmap to 6G: AI Empowered Wireless Networks. *IEEE Communications Magazine*, *57*(8), 84–90.

Li & Avestimehr. (2017). A scalable framework for wireless distributed computing. *IEEE/ACM Transactions on Networking, 25*(5), 2643-2654.

Li & Wu. (2014). Wireless backhaul of dense small cell networks with high dimension MIMO. *IEEE Globecom Workshops (GC Wkshps)*, 1254-1259.

Liang, L., Zheng, K., Sheng, Q., & Huang, X. (2016, December). A denial-of-service attack method for an iot system. In *2016 8th international conference on Information Technology in Medicine and Education (ITME)* (pp. 360-364). IEEE. 10.1109/ITME.2016.0087

Lin, T. (2020, November). Deep Learning for IoT. In *2020 IEEE 39th International Performance Computing and Communications Conference (IPCCC)* (pp. 1-4). IEEE.

Liu & Vucetic. (2018). Backscatter multiplicative multiple-access systems: Fundamental limits and practical design. *IEEE Transactions on Wireless Communications, 17*(9), 5713-5728.

Liu, Y., Wang, C. X., Chang, H., He, Y., & Bian, J. (2021). A Novel Non-Stationary 6G UAV Channel Model for Maritime Communications. *IEEE Journal on Selected Areas in Communications.*

Liu. (2017). Polar Codes for Block Fading Channels. *IEEE Wireless Communications and Networking Conference Workshops (WCNCW)*, 1-6.

Liu, C., Feng, W., Chen, Y., Wang, C. X., & Ge, N. (2020). Cell-free satellite-UAV networks for 6G wide-area Internet of Things. *IEEE Journal on Selected Areas in Communications, 39*(4), 1116–1131.

Liu, R., Liu, A., Qu, Z., & Xiong, N. N. (2021). An UAV-Enabled Intelligent Connected Transportation System With 6G Communications for Internet of Vehicles. *IEEE Transactions on Intelligent Transportation Systems*, 1558–0016.

Liu, Y., Yang, C., Jiang, L., Xie, S., & Zhang, Y. (2019). Intelligent Edge Computing for IoT-Based Energy Management in Smart Cities. *IEEE Network, 33*(2), 111–117. doi:10.1109/MNET.2019.1800254

Liu, Y., Yuan, X., Xiong, Z., Kang, J., Wang, X., & Niyato, D. (2020). Federated learning for 6G communications: Challenges, methods, and future directions. *China Communications, 17*(9), 105–118. doi:10.23919/JCC.2020.09.009

Li, W., Su, Z., Li, R., Zhang, K., & Wang, Y. (2020). Blockchain-based data security for artificial intelligence applications in 6G networks. *IEEE Network, 34*(6), 31–37. doi:10.1109/MNET.021.1900629

Loscri, V., Marchal, C., Mitton, N., Fortino, G., & Vasilakos, A. V. (2014). Security and privacy in molecular communication and networking: Opportunities and challenges. *IEEE Transactions on Nanobioscience, 13*(3), 198–207. doi:10.1109/TNB.2014.2349111 PMID:25148668

Lovén, L., Leppänen, T., Peltonen, E., Partala, J., Harjula, E., Porambage, P., ... Riekki, J. (2019). EdgeAI: A vision for distributed, edge-native artificial intelligence in future 6G networks. *The 1st 6G Wireless Summit*, 1-2.

Luo, C., Miao, W., Ullah, H., McClean, S., Parr, G., & Min, G. (2019). Unmanned aerial vehicles for disaster management. In *Geological disaster monitoring based on sensor networks* (pp. 83–107). Springer.

Luo, J., Fan, L., & Li, H. (2017). Indoor positioning systems based on visible light communication: State of the art. *IEEE Communications Surveys and Tutorials, 19*(4), 2871–2893. doi:10.1109/COMST.2017.2743228

Lu, Y., Higgins, M. D., & Leeson, M. S. (2015). Comparison of channel coding schemes for molecular communications systems. *IEEE Transactions on Communications, 63*(11), 3991–4001. doi:10.1109/TCOMM.2015.2480752

Lv, Z., Qiao, L., & You, I. (2020). 6G-enabled network in box for internet of connected vehicles. *IEEE Transactions on Intelligent Transportation Systems.*

Maddikunta, P. K. R., Hakak, S., Alazab, M., Bhattacharya, S., Gadekallu, T. R., Khan, W. Z., & Pham, Q. V. (2021). Unmanned aerial vehicles in smart agriculture: Applications, requirements, and challenges. *IEEE Sensors Journal.*

Mafamane, R., Ouadou, M., Hassani, A. T. J., & Minaoui, K. (2021, April). Study of the heterogeneity problem in the Internet of Things and Cloud Computing integration. In *2020 10th International Symposium on Signal, Image, Video and Communications (ISIVC)* (pp. 1-6). IEEE. 10.1109/ISIVC49222.2021.9487539

Mahdavinejad, M. S., Rezvan, M., Barekatain, M., Adibi, P., Barnaghi, P., & Sheth, A. P. (2018). Machine learning for internet of things data analysis: A survey. *Digital Communications and Networks, 4*(3), 161–175. doi:10.1016/j.dcan.2017.10.002

Ma, J., Shrestha, R., Adelberg, J., Yeh, C. Y., Hossain, Z., Knightly, E., Jornet, J. M., & Mittleman, D. M. (2018). Security and eavesdropping in terahertz wireless links. *Nature, 563*(7729), 89–93. doi:10.103841586-018-0609-x PMID:30323288

Mao, B., Tang, F., Yuichi, K., & Kato, N. (2021). *AI based Service Management for 6G Green Communications.* arXiv preprint arXiv:2101.01588.

Mao. (2020). AI-based joint optimization of QoS and security for 6G energy harvesting Internet of Things. *IEEE Internet of Things Journal, 7*(8), 7032-7042.

Mao, B., Tang, F., Kawamoto, Y., & Kato, N. (2021). Optimizing Computation Offloading in Satellite-UAV-Served 6G IoT: A Deep Learning Approach. *IEEE Network, 35*(4), 102–108.

Marline, Kumari, & Krishna. (2018). Prognosis of diseases using machine learning algorithms: A survey. *2018 International Conference on Current Trends towards Converging Technologies (ICCTCT)*, 1-9.

Masaracchia, A., Li, Y., Nguyen, K. K., Yin, C., Khosravirad, S. R., Da Costa, D. B., & Duong, T. Q. (2021). UAV-enabled Ultra-Reliable Low-Latency Communications for 6G: A Comprehensive Survey. *IEEE Access.*

Matti, E., Johns, O., Khan, S., Gurtov, A., & Josefsson, B. (2020, October). Aviation Scenarios for 5G and Beyond. In *2020 AIAA/IEEE 39th Digital Avionics Systems Conference (DASC)* (pp. 1-10). IEEE.

Mbarek, B., Ge, M., & Pitner, T. (2020, September). Blockchain-Based Access Control for IoT in Smart Home Systems. In *International Conference on Database and Expert Systems Applications* (pp. 17-32). Springer. 10.1007/978-3-030-59051-2_2

McMahan, B. (2017). Communication-efficient learning of deep networks from decentralized data. *Int'l. Conf. Artificial Intell. Stat., 54*, 1273–1282.

Melamed, T. (n.d.). *An Active Man-In-The-Middle Attack on Bluetooth Smart Devices.* doi:10.1109/GLOBECOM42002.2020.9322070

Merenda, M., Porcaro, C., & Iero, D. (2020). Edge Machine Learning for AI-Enabled IoT Devices: A Review. *Sensors (Basel), 20*(9), 2533. doi:10.339020092533 PMID:32365645

Miraz, M. H., Ali, M., Excell, P. S., & Picking, R. (2015, September). A review on Internet of Things (IoT), Internet of everything (IoE) and Internet of nano things (IoNT). In 2015 Internet Technologies and Applications (ITA) (pp. 219-224). IEEE.

Mostafa, A., & Lampe, L. (2014, June). Physical-layer security for indoor visible light communications. In *2014 IEEE International Conference on Communications (ICC)* (pp. 3342-3347). IEEE. 10.1109/ICC.2014.6883837

Mozaffari & Debbah. (2018). Beyond 5G with UAVs: Foundations of a 3D wireless cellular network. *IEEE Transactions on Wireless Communications, 18*(1), 357-372.

Mucchi. (2020). How 6G Technology Can Change the Future Wireless Healthcare. *2nd 6G Wireless Summit (6G SUMMIT)*, 1-6. doi: . doi:10.1109/6GSUMMIT49458.2020.9083916

Mung, P. S., & Phyu, S. (2020). Effective analytics on healthcare big data using ensemble learning. *2020 IEEE Conference on Computer Applications (ICCA)*, 1-4.

Murphy, K. P. (2012). *Machine Learning: A Probabilistic Perspective*. MIT press.

Nakano, T., Okaie, Y., Kobayashi, S., Hara, T., Hiraoka, Y., & Haraguchi, T. (2019). Methods and applications of mobile molecular communication. *Proceedings of the IEEE*, *107*(7), 1442–1456. doi:10.1109/JPROC.2019.2917625

Naser, S., & Bariah, L. (2021). Towards Federated Learning-Enabled Visible Light Communication in 6G Systems. doi:10.36227/techrxiv.14632551.v2

Nawaz, S. J., Sharma, S. K., Wyne, S., Patwary, M. N., & Asaduzzaman, M. (2019). Quantum machine learning for 6G communication networks: State-of-the-art and vision for the future. *IEEE Access: Practical Innovations, Open Solutions*, *7*, 46317–46350. doi:10.1109/ACCESS.2019.2909490

Nayak, S., & Patgiri, R. (2020). *6G communications: A vision on the potential applications*. arXiv preprint arXiv:2005.07531.

Nayak, S., & Patgiri, R. (2021). 6G communication technology: A vision on intelligent healthcare. In *Health Informatics: A Computational Perspective in Healthcare* (pp. 1–18). Springer.

Ngo, Q. D., Nguyen, H. T., Le, V. H., & Nguyen, D. H. (2020). A survey of IoT malware and detection methods based on static features. *ICT Express*, *6*(4), 280–286. doi:10.1016/j.icte.2020.04.005

Nguyen & Pirttikangas. (2020). Privacy-aware blockchain innovation for 6G: Challenges and opportunities. *6G Wireless Summit (6G SUMMIT)* (pp. 1-5),

Nguyen, D. C., Ding, M., Pathirana, P. N., Seneviratne, A., Li, J., Niyato, D., & Poor, H. V. (2021). *6G Internet of Things: A Comprehensive Survey. IEEE Internet of Things Journal.*

Nguyen, D. C., Pathirana, P. N., Ding, M., & Seneviratne, A. (2020). Blockchain for 5G and beyond networks: A state of the art survey. *Journal of Network and Computer Applications*, *166*, 102693. doi:10.1016/j.jnca.2020.102693

Ni, J., Lin, X., & Shen, X. (2019). Toward privacy-preserving valet parking in autonomous driving era. *IEEE Transactions on Vehicular Technology*, *68*(3), 2893–2905. doi:10.1109/TVT.2019.2894720

Nikam, S. S. (2015). A comparative study of classification techniques in data mining algorithms. *Oriental Journal of Computer Science & Technology*, *8*(1), 13–19.

Noorbakhsh-Sabet, N., Zand, R., Zhang, Y., & Abedi, V. (2019). Artificial intelligence transforms the future of healthcare. *The American Journal of Medicine*.

Oyinloye, D. P., Teh, J. S., Jamil, N., & Alawida, M. (2021). Blockchain Consensus: An Overview of Alternative Protocols. *Symmetry*, *13*(8), 1363.

Pandey. (2017). *Machine Learning and IoT for Prediction and Detection of Stress*. Academic Press.

Pan, L., Yi, J., Yin, C., Yu, J., & Li, X. (2019). Joint 3D UAV placement and resource allocation in software-defined cellular networks with wireless backhaul. *IEEE Access: Practical Innovations, Open Solutions*, *7*, 104279–104293. doi:10.1109/ACCESS.2019.2927521

Papageorgiou, A., Zahn, M., & Kovacs, E. (2014). Efficient auto-configuration of energyrelated parameters in cloud-based iot platforms. *Cloud Networking (CloudNet), IEEE 3rd International Conference on*, 236–241.

Pardo. (2006). Designing learning analytics experiences. In Learning Analytics. From Research to Practice (pp. 15–38). Springer.

Pathak, P. H., Feng, X., Hu, P., & Mohapatra, P. (2015). Visible light communication, networking, and sensing: A survey, potential and challenges. *IEEE Communications Surveys and Tutorials*, *17*(4), 2047–2077. doi:10.1109/COMST.2015.2476474

Peltonen, E., Bennis, M., Capobianco, M., Debbah, M., Ding, A., Gil-Castiñeira, F., . . . Yang, T. (2020). *6G white paper on edge intelligence*. arXiv preprint arXiv:2004.14850.

Peltonen, E., Bennis, M., Capobianco, M., Debbah, M., Ding, A., Gil-Castiñeira, F., Jurmu, M., Karvonen, T., Kelanti, M., Kliks, A., Leppänen, T., Lovén, L., Mikkonen, T., Rao, A., Samarakoon, S., Seppänen, K., Sroka, P., Tarkoma, S., & Yang, T. (2020). *6G White Paper on Edge Intelligence* (White paper) (6G Research Visions, No. 8). University of Oulu. http://urn.fi/urn:isbn:978952622677

Peteiro-Barral, D., & Guijarro-Berdiñas, B. (2012). A survey of methods for distributed machine learning. *Progress in Artificial Intelligence*, *2*(1), 1–11.

Piran, M. J., & Suh, D. Y. (2019). Learning-Driven Wireless Communications, towards 6G. *2019 International Conference on Computing, Electronics & Communications Engineering (iCCECE)*, 219-224. 10.1109/iCCECE46942.2019.8941882

Plastiras, Terzi, Kyrkou, & Theocharides. (2018). *Edge Intelligence: Challenges and Opportunities of Near-Sensor Machine Learning Applications*. IEEE.

Popovski. (2018). 5G wireless network slicing for eMBB, URLLC, and mMTC: A communication-theoretic view. *IEEE Access, 6*, 55765-79.

Porambage, P., Kumar, T., Liyanage, M., Partala, J., Lovén, L., Ylianttila, M., & Seppänen, T. (2019). Sec-EdgeAI: AI for edge security Vs security for edge AI. The 1st 6G Wireless Summit.

Porambage, P., Gür, G., Osorio, D. P. M., Liyanage, M., Gurtov, A., & Ylianttila, M. (2021). The roadmap to 6G security and privacy. *IEEE Open Journal of the Communications Society*, *2*, 1094–1122.

Porambage, P., Gür, G., Osorio, D. P. M., Liyanage, M., Gurtov, A., & Ylianttila, M. (2021). The Roadmap to 6G Security and Privacy. *IEEE Open Journal of the Communications Society*, *2*, 1094–1122. doi:10.1109/OJCOMS.2021.3078081

Prates, N., Vergütz, A., Macedo, R. T., Santos, A., & Nogueira, M. (2020, December). A Defense Mechanism for Timing-based Side-Channel Attacks on IoT Traffic. In *GLOBECOM 2020-2020 IEEE Global Communications Conference* (pp. 1-6). IEEE.

Puthal, D., Malik, N., Mohanty, S. P., Kougianos, E., & Yang, C. (2018). The blockchain as a decentralized security framework. *IEEE Consumer Electronics Magazine*, *7*(2), 18–21. doi:10.1109/MCE.2017.2776459

Qi & Li. (2016). Energy-efficient cooperative network coding with joint relay scheduling and power allocation. *IEEE Transactions on Communications, 64*(11), 4506-19.

Rahman. (2019). Network Intelligentizing for Future 6G Wireless Networks. *Digital Communications and Networks, 6*(3), 281-291. doi:10.1016/j.dcan.2020.07.003

Ramadan, R. A., & Vasilakos, A. V. (2017). Brain computer interface: Control signals review. *Neurocomputing*, *223*, 26–44. doi:10.1016/j.neucom.2016.10.024

Ramakrishnan, D., & Radhakrishnan, K. (2022, March). Applying Deep Convolutional Neural Network (DCNN) Algorithm in the Cloud Autonomous Vehicles Traffic Model. *The International Arab Journal of Information Technology*, *19*(2), 186–194.

Rappaport, T. S., Xing, Y., Kanhere, O., Ju, S., Madanayake, A., Mandal, S., & Trichopoulos, G. C. (2008). Wireless communications and applications above 100 GHz: Opportunities and challenges for 6G and beyond. *IEEE Access: Practical Innovations, Open Solutions*, *7*, 78729–78757. doi:10.1109/ACCESS.2019.2921522

Refaey, A., Hammad, K., Magierowski, S., & Hossain, E. (2019). A blockchain policy and charging control framework for roaming in cellular networks. *IEEE Network*, *34*(3), 170–177. doi:10.1109/MNET.001.1900336

Renn, O., & Klinke, A. (2001). Environmental Risks : Perception, Evaluation and Management Article information. Environmental Risks: Perception, Evaluation and Management.

Romero, C., & Ventura, S. (2010). Educational data mining: A review of the state of the art. *IEEE Transactions on Systems, Man and Cybernetics. Part C, Applications and Reviews*, *40*(6), 601–618.

Rommel, Raddo, & Monroy. (2018). Data center connectivity by 6G wireless systems. In 2018 Photonics in switching and computing (PSC) (pp. 1–3). IEEE.

Saad, C., Bennis, M., & Chen, M. (2019). A vision of 6G wireless systems: Applications, trends, technologies, and open research problems. *IEEE Network*, *34*(3), 134–142. doi:10.1109/MNET.001.1900287

Salman, T., & Jain, R. (2019). *A survey of protocols and standards for internet of things*. arXiv preprint arXiv:1903.11549.

Sangeetha, S. K. B., & Dhaya, R. (2022). Deep Learning Era for Future 6G Wireless Communications—Theory, Applications, and Challenges. *Artificial Intelligent Techniques for Wireless Communication and Networking,* 105-119.

Sangeetha, S. K. B., & Dhaya, R. (2016). Fuzzy integrated gaming approach for relay selection in cooperative communication. *Journal of Applied Sciences Research*, *12*(3), 45–49.

Sangeetha, S. K. B., Dhaya, R., & Kanthavel, R. (2019). Improving performance of cooperative communication in heterogeneous manet environment. *Cluster Computing*, *22*(5), 12389–12395.

Sanjab, A., Saad, W., & Başar, T. (2017, May). Prospect theory for enhanced cyber-physical security of drone delivery systems: A network interdiction game. In 2017 IEEE international conference on communications (ICC) (pp. 1-6). IEEE. doi:10.1109/ICC.2017.7996862

Sathyaraj, R., Kanthavel, R., Leonardo Cavaliere, L. P., Vyas, S., Maheswari, S., Gupta, R. K., Ramkumar Raja, M., Dhaya, R., Gupta, M. K., & Sengan, S. (2022). Analysis on Prediction of Covid-19 with Machine Learning Algorithms. *International Journal of Uncertainty, Fuzziness and Knowledge-based Systems*. Advance online publication. doi:10.1142/S0218488522400049

Sattiraju, R., Kochems, J., & Schotten, H. D. (2017, May). Machine learning based obstacle detection for Automatic Train Pairing. In *2017 IEEE 13th International Workshop on Factory Communication Systems (WFCS)* (pp. 1-4). IEEE. 10.1109/WFCS.2017.7991962

Sattiraju, R., Weinand, A., & Schotten, H. D. (2019). *AI-assisted PHY technologies for 6G and beyond wireless networks.* arXiv preprint arXiv:1908.09523.

Sattiraju, R., Weinand, A., & Schotten, H. D. (2018, December). Performance analysis of deep learning based on recurrent neural networks for channel coding. In *2018 IEEE International Conference on Advanced Networks and Telecommunications Systems (ANTS)* (pp. 1-6). IEEE. 10.1109/ANTS.2018.8710159

Sergey, A., Vitaly, P., Mischa, D., & Halim, Y. (2019). Future of ultra-dense networks beyond 5G: Harnessing heterogeneous moving cells. *IEEE Communications Magazine*, *57*(6), 86–92. doi:10.1109/MCOM.2019.1800056

Shah, P., Arora, M., & Adhvaryu, K. (2020, October). Lightweight Cryptography Algorithms in IoT-A Study. In *2020 Fourth International Conference on I-SMAC (IoT in Social, Mobile, Analytics and Cloud)(I-SMAC)* (pp. 332-336). IEEE.

Shakiba-Herfeh, M., Chorti, A., & Poor, H. V. (2020). *Physical layer security: authentication, integrity and confidentiality.* https://arxiv.org/pdf/2001.07153.pdf

Shanthamallu, Spanias, Tepedelenlioglu, & Stanley. (2020). *A Brief Survey of Machine Learning Methods and their Sensor and IoT Applications.* IEEE.

Shanzhi, C., Ying-Chang, L., Shaohui, S., Shaoli, K., Wenchi, C., & Mugen, P. (2020). Vision, requirements, and technology trend of 6G: How to tackle the challenges of system coverage, capacity, user data-rate and movement speed. *IEEE Wireless Communications, 27*(2), 218–228. doi:10.1109/MWC.001.1900333

Sharma, R. C., Kawachi, P., & Bozkurt, A. (2019). The landscape of artificial intelligence in open, online and distance education: Promises and concerns. *Asian J. Distance Educ., 14*(2), 1–2.

Shea, T. O., & Hoydis, J. (2017). An introduction to deep learning for the physical layer. *IEEE Transactions on Cognitive Communications and Networking, 3*(4), 563–575. doi:10.1109/TCCN.2017.2758370

Sheth, K., Patel, K., Shah, H., Tanwar, S., Gupta, R., & Kumar, N. (2020). A Taxonomy of AI Techniques for 6G Communication Networks. *Computer Communications, 161*, 279–303. doi:10.1016/j.comcom.2020.07.035

Shi, Y., Yang, K., Jiang, T., Zhang, J., & Letaief, K. B. (2020). *Communication-efficient edge ai: Algorithms and systems.* arXiv preprint arXiv:2002.09668.

Shrestha, R., Bajracharya, R., & Kim, S. (2021). 6G Enabled Unmanned Aerial Vehicle Traffic Management: A Perspective. *IEEE Access: Practical Innovations, Open Solutions, 9*, 91119–91136. doi:10.1109/ACCESS.2021.3092039

Shurman, M., Obeidat, A. A. R., & Al-Shurman, S. A. D. (2020, April). Blockchain and smart contract for IoT. In *2020 11th International Conference on Information and Communication Systems (ICICS)* (pp. 361-366). IEEE.

Siriwardhana, Y., Porambage, P., Liyanage, M., & Ylianttila, M. (2021). *AI and 6G Security: Opportunities and Challenges.* 10.1109 / EuCNC / 6GSummit 51104.2021.9482503

Sivaramakrishnan, S. (2021). Terrance Frederick Fernandez; R. G. Babukarthik; S. Premalatha, "4 Forecasting Time Series Data Using ARIMA and Facebook Prophet Models. In *Big Data Management in Sensing: Applications in AI and IoT* (pp. 47–60). River Publishers.

Sivaramakrishnan, S., & Kesavamurthy, T. (2017). Emseep—Effective Modelling of Scalable Energy Efficient Protocol In Wireless Sensor Network. *Advances in Natural and Applied Sciences, 11*(12), 18–26.

Sivaramakrishnan, S., & Kesavamurthy, T. (2017). Identifying Cluster Head and Data Transmission Through Them for Efficient Communication in Wireless Sensor Network. *Journal of Computational and Theoretical Nanoscience, 14*(8), 4014–4020.

Sivaramakrishnan, S., & Kesavamurthy, T. (2017). Optimum Routing of Data through Cluster Head in Wireless Sensor Network. *International Journals of Control Theory and Applications, 10*(31), 27–35.

Soll, M., Hinz, T., Magg, S., & Wermter, S. (2019). Evaluating defensive distillation for defending text processing neural networks against adversarial examples. In I. V. Tetko, V. Kþurková, P. Karpov, & F. Theis (Eds.), Artificial Neural Networks and Machine Learning—ICANN Image Processing (pp. 685–696). Academic Press.

Stamp, M. (2017). *Introduction to Machine Learning with Applications in Information Security.* Chapman and Hall/CRC.

Statista. (2020). *IoT: Number of Connected Devices Worldwide 2012–2025*. Available online: https://www.statista. com/statistics/471264/iot-number-of-connected-devices-worldwide/

Stoica & Freitas de Abreu. (2019). *6G: the Wireless Communications Network for Collaborative and AI Applications.* arXiv. arXiv:1904.03413v1.

Strinati. (2019). 6G: The next frontier: From holographic messaging to artificial intelligence using subterahertz and visible light communication. *IEEE Vehicular Technology Magazine*, *14*(3), 42-50.

Strinati, E. C., Barbarossa, S., Gonzalez-Jimenez, J. L., Ktenas, D., Cassiau, N., Maret, L., & Dehos, C. (2019). 6G: The next frontier: From holographic messaging to artificial intelligence using subterahertz and visible light communication. *IEEE Vehicular Technology Magazine*, *14*(3), 42–50.

Strukov, D., Indiveri, G., Grollier, J. S., & Fusi, S. (2019). Building brain inspired computing. *Nature Communications*, 10.

Su & Chih-Lin. (2019) Energy and spectral efficient frequency reuse of ultra-dense networks. *IEEE Transactions on Wireless Communications*, *15*(8), 5384-98.

Su, Sperli, Moscato, Picariello, Esposito, & Choi. (2019). An Edge Intelligence Empowered Recommender System Enabling Cultural Heritage Applications. IEEE Transactions On Industrial Informatics, 15(7).

Sun, Liu. & Yue. (2019). *AI-Enhanced Offloading in Edge Computing: When Machine Learning Meets Industrial IoT.* IEEE Network.

Sun, H., Hua, S., Zhou, E., Pi, B., Sun, J., & Yamashita, K. (2018, June). Using ethereum blockchain in Internet of Things: A solution for electric vehicle battery refueling. In *International Conference on Blockchain* (pp. 3-17). Springer.

Sun, X., Yang, W., Cai, Y., Ma, R., & Tao, L. (2019). Physical layer security in millimeter wave SWIPT UAV-based relay networks. *IEEE Access: Practical Innovations, Open Solutions*, *7*, 35851–35862. doi:10.1109/ACCESS.2019.2904856

Švogor, I., & Kišasondi, T. (2012, June). Two factor authentication using EEG augmented passwords. In *Proceedings of the ITI 2012 34th International Conference on Information Technology Interfaces* (pp. 373-378). IEEE.

Syed, A. S., Sierra-Sosa, D., Kumar, A., & Elmaghraby, A. (2021). IoT in Smart Cities: A Survey of Technologies, Practices and Challenges. *Smart Cities*, *4*(2), 429–475. doi:10.3390martcities4020024

Szporer, R. (2020). *The Climb to the Top of the Smart Factory Data Pyramid.* https://www.6gworld.com/exclusives/the-climb-to-the-top-of-the-smart-factory-data-pyramid

Tang. (2019). Future intelligent and secure vehicular network toward 6G: Machine-learning approaches. *Proceedings of the IEEE, 108*(2), 292-307.

Tan, X. S., Xu, W. H., Sun, K., Xu, Y., Be'ery, Y., You, X., & Zhang, C. (2020). Improving massive MIMO message passing detectors with deep neural network. *IEEE Transactions on Vehicular Technology*, *69*(2), 1267–1280. doi:10.1109/TVT.2019.2960763

Tariq, F., Khandaker, M. R., Wong, K. K., Imran, M. A., Bennis, M., & Debbah, M. (2020). A speculative study on 6G. *IEEE Wireless Communications*, *27*(4), 118–125. doi:10.1109/MWC.001.1900488

Tomkos, Klonidis, Pikasis, & Theodoridis. (2020). Toward the 6G Network Era: Opportunities and Challenges. *IT Professional, 22*(1), 34-38. . doi:10.1109/MITP.2019.2963491

Tramèr, F., Kurakin, A., Papernot, N., Goodfellow, I., Boneh, D., & McDaniel, P. (2017). Ensemble adversarial training: Attacks and defenses. *Proc. Int. Conf. Learn. Represent.* Available: https://arxiv.org/abs/1705.07204

Tran, N.-T. (2019). Self-Supervised GAN: Analysis And Improvement with Multi-Class Minimax Game. NeurIPS, Vancouver, Canada.

Tzagkarakis, C., Petroulakis, N., & Ioannidis, S. (2019, June). *Botnet attack detection at the IoT edge based on sparse representation. In 2019 Global IoT Summit (GIoTS)*. IEEE.

Ucar, S., Coleri Ergen, S., Ozkasap, O., Tsonev, D., & Burchardt, H. (2016, November). Secvlc: Secure visible light communication for military vehicular networks. In *Proceedings of the 14th ACM International Symposium on Mobility Management and Wireless Access* (pp. 123-129). 10.1145/2989250.2989259

Un, M., Shao, M., Ma, W., & Ching, P. C. (2019). Deep mimo detection using ADMM unfolding. *Proc. of IEEE Data Science Workshop*.

Vanitha, U., Sivaramakrishnan, S., Karthika, S., & Babukarthik, R. G. (2020). Efficient Communication Routing Through WiMax Network During Disaster. In *IOP Conference Series: Materials Science and Engineering* (*vol. 994*, no. 1, p. 012035). IOP Publishing.

Varfolomeev, A. A., Alfarhani, L. H., & Oleiwi, Z. C. (2021). Secure-reliable smart contract applications based blockchain technology in smart cities environment. *Procedia Computer Science*, *186*, 669–676.

Verbeke, W. (2001). Beliefs, attitude and behaviour towards fresh meat revisited after the Belgian dioxin crisis. *Food Quality and Preference*, *12*(8), 489–498. doi:10.1016/S0950-3293(01)00042-8

Vernon, D., Metta, G., & Sandini, G. (2014, April). A survey of artificial cognitive systems: Implications for the autonomous development of mental capabilities in computational agents. *IEEE Transactions on Evolutionary*, *A247*, 529–551.

Wale & Sonawani. (2018). A Survey on Deep Learning Approaches in Healthcare. *International Journal of Innovative Research in Computer and Communication Engineering, 6*(5).

Wang & Hepsaydir. (2014). Cellular architecture and key technologies for 5G wireless communication networks. *IEEE Communications Magazine, 52*(2), 122-130.

Wang, H., Cheng, H., & Hao, H. (2020, October). The use of unmanned aerial vehicle in military operations. In *International Conference on Man-Machine-Environment System Engineering* (pp. 939-945). Springer.

Wang, C. X., Huang, J., Wang, H., Gao, X., You, X., & Hao, Y. (2020). 6G wireless channel measurements and models: Trends and challenges. *IEEE Vehicular Technology Magazine*, *15*(4), 22–32. doi:10.1109/MVT.2020.3018436

Wang, N., Jiang, T., Lv, S., & Xiao, L. (2017, July). Physical-layer authentication based on extreme learning machine. *IEEE Communications Letters*, *21*(7), 1557–1560.

Wang, Q., Zhu, X., Ni, Y., Gu, L., & Zhu, H. (2020). Blockchain for the IoT and industrial IoT: A review. *Internet of Things*, *10*, 100081.

Wara, M. S., & Yu, Q. (2020, December). New Replay Attacks on ZigBee Devices for Internet-of-Things (IoT) Applications. In *2020 IEEE International Conference on Embedded Software and Systems (ICESS)* (pp. 1-6). IEEE.

Watson, C., & Woods, K. (2021). TW: 6G and Artificial Intelligence & Machine Learning. The MITRE Corporation.

Weinand, A., Karrenbauer, M., Lianghai, J., & Schotten, H. D. (2017, June). Physical layer authentication for mission critical machine type communication using Gaussian mixture model based clustering. In *2017 IEEE 85th Vehicular Technology Conference (VTC Spring)* (pp. 1-5). IEEE. 10.1109/VTCSpring.2017.8108527

Wen, Wu, Wang, Chen, & Li. (2016). Learning structured sparsity in deep neural networks. *NIPS*, 2082–2090.

Wen, W. S. C., & Jin, S. (2018). Deep learning for massive mimo csi feedback. *IEEE Wireless Communications Letters, 7*(5), 748–751. doi:10.1109/LWC.2018.2818160

Wolpaw, J. R., Birbaumer, N. D., McFarland, J., Pfurtscheller, G., & Vaughan, T. M. (2002). Brain–Computer Interfaces for Communication and Control. *Clinical Neurophysiology, 113*(6), 767–791.

Wu & Zhang. (2019). Intelligent reflecting surface enhanced wireless network via joint active and passive beam-forming. *IEEE Transactions on Wireless Communications, 18*(11), 5394-09.

Xiao, Y., Shi, G., Li, Y., Saad, W., & Poor, H. V. (2020). Toward Self-Learning Edge Intelligence in 6G. *IEEE Communications Magazine, 58*(12), 34–40. doi:10.1109/MCOM.001.2000388

Xie, J., Yu, F. R., Huang, T., Xie, R., Liu, J., & Liu, Y. (2019). A survey on the scalability of blockchain systems. *IEEE Network, 33*(5), 166–173. doi:10.1109/MNET.001.1800290

Xu, W. H., Wu, Z. Z., & Ueng, Y. L. (2017). Improved polar decoder based on deep learning. *Proceedings of IEEE International Workshop on Signal Processing Systems*, 1–6. 10.1109/SiPS.2017.8109997

Yaacoub & Alouini. (2019). *A key 6G challenge and opportunity--connecting the remaining 4 billions: a survey on rural connectivity.* arXiv preprint arXiv:1906.14.

Yang & Li. (2019). 6G wireless communications: Vision and potential techniques. *IEEE Network, 33*(4), 70-75.

Yang, H., Zhao, J., Nie, J., Kumar, N., Lam, K. Y., & Xiong, Z. (2021, May). UAV-Assisted 5G/6G Networks: Joint Scheduling and Resource Allocation Based on Asynchronous Reinforcement Learning. In *IEEE INFOCOM 2021-IEEE Conference on Computer Communications Workshops (INFOCOM WKSHPS)* (pp. 1-6). IEEE.

Yang, Z., Chen, M., Wong, K. K., Poor, H. V., & Cui, S. (2021). *Federated learning for 6G: Applications, challenges, and opportunities*. arXiv preprint arXiv:2101.01338.

Yang, H., Alphones, A., Xiong, Z., Niyato, D., Zhao, J., & Wu, K. (2020). Artificial-Intelligence-Enabled Intelligent 6G Networks. *IEEE Network, 34*(6), 272–280. doi:10.1109/MNET.011.2000195

Yang, P., Xiao, Y., Xiao, M., & Li, S. (2019). 6G wireless communications: Vision and potential techniques. *IEEE Network, 33*(4), 70–75.

Yastrebova & Koucheryavy. (2018). Future Networks 2030: Architecture & Requirements. *Proceedings of 10th International Congress on Ultra-Modern Telecommunications and Control Systems and Workshops (ICUMT),* 1–8.

Yastrebova, A., Kirichek, R., Koucheryavy, Y., Borodin, A., & Koucheryavy, A. (2018). Future networks 2030: Architecture & requirements. In *2018 10th International Congress on Ultra Modern Telecommunications and Control Systems and Workshops (ICUMT)* (pp. 1-8). IEEE.

Ye, B. F. J. H., Li, G. Y., & Sivanesan, K. (2018). Channel agnostic end-to-end learning based communication systems with conditional gan. *IEEE Globecom Workshops, 16*, 1–5. doi:10.1109/GLOCOMW.2018.8644250

Ye, H., Li, G. Y., & Juang, B. H. (2018). Power of deep learning for channel estimation and signal detection in OFDM systems. *IEEE Wireless Communications Letters, 7*(1), 114–117. doi:10.1109/LWC.2017.2757490

You, X. H., Zhang, C., Tan, X. S., Jin, S., & Wu, H. (2019). AI for 5G: Research directions and paradigms. *Science China. Information Sciences, 62*(2), 021301. doi:10.100711432-018-9596-5

You, X., Wang, C. X., Huang, J., Gao, X., Zhang, Z., Wang, M., Huang, Y., Zhang, C., Jiang, Y., Wang, J., Zhu, M., Sheng, B., Wang, D., Pan, Z., Zhu, P., Yang, Y., Liu, Z., Zhang, P., Tao, X., ... Liang, Y.-C. (2021). Towards 6G wireless communication networks: Vision, enabling technologies, and new paradigm shifts. *Science China. Information Sciences*, *64*(1), 110301. doi:10.100711432-020-2955-6

Yueyue, D. (2019). Blockchain and deep reinforcement learning empowered intelligent 5G beyond. *IEEE Network*, *33*(3), 10–17. doi:10.1109/MNET.2019.1800376

Yu, K.-H., Beam, A. L., & Kohane, I. S. (2018). Artificial intelligence in healthcare. *Nature Biomedical Engineering*, *10*, 719–731.

Zantalis, F., Koulouras, G., Karabetsos, S., & Kandris, D. (2019). A Review of Machine Learning and IoT in Smart Transportation. *Future Internet*, *11*(4), 94. doi:10.3390/fi11040094

Zappone, A., Debbah, M., & Altman, Z. (2018). Online energy efficient power control in wireless networks by deep neural networks. *IEEE 19th Int. Workshop Signal Process. Adv. Wireless Commun.*, 1–5. 10.1109/SPAWC.2018.8445857

Zappone, M. D. R., & Debbah, M. (2019). Wireless Networks Design in the Era of Deep Learning: Model-Based, AI-Based, or Both? *IEEE Transactions on Communications*, *67*(5), 7331–7376. doi:10.1109/TCOMM.2019.2924010

Zero-Touch Network and Service Management (ZSM). (2020). *Landscape*. ETSI Standard GR ZSM 004.

Zhang & Leung. (2016). Fronthauling for 5G LTE-U ultra-dense cloud small cell networks. *IEEE Wireless Communications, 23*(6), 48-53.

Zhang & Nallanathan. (2019). Artificial intelligence-based resource allocation in ultradense networks: Applying event-triggered Q-learning algorithms. *IEEE Vehicular Technology Magazine, 14*(4), 56-63.

Zhang & Song. (2020). HetMEC: Heterogeneous multi-layer mobile edge computing in the 6 G era. *IEEE Transactions on Vehicular Technology, 69*(4), 4388-00.

Zhang, Q., Ferdowsi, A., Saad, W., & Bennis, M. (2021) *Distributed conditional generative adversarial networks (GANs) for data-driven millimeter wave communications in UAV networks arXiv preprint*. arXiv:2102.01751.

Zhang, S., & Zhu, D. (2020). Towards artificial intelligence enabled 6G: State of the art, challenges, and opportunities. *Computer Networks, 183*. doi:10.1016/j.comnet.2020.107556

Zhang, X., Kunz, A. S., & Schroder, S. (2017). Overview of 5G security in 3GPP. *2017 IEEE Conference on Standards for Communications and Networking (CSCN)*, 181-186.

Zhang, Z. K., Cho, M. C. Y., Wang, C. W., Hsu, C. W., Chen, C. K., & Shieh, S. (2014, November). IoT security: Ongoing challenges and research opportunities. In *2014 IEEE 7th international conference on service-oriented computing and applications* (pp. 230-234). IEEE.

Zhang. (2019). 6G Visions: Mobile ultra-broadband, super internet-of-things, and artificial intelligence. *China Communications, 16*(8), 1-14.

Zhang. (2019). 6G wireless networks: Vision, requirements, architecture, and key technologies. *IEEE Vehicular Technology Magazine, 14*(3), 28-41.

Zhang, C., & Ma, Y. (2012). *Ensemble Machine Learning: Methods and Applications*. Springer.

Zhang, J. J., Huang, Y. M., & Wang, J. H. (2019) Intelligent beam training for millimeter-wave communications via deep reinforcement learning. *Proceedings of IEEE Global Communications Conference*, 1–7. 10.1109/GLOBECOM38437.2019.9014113

Zhang, J., Kang, K., Huang, Y., Shafi, M., & Molisch, A. F. (2019). Millimeter and THz wave for 5G and beyond. *China Communications, 16*(2), iii–vi.

Zhang, L., Yan, Z., & Kantola, R. (2017). *Privacy-preserving trust management for unwanted traffic control.* Elsevier Future Generation Computer Systems.

Zhang, W., Ding, D. S., Sheng, Y. B., Zhou, L., Shi, B. S., & Guo, G. C. (2017). Quantum secure direct communication with quantum memory. *Physical Review Letters, 118*(22), 220501. doi:10.1103/PhysRevLett.118.220501 PMID:28621985

Zhang, Z. M., Hua, M., Li, C. G., Huang, Y., & Yang, L. (2019). Placement delivery array design via attention-based sequence-to-sequence model with deep neural network. *IEEE Wireless Communications Letters, 8*(2), 372–375. doi:10.1109/LWC.2018.2873334

Zhang, Z., Xiao, Y., Ma, Z., Xiao, M., Ding, Z., Lei, X., Karagiannidis, G. K., & Fan, P. (2019). 6G Wireless Networks: Vision, Requirements, Architecture, and Key Technologies. *IEEE Vehicular Technology Magazine, 14*(3), 28–41. doi:10.1109/MVT.2019.2921208

Zhao. (2019). *A survey of intelligent reflecting surfaces (IRSs): Towards 6G wireless communication networks.* arXiv preprint arXiv:1907.04789.

Zheng, Z., Xie, S., Dai, H., Chen, X., & Wang, H. (2017, June). An overview of blockchain technology: Architecture, consensus, and future trends. In *2017 IEEE international congress on big data (BigData congress)* (pp. 557-564). IEEE.

Zhou, C., & Li, Z., Luo, & Zhang. (2019). Edge Intelligence: Paving the Last Mile of Artificial Intelligence With Edge Computing. *Proceedings of the IEEE, 7*(8).

Zhu, T., & Philip, S. Y. (2019, July). Applying differential privacy mechanism in artificial intelligence. In *2019 IEEE 39th International Conference on Distributed Computing Systems (ICDCS)* (pp. 1601-1609). IEEE. 10.1109/ICDCS.2019.00159

Ziegler, V., Viswanathan, H., Flinck, H., Hoffmann, M., Räisänen, V., & Hätönen, K. (2020). 6G architecture to connect the worlds. *IEEE Access: Practical Innovations, Open Solutions, 8*, 173508–173520.

Zong. (2019). 6G technologies: Key drivers, core requirements, system architectures, and enabling technologies. *IEEE Vehicular Technology Magazine, 14*(3), 18-27.

Zuhair, M., Patel, F., Navapara, D., Bhattacharya, P., & Saraswat, D. (2021, April). BloCoV6: A blockchain-based 6G-assisted UAV contact tracing scheme for COVID-19 pandemic. In *2021 2nd International Conference on Intelligent Engineering and Management (ICIEM)* (pp. 271-276). IEEE.

Zuo. (2019). Improving the tolerance of stochastic LDPC decoders to overclocking-induced timing errors: A tutorial and a design example. *IEEE Access, 4*, 1607-1629.

About the Contributors

A. M. Viswa Bharathy completed his B.E. and M.E. in CSE in 2008 and 2011 respectively from Anna University, India. He completed his Ph.D. in the area of Computer Networks and Security from the same University in 2017 with highly commendable peformance. He has a total of 13+ years of experience in academic teaching. His interests are Computer Networks, Wireless Networks, Adhoc and Sensor Networks, Mobile Communications and Data Science. He has been honoured with awards two times. He has delivered guest lectures and keynote addresses in the national and international arena many a times. He is reviewer and editorial member of many WOS and Scopus journals. He has presented and published more than 30 papers in international conferences and journals.

* * *

Harshitha B. is an Assistant Professor working in the Department of Electronics and Communication Engineering in B M S College of Engineering, Bangalore, Karnataka, India. She has one Indian Patent, three Scopus indexed journal papers in her credit. Her post-graduation was in Electronics and communication engineering in July 2003 and her Graduation in Electronics Engineering in August 1997. Her areas of interest are Networking, Internet of Things, Data Security and Analog electronics. Having almost three years of Industry Experience, Harshitha has around 14 years of teaching experience and has a passion to teach. Before joining B.M.S. college of Engineering, Harshitha worked at Juniper Networks India pvt ltd., as senior technical trainer, where she was involved in imparting training on Junos operating system to the software and hardware developers enabling them to add new features to the Juniper products. After joining the noble teaching profession, Harshitha emphasized on being a student facilitator rather being a knowledge repository. Guided more than four post graduates and thirty two undergraduate students for academic projects under different verticals. She is keen in teaching subjects like Data communication Networks , Computer communication Networks, Analog Electronic circuits , Mixed Signal Design Basic electronics, Network Security and Cryptography, Microcontroller, operating system. She has coordinated workshops, attended more than ten workshops and faculty development programs. She has Two Technical Memberships: 1. Life Member of the Indian Society for the Technical Education. 2. Member of International Association of Engineers (MIAENG).

Abhinav Bhandari is an Assistant Professor in department of Computer Science & Engineering at Punjabi University, Patiala, Punjab, India. He has done his B. Tech. Computer Science & Engineering from GTBKIET, Malout, Punjab, India in 2001. He has completed his Master in Technology from DAVIET, Jalandhar,Pujjab, India in 2008. He finished his Ph.D. from DR. B.R Ambedkar National Institute of Technolgy, jalandhar, Punjab, India in September, 2017. He has more than 20 years of teaching, research and administrative experience. His general research interests are in the areas of Network Security and Computer Networks. Specific research interests include DDoS detection and Security aspects of Software Defined Networks. He has published 3 Books in the field of Computer Science & Network Security. He has published more than 50 papers in national / International peer reviewed / Indexed / impact factor Journals and IEEE, Elsevier and Springer proceedings.

Saravana Kumar G. is presently associated with Department of ECE, Karpagam College of Engineering as a Professor. He had completed his B.E., (ECE), M.Tech (Applied Electronics) and Ph.D., . His area of research interests are Brain Computer Interface, Bio Signal Processing, Artificial Intelligence and Deep Learning. He has 24 years of Teaching Experience in various institutions. He had published 34 international journals and had received grants from AICTE. He worked as Dean - Internal Quality Assurance Cell for eights and has experience in accreditation activities. He had completed B.E., (Electronics and Communication Engineering), M. Tech(Applied Electronics) and Ph.D in Bio Signal Processing.

Shikha Garg is pursuing a post-doctoral degree from Punjabi University, Patiala, in electronics and communication engineering. Her research interest is wireless communication and Network Security.

Sonia Goyal has received the B.Tech and M.E. degree in engineering from Punjab Technical University, Jalandhar, India and Punjab Engineering College PEC, Chandigarh, India respectively. She did her Ph.D. from Punjabi University Patiala. She has been in teaching for the last 14 years. Presently she is Assistant Professor in Department of Electronics and Communication Engineering, Punjabi University, Patiala. She has to her credit many research papers in international journals and national and international conferences. Her current interests are Wireless Sensor Networks, Optimization Techniques and Antenna System.

Surendra H. H. has been working at BMS college of engineering with total teaching and research experience of 17 years. His research areas include artificial intelligence, data analytics and renewable energy.

Janani K. is a Research Scholar in the Department of Computing Technologies in SRM Institute of Science and Technology, Kattankulathur. She Received her MTech degree in Computer Science and Engineering from SRM Institute of Science and Technology, Kattankulathur, Chennai, India. She has 5 Years' experience in Healthcare Industry as Quality Lead and 2 Years' experience as Teaching Assistant in SRM Institute of Science and Technology, Kattankulathur Her Research Interest Include IoT, BlockChain and Deep Learning. She has presented and published more than 10 papers in international conferences and journals..

Madhusudhan K. N. is an Assistant Professor working in the Department of Electronics and Communication Engineering in B M S College of Engineering, Bangalore, Karnataka, India. He has five International Patents, three Science Citation indexed journal Papers, and four Scopus indexed journal papers in his credit. He received his Doctor of Philosophy from Anna University Chennai for his Image security topic "Contemporary Attack Resilient Hybrid Crypto-Stegano-Model for Secure Healthcare Data Communication" in the faculty of Information and Communication Technology in July 2021. His post-graduation was in Digital Communication and Networking in April 2006 and his Graduation in

Telecommunication Engineering in May 2004. His areas of interest are Image Security, Data Security, Networking, and Internet of Things. Having almost five years of Industry Experience, Dr. Madhusudhan has around 12 years of teaching experience and has a passion to teach. Before joining the teaching profession, Dr.Madhusudhan was a software Engineer in Adamya Technologies and Delphi Automotives. Experienced in Bluetooth technology, worked in application development, validation and verification of different Bluetooth Profiles. Validation and Verification of different vehicular cluster dashboards. After joining the noble teaching profession, Dr.Madhusudhan emphasized on being a student facilitator rather being a knowledge repository. Guided more than ten post graduates and thirty two undergraduate students for academic projects under different verticals. He is keen in teaching subjects like electromagnetic wave theory, Microwave Engineering, Antenna Theory, Basic electronics, Satellite Communication, Computer Communication Networks, Network Security and Cryptography. He has coordinated more than ten workshops, attended more than thirty workshops and faculty development programs. He has five papers presented in technical paper presentation in national and international conferences. He has three Technical Life Memberships: 1. Life Member of the Institution of Electronics and Telecommunication Engineering (MIETE- M27195) 2. Life Member of the Indian Society for the Technical Education (MISTE-LM 131072). 3. Member of International Association of Engineers (MIAENG-270222) Also he is a Microsoft Certified Professional.

Kayalvizhi M. completed M.tech degree from the Information Security Department, Pondicherry Engineering College, India and she is currently pursuing her Ph.D. degree from Computing Technology Department, SRM University, India. Her research interests include wireless sensor network, UAV and Cyber security Attacks.

Rekha M., M.E., is currently working as an Assistant Professor in the Department of Electronics and Communication Engineering at Dhanalakshmi Srinivasan Engineering College (Autonomous), Perambalur-621212, Tamil Nadu. She has a decade of teaching experience. Her research work encircles in the domain of Heterogeneous networks and Next Generation mobile computing.

Supreetha M. A. is an undergraduate Engineering student of B M S College Of Engineering, Bengaluru, Karnataka, India. She has worked in various research domains and has made contribution in inter-domain fields combining Communication, Networking and VLSI domains as part of her academic projects.

Anantha Sunil Maligi is an Assistant Professor working in the Department of Electronics and Communication Engineering in B M S College of Engineering, Bangalore, Karnataka, India. He is a passionate teacher as well as an enthusiastic researcher to guide and learn new technologies. He has completed my doctoral studies in the area of "Solar cell device fabrication using physical and chemical deposition technique" at the Indian Institute of Science (IISc, Bangalore). He has extensive experience in handling various material deposition and characterization tools which include Sputtering, Thermal Evaporation, XRD, SEM, Hall Measurement System, Raman Spectrophotometer, etc. He has published research work in international journals as well as conference proceedings. He has been Awarded Teachers Associateship for Research Excellence (TARE): TAR/2020/000293 by Science and Engineering Research Board for a duration of three years from 2020 to 2023. The host institute is Indian Institute of Science (IISc), Bangalore. He is a member of the Institute for Smart structures and systems (ISSS).

Bhuvaneswari Mariappan, M.E., Ph.D., is currently working as Assistant Professor (Senior Grade) in the Department of Electronics and Communication Engineering, BIT Campus Trichy-24, Anna University Chennai. She has completed her PhD in the domain of Heterogeneous Networks. She has nearly twenty five years of teaching and research experience. She has more than 10 National and International research publications, presented more than 15 conference papers and published book chapter in International Publishers. Her primary area of research is in Heterogeneous Networks, Mobile Networks and Next Generation Networks.

Arunachalam Murugesan is a Professor in Department of Information technology at Sri Krishna College of Engineering and technology, Coimbatore, TamilNadu, India. He is an accomplished academician with sixteen years of experience. He completed his Ph.D at Anna University, India and Post graduate studies at National Engineering College affiliated to Anna University. His research area lie in the area of Networks, Data Science and Internet-of-things. He has collaborated actively with researchers in several other disciplines of computer science, particularly in the area of Networks and IOT and published research articles and a patent in reputed journals and conferences. Arunachalam has served on various conferences and workshop committee members and served as Program Chair cum Reviewer for Conferences & Journals.

Karthikeyan P. obtained his the Bachelor of Engineering (B.E.) in Computer Science and Engineering from Anna University, Chennai, and Tamil Nadu, India in 2005 and received his Master of Engineering (M.E.) in Computer Science and Engineering from Anna University, Coimbatore India in 2009. He has completed Ph.D. degree in Anna University, Chennai in 2018. Skilled in developing projects and carrying out research in the area of Cloud computing and Data science with the programming skill in Java, Python, R and C. He published more than 20 International journals with good impact factor and presented more than 10 International conferences. He was the reviewer of Elsevier, Springer, Inderscience and reputed Scopus indexed journals. He is acting as editorial board members in EAI Endorsed Transactions on Energy Web, The International Arab Journal of Information Technology and Blue Eyes Intelligence Engineering and Sciences Publication journal.

V. Parthasarathy is presently working as a Professor & Dean (R&D and Industry Relations) at Karpagam Academy of Higher Education (Deemed to be University), Tamilnadu, India. He has graduated with a Bachelor of Engineering in Electrical and Electronics Engineering from Government College of Technology, Coimbatore. Further, he has completed his post-graduation for Master of Engineering in Computer Science and Engineering from College of Engineering, Guindy, Anna University and Ph.D. in Information and Communication engineering from Anna University, Tamilnadu, India. He has published over 100 scientific papers in reputed journals and conferences, He has hosted many conferences as Chair and Co-Chair, and delivered keynotes. Besides, he is a reviewer in several peer-reviewed journals. His research interests include the Internet of Things, Wireless Sensor Network, Biomedical Informatics, Next-Generation Communication Networks, and Artificial Intelligence. He is a member of many professional societies such as IEEE, ISTE, ACM, OSA, etc. He is the recipient of the Indian Government's Fund for Improvement of Science and Technology (FIST) project and Co-investigator for the project grant from Defence Research and Development Organization (DRDO).

Dhaya R. has 16 years of experience in teaching and research in the field of Computer Science and Engineering. She published more than 80 research papers in peer-reviewed international Journals. She was the recipient of the IEI Young women's Engineer award. Her areas of interest are wireless sensor networks, embedded systems, Machine Learning, Communication Systems.

Kanthavel R. has 22 years of experience in teaching and research in the field of information and Communication Engineering. He has the credit of more than 100 research articles in peer-reviewed international Journals. His areas of interest are computer networking, Machine Learning, and AI, Co-operative communication, computing, and mobile networks.

Rajagopal R. is Associate Professor in the Department of Computer Science & Engineering at Narsimha Reddy Engineering College (Autonomous), Hyderabad, India. He received the Bachelor of Engineering (B.E) in Computer Science and Engineering from Anna University, Chennai, Tamilnadu, India, in 2005 and received his Master of Engineering (M.E,) in Computer Science and Engineering from Anna University, Coimbatore, India in 2009 and Ph.D degree in Anna University, Chennai, in 2018. His areas of research cover the fields of cryptography and network security; cloud computing, machine learning, He acts as an editorial board member for the International journal of mathematical science and Engineering, and Sciences Publication journal. He is a member of CSI, IAENG, ISSE, and CSTA. He published more than 22 International journals and presented more than 20 International conferences.

Ramya R is a Research Scholar in the department of Computer science and Technology in SRM Institute of Science and Technology, Kattankulathur. She completed her B.E Computer science and Engineering from Madras University and M.E Computer science and Engineering from Vinayaka Mission University. She has total of 14+ years of experience in academic teaching from Engineering colleges. Her interests are Computer networks , Internet of Things(IoT), Machine Learning and Deep Learning.

Lalitha S. is currently working as an Assistant Professor in the department of Electronics & Communication, BMSCE Bangalore Karnataka holds, M.Tech (bio-medical signal processing) from Visvesvaraya Technological University(2006) and MBA(HR) from Pondicherry University(2009). She has over 15+ years of industrial experience with strong exposure in consumer and industrial automation domain, along with 15 years of teaching experience in the field of Signal and Image processing domain. She has journal publications in the field of Image classification and analysis as well as in education domain. she holds a Innovation Australian Patent, along with book chapter published in CRC Press. Her research interests include Image analysis, Pattern recognition, ANN, AI / ML and computer vision.

Ramamoorthy S. obtained his under graduate and Master degree and PhD from Anna University, Chennai. Presently working as an Associate Professor in SRM Institute of Science and Technology, Kattankulathur, Chennai. Having 16 Years of Overall Teaching Experience and 10 Years of Research Experience in Computer Science and Engineering discipline. More than 50 research papers Published in International Journals with High Impact factor which includes SCI, Scopus, Web of Science and so forth. Over 23 research papers are published in International Conferences like IEEE, Springer conferences and so on. Received the Travel Grant from TNSCST to present a paper in International Conference held in Malaysia. He has Published More than 5 Indian Patents and acting as an Editorial board member and reviewer for many international research forums.

Sivaramakrishnan S. is an Assistant Professor in the Department of Electronics and Communication Engineering, Dayananda Sagar University, Bangalore, India. He has received his Ph.D from Anna University, Chennai. He has completed his Bachelor of Engineering in Electronics and Communication Engineering and Masters of Engineering with specialization in Communication Systems from Anna University, Chennai. His area of interest includes machine learning, optimization of resources in wireless sensor networks and analysis on data using machine learning algorithms. He has 13 years of teaching experience. He has presented 5 papers in national conferences and published 5 papers in various indexed international journals.

Sanjana T. received M.Tech degree in Digital Communication from BMS College of Engineering, Bangalore in 2014, and B.E degree from Siddaganga Institute of Technology, Tumkur, Karnataka in 2012. She is currently working as Assistant professor in the Department of Electronics and Communication, BMS College of Engineering, Bangalore, Karnataka, India. She has total teaching experience of 7 years. Her current areas of interest include Multiple access schemes and Wireless communication. She has published her research work in several International Journals and presented in Conferences. Her current area of research is different multiple access schemes and signal processing operations which optimizes the performance of physical layer of future generation networks.

Index

K

L

M

N

P

Q

S

T

U

V

W

www.ingramcontent.com/pod-product-compliance
Lightning Source LLC
LaVergne TN
LVHW062155300926
849161LV00010B/2673